POLYMER SCIENCE AND TECHNOLOGY
Volume 19

COORDINATION
POLYMERIZATION

POLYMER SCIENCE AND TECHNOLOGY

Recent volumes in the series:

A Continuation Order Plan is available for this series. A continuation order will bring delivery of each new volume immediately upon publication. Volumes are billed only upon actual shipment. For further information please contact the publisher.

POLYMER SCIENCE AND TECHNOLOGY
Volume 19

COORDINATION POLYMERIZATION

Edited by

Charles C. Price
University of Pennsylvania
Philadelphia, Pennsylvania

and

Edwin J. Vandenberg
Hercules Incorporated Research Center
Wilmington, Delaware

PLENUM PRESS ● NEW YORK AND LONDON

Library of Congress Cataloging in Publication Data
Main entry under title:

Coordination polymerization.

(Polymer science and technology; v. 19)
"Proceedings of a symposium on coordination polymerization held March 30, 1981,
at the American Chemical Society meeting, in Atlanta, Georgia"—Copr. p.
Includes bibliographical references and index.
1. Polymers and polymerization—Congresses. 2. Coordination compounds—Congresses. I. Price, Charles C. (Charles Coale), 1913–   . II. Vandenberg, Edwin J.,
1918–   . III. American Chemical Society. IV. Series.
QC380.C658 1983   547′.28   82-19081

ISBN-13:978-1-4613-3577-1   e-ISBN-13: 978-1-4613-3575-7
DOI: 10.1007/978-1-4613-3575-7

Proceedings of a symposium on Coordination Polymerization held March 30, 1981, at
the American Chemical Society meeting, in Atlanta, Georgia

© 1983 Plenum Press, New York
Softcover reprint of the hard cover 1st edition 1983
A Division of Plenum Publishing Corporation
233 Spring Street, New York, N.Y. 10013

PREFACE

Edwin J. Vandenberg, a long time researcher at the Research Center of Hercules Incorporated, was the 1981 winner of the American Chemical Society Award in Polymer Chemistry, sponsored by the Witco Chemical Corp. Following is the citation of the accomplishments which led to this award.

"In recognition of his pioneering research that advanced polymer science and that led to the discovery and development of isotactic polypropylene, epichlorohydrin and propylene oxide elastomers, new polymerization catalysts, and the hydrogen method of controlling molecular weights of Ziegler polyolefins."

It was my pleasure to arrange a symposium to celebrate this award at the Atlanta Meeting of the American Chemical Society on March 30, 1981. In considering the broad range of Vandenberg's contributions to polymer chemistry, it was decided to choose the subject of "Coordination Polymerization" for the symposium. This area is both one to which Vandenberg has made major contributions and one of great industrial and scientific interest.

Since Vandenberg has been involved in coordination polymerization in both ring-opening and α-olefin type polymerizations, both were covered in the symposium, whose program follows.

1. "The Polymerization of 1,2-Epoxides Catalyzed by the Condensation Products of Metal-Containing Compounds with

   Alkylphosphates", T. Nakata, Research Laboratories, Osaka Soda Co., Ltd., 9 Otakasu-cho, Amagasaki City, Hyogo Pref. 660, Japan.

2. "Stereochemical Aspects of the Cationic Polymerization of Chiral Alkyl Vinyl Ethers", Emo Chiellini, Roberto Solaro and Francesco Masi, Instituto di Chimica Organica Industriale, Universita di Pisa, 56100 Pisa, Italy.

3. "Stereochemical Aspects of the Polymerization of Cis and Trans
    Thiiranes Using Chiral Initiators", N. Spassky, A. Momtaz and
    P. Sigwalt, Laboratory of Macromolecular Chemistry, University
    of Paris VI, 4, Place Jussieu, 75230 Paris Cedex, France.

4. "Coordination Polymerization", E. J. Vandenberg, Hercules
    Research Center, Wilmington, Delaware 19899.

5. "Stereoselective and Stereoelective Polymerization of Racemic
    α-Olefins with Supported Titanium Catalysts", P. Pino, G.
    Fochi, R. Mulhaupt, O. Piccolo, U. Giannini and A. Oschwald,
    Swiss Federal Institute of Technology, Zurich, Switzerland.

6. "1,4 Polybutadiene Block Copolymers from "Living" Coordination
    Catalysts", Ph. Teyssie, P. Hadjiandreou and M. Julemont,
    Laboratory of Macromolecular Chemistry and Organic Catalysis,
    University of Liege, Sart-Tilman, 4000 Liege, Belgium.

7. "Supported Catalysts for Polypropylene: Aluminum Alkyl-Ester
    Chemistry", A. W. Langer, T. J. Burkhardt and J. J. Steger,
    Exxon Research & Engineering Company, Corporate Research -
    Science Laboratories, P.O. Box 45, Linden, NJ 07076.

8. "Tetra (Neophyl) Zirconium and Its Use in the Polymerization
    of Olefins, R. A. Setterquist and F. N. Tebbe, E. I. du Pont
    de Nemours & Company, Central Research & Development
    Department, Experimental Station, Wilmington, Delaware 19898.

9. "Polymerization of Ethylene Using Organoscandium Compounds",
    Glenn A. Moser, Hercules Research Center, Wilmington, DE
    19899.

        There were other investigators whose contributions would
    have been appropriate but who, for one reason or another,
    could not be fitted into the one day symposium in Atlanta.
    When we decided the material was of suitable interest to
    warrant publication in book form, we invited others to submit
    chapters for this purpose. We are pleased that the following
    were able to participate in this way.

10. " 'Living' Coordination Polymerization of Propylene and Its
    Application to Block Copolymer Synthesis", T. Keii, Tokyo
    Institute of Technology, Ookyama, Meguro-ku, Tokyo, Japan.

11. "Some aspects of coordination homo- and copolymerization of
    high α-olefins cyclic and allene hydrocarbons," B. A.
    Krentsel, V. I. Kleiner, E. A. Mushina and L. L. Stotskaya,
    The Topchiev Institute of Petrochemical Synthesis, Academy of
    Sciences of the U.S.S. R, Moscow, U.S.S.R.

12. "Novel Catalyst Systems, Based on Organozine Compounds and Group IVB Organometallic Compounds, for the Stereospecific Polymerization of Oxiranes", H. C. W. M. Buys, H. G. J. Overmars, and J. G. Noltes, Institute for Organic Chemistry TNO, Utrecht, The Netherlands.

13. "Molecular Level Elucidation of Stereospecific Polymerization of Methyloxirane Using a Well-Defined Organozinc Complex", T. Tsuruta, Science University of Tokyo, Kagurazaka, Shinjuku-ku, Tokyo 162, Japan

14. "Polymers and Copolymers of Methyl ω-Epoxyalkanoates", O. Vogl, P. Loeffler, D. Bansleben and J. Muggee, University of Massachusetts, Amherst, Massachusetts.

15. "Mechanism of Ziegler-Natta Polymerization on the Basis of Data on the Number of Active Centers and their Reactivity," V. A. Zakharov, G. D. Bukatov and Yu. I. Yermakov, Institute of Catalysis, Novosibirsk, 630090 USSR

An introductory chapter on "Selectivity in Addition Polymerization" was also added to provide background for this aspect of coordination polymerization.

All of us who have participated in this endeavour have done so gladly to express our appreciation to Ed Vandenberg, both for his friendship as an esteemed colleague in the forefront of Polymer Chemistry and for his many important scientific and industrial contributions to our chosen field of research.

Charles C. Price
Benjamin Franklin Professor
    of Chemistry (Emeritus)
University of Pennsylvania,
Philadelphia, PA 19104

CONTENTS

SELECTIVITY IN ADDITION POLYMERIZATION

Charles C. Price

Benjamin Franklin Professor of Chemistry (Emeritus)
University of Pennsylvania
Philadelphia, Pa 19104

INTRODUCTION

The development of methods to control selectivity in polymerization processes has been a major accomplishment of polymer chemists during the past quarter of a century. Despite the immense amount of industrial and academic effort expended, and despite much broad understanding of the factors involved, a full understanding of much of the detail of many aspects of selectivity in polymerization remains as a challenge to chemists.

It is the purpose of this symposium herein to report on recent advances in selectivity in vinyl and ring-opening addition polymerizations. This introductory chapter provides historical perspective for these reports.

VINYL POLYMERIZATION

Head-to-Tail Selectivity

Normally, structural selectivity for head-to-tail units in

vinyl polymerization has been very high. Whether by radical,
anionic or cationic propagation, at least two factors are
involved. Especially for monomers unsubstituted at one carbon of
the double bond, one factor must be the sterically favored access
to the unsubstituted carbon in the propagation step.

$$\text{\raisebox{0.2em}{$\sim$}CH}_2-\underset{\underset{Y}{|}}{\overset{\overset{X}{|}}{C}}* \; + \; \text{CH}_2=\underset{\backslash Y}{\overset{/X}{C}} \;\; \xrightarrow{\text{kp}} \;\; \text{\raisebox{0.2em}{$\sim$}CH}_2\underset{\backslash Y}{\overset{/X}{C}}\text{CH}_2\underset{\backslash Y}{\overset{/X}{C}}*$$

The second factor is the capability of the groups X or Y to
stabilize the reactive center. Some groups, such as X=phenyl or
vinyl, can stabilize radical, anion or cation centers. Some, such
as X=nitrile or carbonyl, are especially effective for anionic
centers, others, such as X=ether, are particularly effective for
stabilizing cationic centers.

     One of the polymers for which unequivocal quantitative
evidence exists for head-to-head units is vinyl acetate by free
radical polymerization. On hydrolysis to poly(vinyl alcohol),
head-to-head units will provide 1,2-glycol links subject to rapid
quantitative cleavage by periodic acid. Marvel and Denoon[1]
found small but significant amounts (1-2%) of such groupings. In
this particular case, both the steric and resonance factors would
be minimal. The Van der Waals radius of oxygen (1.4Å) is only
slightly greater than hydrogen (1.2Å) and much less than a
methylene group (2.0Å). Since the oxygen atom attached to a
double bond (or the intermediate free radical) has no readily
available orbitals to interact with an odd electron, the resonance
stabilization factor (Q) is also very low.

Copolymerization Selectivity

     The question of how selective monomers might be in
copolymerization was an early question of importance, since the
properties of copolymers are influenced by the sequencing of the
units. It was early recognized that selectivity in
copolymerization could be very different for free radical, anionic
or cationic copolymerization. Electrical donor or acceptor groups
attached to the double bond are an overwhelming influence for
cationic and anionic propagation. For free radical propagation it
was necessary to consider both electrical and resonance
stabilization factors, which were expressed quantitatively in the
Alfrey-Price equation[2].

$$r_1 = (Q_1/Q_2)\exp-e_1(e_1-e_2); \quad r_2 = (Q_2/Q_1)\exp-e_2(e_2-e_1)$$

Q_1 and Q_2 are measures of the resonance stabilization of a carbon free radical provided by substituents 1 and 2, while e_1 and e_2 are measures of the electrical effects of these substituents.

This work gave some understanding of the factors influencing selectivity in copolymerization, but provided control only by varying the nature of the substituents on the double bond. A discovery which provides some experimental control of selectivity was that of the influence of Lewis acids on the copolymerizability of certain monomers[3]. For example, the copolymerization characteristics of monomers like acrylates could be dramatically altered by the presence of a Lewis acid capable of coordinating with the oxygen.

The electron-accepting ability of the ester group (or positive field generated at the carbon-carbon) is markedly increased in the complex thereby enhancing copolymerization with monomers with a negative field at the carbon-carbon double bond. This provides for a far greater tendency for regular alternation of the two monomer units.

Stereoselection

With the report by Natta (1955), that the Ziegler type coordination catalyst gave a crystalline polymer of propylene, deemed isotactic based on its X-ray pattern, there was an explosion of research leading to the discovery of a wide spectrum of catalysts capable of such steric control of α-olefin polymerizations. Price and Osgan (1956)[4] suggested the now generally accepted coordination-rearrangement mechanism for this process. After coordination of the monomer to the Lewis acid center (M), the propagation step is completed by migration of the δ^- carbon of the growing chain to the δ^+ carbon of the coordinated monomer.

$$M \diagup \overset{\diagup}{\underset{}{CH_2 - \overset{\diagup}{C}HR}} \qquad + \; CH_2 = CHR \; \longrightarrow \; M \diagup \overset{\delta^-}{\underset{CH_2 \diagup\diagup^{\; \delta^+}}{CH_2 - \overset{\diagup}{C}HR}}$$

(M = Lewis acid metal)

$$M \diagdown \\ \downarrow \\ CH_2CH - CH_2\overset{\diagup}{C}H \\ \quad | \qquad \qquad | \\ \quad R \qquad \qquad R$$

The exact nature of the active catalyst site for most of the catalysts has not yet been clearly established. Changes in the catalyst can have a remarkable influence on the course of the reaction. For example, it has been reported that butadiene can be polymerized selectively to give either the isotactic or syndiotactic 1,2-polymer, or the all-<u>cis</u> or all-<u>trans</u>-1,4-polymer.

Before going on to ring-opening polymerization, perhaps a few words about the stereochemistry of isotactic and syndiotactic chains is in order.

Isotactic   Syndiotactic

It is important to recognize that the stereochemical difference between these two sequences does not produce <u>asymmetric</u> or <u>chiral</u> molecules. These three centers are like the "asymmetric" centers in meso$_1$ and meso$_2$ trihydroxyglutaric acid.

meso$_2$, mp 152°   meso$_1$, mp 170°

In the meso$_1$ and meso$_2$ forms, the central atom has four different groups attached to it, although two differ only in

configuration, being mirror images of each other*. For this
reason, both meso$_1$ and meso$_2$ are symmetrical molecules, with a
plane of symmetry through the central carbon atom. They are
however, clearly of different geometry, and the different
molecular architecture gives them quite distinct chemical and
physical properties. The central atoms of meso$_1$ and meso$_2$
have been termed pseudoasymmetric, since a change in their
configuration gives a new molecular architecture but does not make
the molecules asymmetric or chiral.

The atoms in isotactic or syndiotactic polymers are similarly
to be considered as pseudoasymmetric. For most of the atoms in
the polymer chain, the groups on one side are mirror images of the
groups on the other side, just as in the meso$_1$ and meso$_2$ forms
above. Strictly speaking, only the center carbon could be truly
pseudoasymmetric because the length of the chain in one direction
will differ from the length of the chain in the other direction.
Since a carbon atom can be significantly influenced only by other
atoms relatively nearby, only the potentially chiral atoms at the
ends of the chains should be able to contribute optical activity.
Perhaps one day a suitable method to demonstrate this will be
developed.

RING–OPENING POLYMERIZATION

An early example of stereoselectivity in ring–opening
polymerization was the patent issued to Pruitt and Baggett (1955)
disclosing that certain modified ferric chloride catalysts
converted racemic propylene oxide to an exceptionally high
molecular weight crystalline polymer. Price and Osgan (1956)[4]
showed that this polymer had the same X-ray pattern as the
optically-active polymers prepared from pure R-(or S-)-propylene
oxide. This proved that the Pruitt and Baggett catalysts (and
many others discovered since) were stereoselective in that some
catalyst sites selectively polymerized the R-monomer, others (in
equal numbers) the S-monomer.

An interesting facet of these studies was the revelation that
R-, S- and isotactic RS-poly(propylene oxide) of equivalent
molecular weights had identical X-ray patterns and melting
points. This means that R- and S-poly(propylene oxide) form what
is known as a racemic solid solutions, i.e., the chains are

---

*It is important to recognize that, while from our view from
outside the molecule, we consider all the "asymmetric" centers in
an isotactic polymer to have the same configuration, from the
viewpoint of the atoms in the polymer chain, the "asymmetric"
centers on either side have opposite configurations.

isomorphous and either the R- or the S-chain can fit equally well
into the crystal lattice. X-ray spacing has suggested that the
crystal structure involves an extended zig-zag chain with the
methyl groups on alternate sides.

A chain of opposite configuration arranged in the same backbone
sequence would have the methyl groups on opposite sides. In order
for the methyl groups to fit in the same positions in the crystal
lattice, the chains of opposite configuration could accomodate
such a requirement by having the backbone sequence opposite, i.e.,
they would interchange a methylene group and an ether oxygen.

In the analogous case of propylene sulfide polymers,
Spassky[5] has found that the R- and S-polymers have a
significantly different melting point from their 50-50 mixture.
This behavior is expected from the formation of a racemic
compound, which may melt either higher or lower than the pure R-
or S- components.

In contrast to propylene polymerization, where no asymmetric
center exists in the monomer and those in the polymer are
pseudoasymmetric, for propylene oxide a genuine chiral center
exists in both monomer and polymer. This posed the question as to
whether the ring-opening reaction proceeds with inversion or
retention of configuration. Investigations by Vandenberg with
cis- and trans-2-butene oxides[6] and by Price and Spector with
cis- and trans-1,2-di-deuteroethylene oxides[7] established
unequivocally that ring-opening always occurred with inversion,
whether the polymerization proceeded by anionic, cationic or
coordination mechanisms. In other words, the ring opening
propagation step proceeds by the S_N2 mechanism at the carbon
atom to which the growing chain becomes attached.

This led to a question of the nature of the process in
coordination polymerization that gave the amorphous fraction of
the polymer, even from R- or S-propylene oxide. By degrading such

polymers to dimer glycol, it was shown that significant amounts (20-40%) of diprimary and disecondary glycols were formed. These could arise only from head-to-head units in the polymer. Quantitative correlation of the optical activity of the amorphous polymer with the head-to-head fraction showed that every ring-opening at the asymmetric carbon proceeded with inversion of configuration.

The conclusion is that the iron and zinc catalysts studied had two very different and separate catalyst sites. At one, there was a high degree of selectivity for coordinating only with R- (or with S-) monomer and of propagating strictly by ring-opening exclusively at the primary carbon, leaving the chiral center in its original configuration. At the other, coordination occurs relatively randomly with either R- or S-monomer and ring-opening can occur to a considerable extent at the secondary carbon as well as the primary. The enhanced reactivity at the hindered secondary carbon suggests that this site may have much more cationic character in the oxonium ring which is the reactive growing end of the polymer chain in cationic polymerization.

It is clear that cationic epoxide polymerization is promoted by methyl substitution. Ethylene oxide polymerizes very poorly with cationic catalysts, producing low molecular weight oils. 2-Butene oxide is readily polymerized by cationic catalysts, as is tetramethylethylene oxide. It is, of course, well known that methyl groups tend to stabilize cationic intermediates.

It should be made clear that the intermediate in cationic epoxide polymerization is <u>not</u> a carbocation but an oxonium ion.

However, any tendency to increase the cationic character at the carbon atom undergoing nucleophilic attack by monomer, would facilitate such attack. Certainly, putting a formal positive charge on the oxygen serves this purpose.

An interesting example of a similar oxonium intermediate occurs in the case of the copolymerization of benzenediazooxide and tetrahydrofuran[8].

(A)   (B) $\xrightarrow[h\nu]{k_i}$ $+N_2$

k_{pA} $+A$

$\xleftarrow[+B,-N_2]{k_{pB}}$ A*

The considerable resonance stabilization of N_2^+ by O^- in (A)
is not present in A* so $k_{pB} \gg k_i$.

Selectivity in the polymerization of t-butylethylene oxide
using potassium t-butoxide as catalyst proved to be surprising.
Whereas propylene oxide gave atactic polymer and aryl glycidyl
ethers gave isotactic polymers, t-butylethylene oxide polymerized
reluctantly to give a crystalline polymer different from the
isotactic polymer formed with coordination catalysis. Our first
guess that it was syndiotactic proved to be incorrect when
degradation to dimer glycol produced roughly equal amounts of
erythro- and threo-1°,2°-glycol. We proposed a mechanism[9]
involving chelation of the potassium ion not only with the end
oxyanion but with the two adjacent ethers rationalizing the
formation of alternate isotactic and syndiotactic sequences (which
we christened isosyn sequences). These studies also showed that
ring-opening of t-butylethylene oxide by t-butyloxyanion occurred
much more rapidly than the subsequent propagation reaction,
indicating that either the steric hindrance of the neopentyl-like
oxyanion or the postulated coordination of the potassium cation
with the adjacent ether links (or both) slowed down propagation.

It seems evident that a similar slowdown in propagation in
coordination polymerization at a severely hindered oxygen atom may
be responsible for the useful examples of chiral selectivity in
epoxide polymerization by optically active coordination-
catalysts. Tsuruta[10], using zinc-borneol catalysts, and
Spassky[5], using zinc t-butylethylene glycolate catalysts, have

shown that RS-propylene oxide will give isotactic polymer with
excess of one optical antipode, leaving an excess of the other in
the unreacted monomer. Since many less-hindered chiral alcohols
gave no such stereoelective selection, it suggests that perhaps
the hindered chiral alcoholates, which do not readily propagate,
keep their chiral center at the active site, thus influencing
preferential selection of one antipode of the monomer over the
other.

REFERENCES

1) C. S. Marvel and C. E. Denoon, Jr., J. Amer. Chem. Soc., 62,
   3499 (1938).
2) T. Alfrey, Jr., and C. C. Price, J. Poly. Sci., 2, 101 (1947).
3) T. Otsu and B. Yamada, Kobunshi, 15, 141 (1965); N. Gaylord
   and A. Takahashi, Polymer Letters, 7, 443 (1969); H. G. Elias,
   "Macromolecules", pp 782-783, Plenum Press, New York 1977.
4) C. C. Price and M. Osgan, J. Amer. Chem. Soc., 78, 4787 (1956).
5) N. Spassky, P. Dumas, M. Sepulchre and P. Sigwalt, J. Poly.
   Sci. Polymer Symposia, 52, 327 (1975). See also H. Tadokoro,
   "Structural Studies of Optically Active and Racemic Isotactic
   Polymers", presented at IUPAC Macromolecular Symposium,
   Florence, Italy (September 1980), Preprints, 1, 58.
6) E. J. Vandenberg, J. Amer. Chem. Soc., 83, 3538 (1961);
   J. Polymer Sci., B2, 1085 (1964).
7) C. C. Price and R. Spector, J. Amer. Chem. Soc., 88, 4171
   (1966).
8) T. Kunitake and C. C. Price, J. Amer. Chem. Soc., 85, 761
   (1963); J. K. Stille, P. Cassidy and L. Plummer, J. Amer.
   Chem. Soc., 85, 1318 (1963).
9) C. C. Price, M. K. Akkapeddi, B. T. DeBona and B. C. Furie,
   J. Amer. Chem. Soc., 94, 3964 (1972).
10) S. Inoue, T. Tsuruta and J. Furukawa, Makromol. Chem., 53, 215
    (1962); 79, 34 (1964).

COORDINATION POLYMERIZATION*

E. J. Vandenberg

Research Center
Hercules Incorporated
Wilmington, Delaware 19899

SYNOPSIS

A new, broad-based method of polymerization, commonly referred
to as "Coordination Polymerization" was first recognized about 25
years ago as the probable mechanism of the newly-discovered Ziegler,
transition metal-based, low pressure ethylene polymerization
catalyst. Rapidly this system was found to apply to other olefins,
diolefins, epoxides, and vinylethers. The author's involvement in
this area from the inception is reviewed, including some unpublished
aspects of his work as well as his extension of the area to cyclic
ethers with non-transition metals such as Al and Mg. Some of the
commercial aspects of this work are discussed. The limited
knowledge of the mechanism aspects of this unusual polymerization
method is reviewed and directions indicated for future investigations. Some of the author's recent unpublished work in the
epoxide area is presented, including the first preparation and
description of the properties of high molecular weight, isotactic
polyglycidol, an unusual water-soluble polymer.

INTRODUCTION

I am greatly honored to have been selected as the 1981 recipient
of the American Chemical Society award in polymer chemistry.

---

*Hercules Research Center Contribution No. 1737

11

I want to make it clear at the outset that my modest achieve-
ments have depended on important contributions by many others at
Hercules in many areas - Research, Development, Engineering, Patent,
Analytical, especially my direct associates, technicians and
secretaries. To some extent then, this award is a recognition of my
company for providing a favorable research atmosphere for myself and
my associates over the years.

Most of my work over the last 35 years has been in the polymer
field, especially in the field of addition polymerization, and
mostly directed toward either improved processes, new products, or
developing an understanding of some of the known polymerization
methods. I have also done some important work in the polymer
modification area and also some even in the organic synthesis area.
However, for the topic of this symposium I picked the specific field
of coordination polymerization. This new addition polymerization
method was first recognized about 25 years ago as the probable
mechanism of the newly-discovered, revolutionary, Ziegler,
transition metal-based, low pressure ethylene polymerization
catalyst. Rapidly this system was applied to other olefins, and
diolefins[1,2] to yield stereoregular polymers, many which had not
been made before. Ultimately this development was found useful for
the synthesis of small molecules via cyclization reactions,
metathesis, etc. However, I will limit my comments to addition
polymerization.

Professor C. C. Price[3], organizer of this Symposium,
proposed the coordination polymerization mechanism for Ziegler
catalyzed olefin polymerization and, at the same time, for the
earlier patent method of Pruitt and Baggett[4] for propylene oxide
polymerization with an iron catalyst. I subsequently reported
numerous non-transition metal coordination catalysts for epoxide
polymerization[5] and for vinyl ether polymerization[6]. Thus,
the coordination polymerization method is a broad one - not just
limited to Ziegler catalysts with olefins and diolefins - and is
applicable to cyclic polar monomers and some polar vinyl-type
monomers. It is then these broader aspects of coordination
polymerization that this Symposium is directed to.

Since I was involved in this area almost from its inception
and since some of Hercules' commercial developments involve this
type of polymerization, it appeared particularly appropriate to
have this Symposium on the topic of coordination polymerization.

This paper will review some of our earlier polymer studies which preceded our work on coordination polymerization, then cover the general aspects of the coordination polymerization method, with emphasis on our own contributions, and then present some of our recent work on epoxide coordination polymerization.

EARLY POLYMER STUDIES

In 1947, I discovered some very unusual redox systems for emulsion polymerization which consisted of organic hydroperoxides such as cumene hydroperoxide in combination with a reducing agent such as a reducing sugar and a trace of an iron compound[7]. Polymerizations could be done in 15 minutes which previously took 15 hours to do. Hercules was involved in selling a special disproportionated rosin soap for the emulsion polymerization of butadiene and styrene. These new redox systems were much less affected by adventitious impurities and inhibitors. Indeed, one

$$ROOH + Fe^{++} \longrightarrow RO\bullet + OH^- + Fe^{+++}$$

$$Fe^{+++} + \text{Reducing Sugar} \longrightarrow Fe^{++}$$

Fig. 1. Hydroperoxide Redox Systems

could even use ordinary rosin soaps which contained a variety of conjugated resin acids which inhibited and/or retarded the ordinary persulfate type systems. We filed for patents on these new highly active redox systems and were subsequently awarded patent coverage[8,9]. We speculated at the time that this unusual behavior was due to two factors. First, as shown in Figure 1, the asymmetry of the hydroperoxide causes it to react with a reducing agent to reduce just the OH half of the molecule to OH^- and give an RO• radical. Second, this RO• radical is apparently fairly unreactive so that it does not react with impurities as readily as say a persulfate radical, hydroxyl radical, or perhaps some other peroxide radicals such as from benzoyl peroxide.

Independently, Kolthoff and Medalia at the University of Minnesota[10] and Fryling, et al at Phillips Petroleum [11] contributed importantly to these new redox systems. The Phillips workers, of course, had a strong commercial interest in this area of polymerization and recognized the value of this to butadiene-styrene polymerization. These higher activity initiating systems permitted them to lower the temperature of polymerization from 50°C to 5°C and this gave a better rubber[12]. Phillips and ultimately other rubber producers adapted these systems for this use. Indeed, organic hydroperoxides have been used and are still being used in making SBR by low temperature emulsion polymerization. Hercules developed and still has a substantial business selling organic hydroperoxides to the emulsion polymerization industry.

In the course of trying to find applications for our new hydroperoxide redox initiating systems, in 1951 we discovered systems which polymerized ethylene at low temperature (5°C) and low pressure (40 atm.), i.e., below the critical temperature (9.9°C). The best system found utilized an insoluble ferrous complex of ethylene diamine tetracetic acid with cumene hydroperoxide in 97.5% tert-butyl alcohol (2.5% H_2O), but rates (0.2%/hr.), conversion (2%) and molecular weights (M_w = 8,000) were low. The product, however, was more linear than the commercial high pressure product, since it contained much fewer CH_3 groups by IR, was higher melting (125°C), and was more crystalline by x-ray. In addition, the product was a brittle wax even though we did get the molecular weight up to 20,000 by raising the pressure to 300 atm. This is obviously the linear high density polyethylene which was unknown at that time but which was very easily made a few years later at much higher molecular weight from the Ziegler development and the Phillips chromium oxide on silica catalyst[2]. This polymer has become a very important, large volume commercial product. Our molecular weights were too low for us to recognize its value and, in any event, our process was too poor to be useful. In the course of this work we considered trying a combination of an organometallic such as a Grignard reagent with an organic chloride and a transition metal compound, such as $CoCl_2$, $NiCl_2$, $CrCl_3$, or $FeCl_3$, as a source of free radicals for initiating polymerization. This combination had been reported by Kharasch in his work on free-radical reactions[13], and it falls within the scope of the subsequently discovered Ziegler catalyst. But, the organometallic was not readily available, and we never tried it. So we missed what might have been a logical development of Ziegler catalysis. Indeed, there are examples in the literature where some element of this new coordination polymerization method was used, but in a very imperfect way, and the authors never recognized that it was a new and different method of polymerization[1].

ZIEGLER POLYMERIZATION

General

In mid-1954, Hercules acquired a license to Ziegler's new low pressure ethylene polymerization method which involved a new type of catalyst consisting of organometallics, such as alkyl aluminums, combined with transition metal compounds, such as $TiCl_4$. In October of 1954 I was assigned to do scouting work in this new area of catalysis. Of course, we recognized the importance of Ziegler's discovery. Needless to say, these were very exciting times in the laboratory. We did not know very much about the mechanism of the polymerization method, although it was obvious that some new type of polymerization method was involved. We did speculate that growth of the polymer on the transition metal might occur, although there was considerable debate in the next ten years as to whether the polymer chain was growing on the transition metal or the aluminum metal[1].

Propylene Polymerization

Just one week after I began to work in the field, I tried, on the same day, polymerizing propylene and also using hydrogen in an ethylene polymerization. In the case of propylene I obtained 92 milligrams of an unusual, insoluble, crystalline polymer which I found to be a fiber former. I concluded that it would be attractive for a further study. In my experiments with hydrogen we ran three experiments plus a control and found that it definitely affected the molecular weight (Table 1). The effect was really quite small and I suppose that a lot of people would have considered this inconsequential. However, we did pursue both of these leads.

In the case of propylene, ordinary Ziegler catalysts as described by Ziegler gave very low yields of crystalline polymer; indeed, most of the polymer was an amorphous rubber or even a viscous liquid. Very quickly in 1955, we found catalysts and

Table 1. First Experiments on the Effect of Hydrogen on the Polymerization of Ethylene with a Ziegler Catalyst

Mole % H_2	$\dfrac{\eta_{sp}}{c}$	M_w x 10^{-6}
0	21.0	1.33
0.6	17.4	1.11
1.1	16.4	1.05
2.3	15.8	1.01

conditions which gave high yields of the crystalline polymer
(Figures 2 and 3). Shortly after, Natta and co-workers reported
that the polymer was stereoregular and coined the term
"isotactic"[14]. We did file a patent application on crystalline
polypropylene as a new composition of matter, and were involved in
a subsequent interference[2]. However, our dates were not early
enough and ultimately we were removed from the interference. In
the case of hydrogen, we tried it on propylene and found it to be
a very effective way of controlling molecular weight. We filed a
patent application which also got involved in an interference. In
this case, we prevailed and obtained a patent on our discovery[15].
This method of controlling molecular weight is the major method
used in the world today to control the molecular weight of
isotactic polypropylene. The production capacity for
polypropylene has grown rapidly so that it is now one of the major
synthetic polymers with a worldwide capacity of about 14 billion
lbs./yr.[16].

Catalysts Giving High Stereoregularity – $TiCl_3 \cdot nAlCl_3$.
In the case of propylene polymerization, our new catalysts
utilized crystalline $TiCl_3$ solids which contained large amounts
of $AlCl_3$, up to one mole per Ti. These $TiCl_3$ solids were
obviously mixed crystals of $TiCl_3$ and $AlCl_3$ since the x-ray
pattern gave no evidence for $AlCl_3$ per se. Both the $TiCl_3$-1.0
$AlCl_3$ and the $TiCl_3$-0.33 $AlCl_3$ mixed crystals gave new and
different x-ray patterns which appeared related to the known

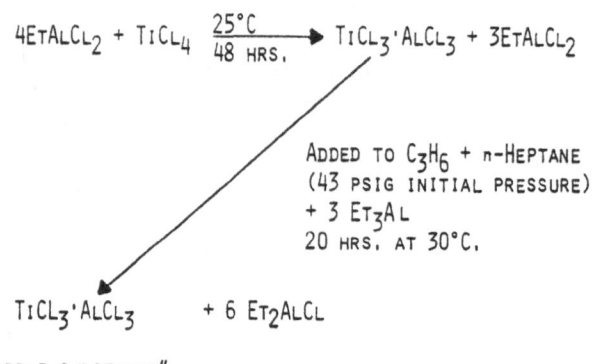

$$4 EtAlCl_2 + TiCl_4 \xrightarrow[\text{48 HRS.}]{25°C} TiCl_3 \cdot AlCl_3 + 3 EtAlCl_2$$

ADDED TO C_3H_6 + n-HEPTANE
(43 PSIG INITIAL PRESSURE)
+ 3 Et_3Al
20 HRS. AT 30°C.

$TiCl_3 \cdot AlCl_3$ + 6 Et_2AlCl

"PURPLE SUBSTANCE"
δ FORM

-95% YIELD, 86% CONV. OF HEPTANE-INSOLUBLE POLYPROPYLENE,
$\frac{\eta_{SP}}{C}$ = 29

Fig. 2.   First High Yield Method for
Isotactic Polypropylene[17,18]

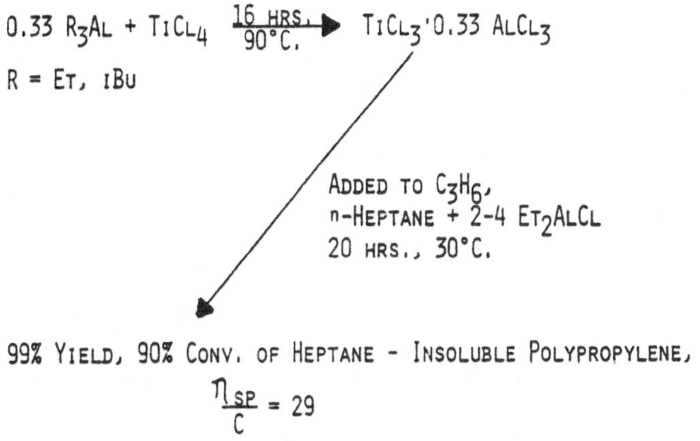

$$0.33\ R_3Al + TiCl_4 \xrightarrow[90°C.]{16\ HRS.} TiCl_3·0.33\ AlCl_3$$

R = Et, iBu

ADDED TO C_3H_6,
n-HEPTANE + 2-4 Et_2AlCl
20 HRS., 30°C.

99% YIELD, 90% CONV. OF HEPTANE - INSOLUBLE POLYPROPYLENE,

$$\frac{\eta_{sp}}{C} = 29$$

Fig. 3. Second High Yield Method for Isotactic
Polypropylene[21,22] - Stoichiometric Catalyst

purple α-TiCl$_3$ rather than the known brown β-TiCl$_3$. Color
varied from reddish purple to violet. The TiCl$_3$-1.0 AlCl$_3$
product from EtAlCl$_2$-TiCl$_4$ reaction was called "purple
substance" by our x-ray specialist to differentiate it from the
known α-TiCl$_3$. This type proved important in the commercial
manufacture of polypropylene. We initially made this new form by
the room temperature reaction of TiCl$_4$ with an excess (2-4
mole/Ti) of EtAlCl$_2$. By using this pre-reaction product with
triethyl aluminum in certain ratios where the triethyl aluminum
reacted with the excess ethyl aluminum dichloride to form diethyl
aluminum chloride (Figure 2), we obtained high yields of
stereoregular polypropylene[17,18]. This was our first high yield
method of making isotactic polypropylene and may indeed be the
first high rate, high yield method discovered by anyone. This
method did involve the new so-called "purple substance" form of
TiCl$_3$. Hercules obtained a composition of matter patent on this
form[19]. Years later, Natta and co-workers[20] reported their
excellent crystallographic studies of the various modifications of
TiCl$_3$ and called this form δ. It turns out that this δ form
was the basis of all the early methods of making polypropylene
commercially[16].

Our second and even higher yield (99%) and higher rate method
of making isotactic polypropylene was discovered in 1955 as a
logical development from our first high yield catalyst. If
TiCl$_3$·1.0 AlCl$_3$ gave this excellent result, how would lower
levels of AlCl$_3$ work? We prepared such a lower AlCl$_3$ catalyst
by the stoichiometric reaction of trialkyl aluminum with TiCl$_4$,
i.e., one alkyl aluminum bond per TiCl$_4$ to form TiCl$_3$-0.33
AlCl$_3$ as the sole product and then using this crystalline solid
with Et$_2$AlCl as an activator (Figure 3)[21,22]. This TiCl$_3$

solid was also a mixed crystal of $TiCl_3$ and $AlCl_3$ (no x-ray
evidence for $AlCl_3$) and had a different x-ray pattern from the
known "violet" α-$TiCl_3$ but is identical to that finally
reported as the γ form by Natta[20], especially when R = iBu in
Figure 3. This method for making isotactic polypropylene was very
effective but the physical form of the product was less favorable
and thus was not developed commercially by Hercules. Similar
processes have been described in the patent literature[23].

However, this stoichiometic type of $TiCl_3$ has become of
major importance in the commercial manufacture of
polypropylene[1], based on the reduction of $TiCl_4$ by aluminum
metal, as first reported by Ruff and Neumann[24]. Improved
methods of making this type of $TiCl_3$ were found and developed by
Exxon[25], Stauffer[26], and others[1]. In particular, Tornqvist
et al[26] found that ballmilling was necessary for good
polymerization rates. In addition, ballmilling converted the
$TiCl_3$ to the δ form[20,27]. Also, the polypropylene was
obtained in better physical form. Tornqvist et al also recognized
the higher activity and other desirable aspects of the
$TiCl_3 \cdot nAlCl_3$ mixed crystals, particularly $TiCl_3 \cdot 0.33$
$AlCl_3$, over $TiCl_3$, and ascribed this to a specific effect of
$AlCl_3$. This was my belief too. This conclusion is somewhat
suspect today since Solvay has reported $TiCl_3$ catalysts of four
fold higher activity and with higher stereospecificity than with
prior $TiCl_3 \cdot nAlCl_3$ catalysts using especially pure δ-type
$TiCl_3$, prepared by special ether extraction and $TiCl_4$
treatment of a Et_2AlCl-$TiCl_4$ reaction product[28]. We were
not able to publish, except by patent, in this area of great
commercial importance.

Commercial Development Aspects. We went on to make large
quantities of isotactic polypropylene and found that it had an
interesting combination of properties, although it was not
immediately taken up by Hercules since much of our effort was
directed toward developing a Ziegler-type process for high density
polyethylene. I personally felt that polypropylene would
ultimately be an important polymer based on the low cost of
propylene. Finally, after about a year or two, Hercules developed
a commercial process in collaboration with Hoechst from our
lead[29]. As you know, Hercules in late 1957 became the first
U.S. producer and is presently the world's largest producer of
polypropylene.

This new Ziegler method of polymerization has proved to be
very useful commercially and is used on a very large scale for
olefin (ethylene and propylene) and diolefin (butadiene and
isoprene) polymerization. There has been a tremendous amount of
work done on olefin and diolefin catalysts from both the practical
and mechanistic point of view and yet we really do not understand

the intimate details of this new type of polymerization very
well. However, this area is still of great interest. There is
much to be learned from a mechanism point of view and there are
new products and improved processes of importance to be discovered
and developed.

Mechanism Aspects

Let us consider now some of the general and mechanism aspects
of the Ziegler olefin polymerization method.

The general concepts involved in coordination polymerization
are indicated schematically in Figure 4 as compared with the
previously known methods of addition polymerization, that is, free
radical, anionic, and cationic. I have shown this for ethylene
polymerization on a titanium metal center. The key point here is
that the polymer chain is covalently attached to the transition
metal. The monomer and/or polymer chain end is activated in some
way to permit the monomer to insert at the metal-carbon bond and
propagate the chain. A favorite hypothesis is that the olefin
coordinates with the metal at a vacant coordination site and this
step in some way enables the olefin to insert in the polymer
chain. In the case of the well-known free radical, cationic, and
anionic polymerization methods, the chain end is relatively free,
although in the field of ionic polymerization one can obviously
have relatively tight ion pairs. Admittedly, there is a big
debate over whether the coordination step does occur and indeed
there are a number of other mechanisms proposed for this

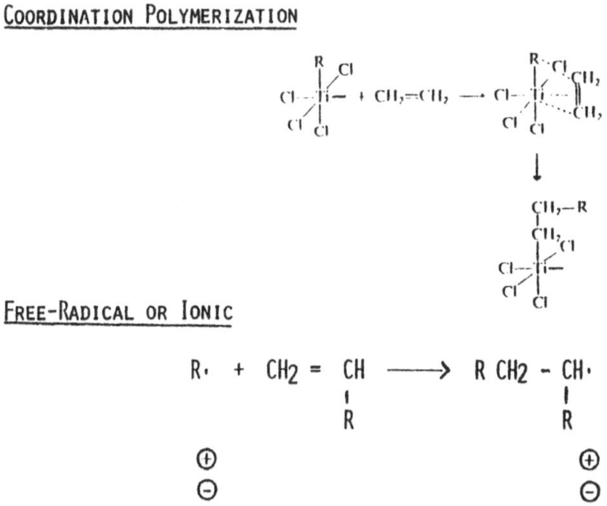

COORDINATION POLYMERIZATION

FREE-RADICAL OR IONIC

$$R\cdot \; + \; CH_2 = CH \; \longrightarrow \; R\,CH_2 - CH\cdot$$
$$\qquad\qquad\quad R \qquad\qquad\qquad R$$

⊕   ⊕

⊖   ⊖

Fig. 4.   Coordination Polymerization Compared to
           Prior Addition Polymerization Methods

polymerization[1,30-32]. But in lieu of any real firm evidence, I believe it is still worthwhile to consider this polymerization as involving some type of coordination mechanism as indicated.

Some of the characteristics of olefin coordination polymerization appear reasonable based on the proposed mechanism with a relatively hindered propagation site due to the covalent attachment of the growing chain to the transition metal (Figure 5). Thus, ethylene with the least steric hindrance to coordination propagates 10-100x faster than propylene[33] even though propylene with its electron donating methyl group should have a higher electron density on its double bond, a factor which should enhance coordination. Also, the copolymerization of ethylene (M_1) with propylene is quite unfavorable with a soluble V catalyst with ethylene entering the copolymer 20x faster than propylene, $r_1 = 20$[34]. Contrast this with the favorable free radical copolymerization of methyl acrylate (M_1) with methyl methacrylate where $r_1 = 0.3$ and the methyl group substituent favors copolymerization[35]. Larger side group olefins copolymerize even much less readily with ethylene than propylene, e.g., butene-1 being 4x less favorable than propylene[36]. Copolymerization reactivity ratios vary with the transition metal, supporting the hypothesis that chain growth is on the transition metal[1]. Also, stereoregular polymerization can go very well

PROPAGATION

ETHYLENE 10-100X FASTER THAN PROPYLENE.

COPOLYMERIZATION

C = C WITH C = C r_1 = 20 FOR V
(M1)   C

VARIES WITH TRANSITION METAL

IN FREE-RADICAL SYSTEM:

C = C   WITH   C = C r_1 = 0.3
    COOME   COOME
(M1)

STEREOREGULAR POLYMERIZATION - READILY FEASIBLE WITH C = C
                                                          R

MOLECULAR WEIGHT - VERY HIGH - USUALLY IN THE MILLIONS

Fig. 5.   Characteristics of Olefin Coordination Polymerization

with 1-olefins and diolefins, although part of the needed steric
hindrance, in this case, comes from having the transition metal
propagation site as part of a solid surface. Thus, one of the
most important properties of the growing site in olefin and other
coordination polymerizations is the steric hindrance at the
growing chain end.

Polar Monomers

    Early in scouting efforts on Ziegler catalysts, other monomers
were studied, particularly polar monomers, to see whether we could
obtain other unusual products. We had some success with vinyl
chloride[37] and methyl methacrylate[38] and this work was
explored by others at our laboratory[39,40]. However, we had most
success with vinyl ethers and found a variety of new catalysts
which proved especially effective for making stereoregular,
isotactic polymers from vinyl ethers[6]. Some of these catalyst
are shown in Figure 6. Initially, we devised a vanadium based
catalyst used for vinyl methyl ether, consisting of the
stoichiometric reduction product from 0.33 Et_3Al plus $V Cl_4$
(SV catalyst) pretreated with iBu_3Al (PSV catalyst) used with
additional R_3Al[41]. Subsequently, we found better, new,
non-transition metal catalysts[42-44] such as the combination of
an aluminum alkoxide or alkyl with sulfuric acid[42]. With these
new catalysts, we were able to obtain much more highly
stereoregular polymers than Schildknecht had obtained some ten
years before with ordinary Lewis acid type catalysts. We
concluded that these new catalysts were examples of cationic
insertion or coordination catalysts as contrasted with Ziegler
catalysts which by that time were considered as examples of

TRANSITION METAL BASED

    PRE-TREATED STOICHIOMETRIC VANADIUM (PSV) CATALYST

$$\left[2\ iBu_3Al\ +\ VCl_3 \cdot 0.33\ AlCl_3 \right]\ +\ R_3Al$$

AL BASED

$$\left[R_3Al\ +\ H_2SO_4 \right]\ +\ R_3Al$$

R = ALKYL OR RO-

Fig. 6. Vinyl Ether Coordination Catalysts

anionic coordination catalysts. Figure 7 shows the type of
mechanism that we proposed in a paper on that subject in 1963[6].
As you can see, the mechanism postulated is different than
considered with olefins in that coordination involves the ether
group, not the double bond, of the monomer and also the ether
group of the last unit in the growing chain. We concluded that
this was an example of coordination polymerization because the
polymerizations were many orders of magnitude slower, even at or
near ambient temperature, than typical cationic polymerizations,
molecular weights were unusually high and stereoregularity could
be very high. Although this work was very interesting
scientifically, it has not yet led to any important commercial
developments.

EPOXIDE POLYMERIZATION

New Catalysts and Polymers

In 1957, we noted that some of our newly developed transition
metal-based vinyl ether catalysts were somewhat similar to a new
epoxide polymerization catalyst, an aluminum isopropoxide-$ZnCl_2$
combination, which Price and Osgan[45] had just found for
polymerizing propylene oxide. Thus, we tried polymerizing an
epoxide, specifically epichlorohydrin, with some of our new vinyl
ether catalysts and some Ziegler catalysts and obtained a small
amount, 9 mg., of crystalline polyepichlorohydrin, a new polymer,
at least to us. Then through exploring this development, with a
catalyst consisting of triisobutyl aluminum and vanadium
trichloride, we found that if we left out the transition metal
component we got even better results. In addition, the results

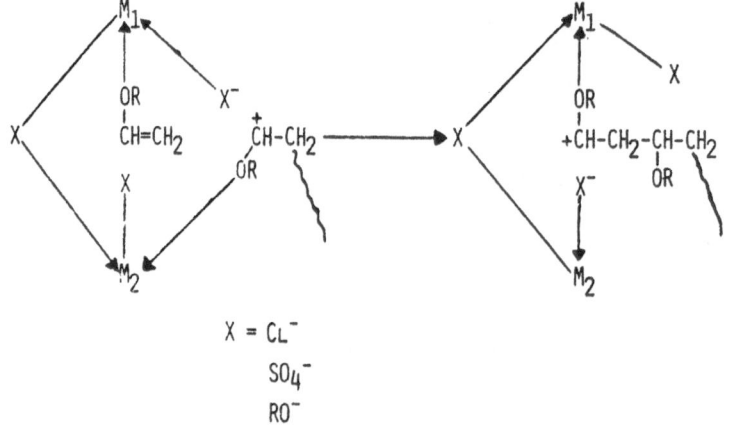

Fig. 7.   Mechanism of Vinyl Ether Coordination Polymerization

were different in that the major product was now an amorphous rubber. This turned out to be another new polymer. In a course of exploring this lead, we found that we could not repeat our previous work with a new batch of monomer. Indeed in our first work, we had used a nearly empty bottle of epichlorohydrin that had been around the laboratory for some time. We immediately suspected that there was probably water in the epichlorohydrin. Thus, we reacted triisobutyl aluminum with water and found it to be very effective catalytically. In this way, we discovered this very important class of new catalysts based on the reaction of aluminum alkyls with water, as shown in Figure 8[46]. In the course of studying the mechanism of polymerization with this new catalyst, we speculated that it was a coordination catalyst, i.e., the epoxide coordinated with the metal prior to its insertion in the chain. We thought that we could prove this hypothesis by adding a chelating agent such as acetyl acetone to block the coordination site on the aluminum and thus prevent polymerization. Much to our surprise, when we tried this, we obtained an even better catalyst. How this improved catalyst works is still not thoroughly understood even today, except to propose that aluminum becomes 5 or 6 coordinate in the propagation step. This catalyst which I will refer to as the chelate catalyst has indeed proven to be one of the most versatile catalysts in the field of epoxide polymerization as well as for the polymerization of some other monomers[47,48]. At the same time, we did discover a wide variety of other useful catalysts for epoxides. Many of these were based on the reaction of organometallics such as zinc

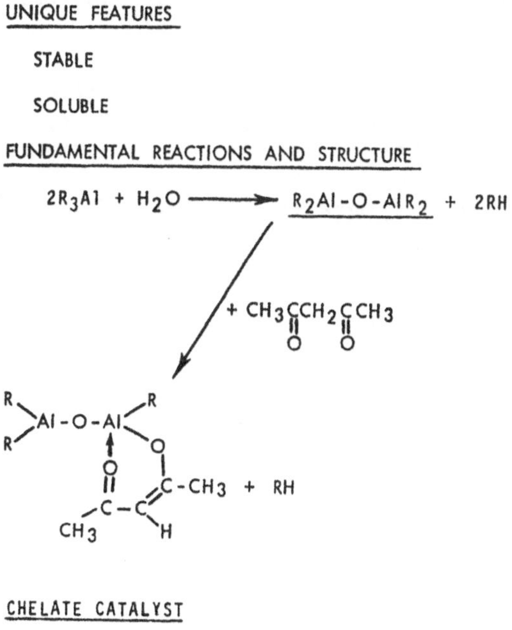
UNIQUE FEATURES
STABLE
SOLUBLE
FUNDAMENTAL REACTIONS AND STRUCTURE
2R3Al + H2O → R2Al-O-AlR2 + 2RH
+ CH3CCH2CCH3
C-CH3 + RH
CH3
H
CHELATE CATALYST

Fig. 8. R_3Al-H_2O Catalysts

or magnesium with di or polyfunctional additives such as water, polyols, ammonia, amines, diacids, diketones[5], etc. As a result of the discovery of these catalysts, particularly the aluminum alkyl-water-chelate catalyst, we were able to make a large number of new high molecular weight polymers, particularly in the epoxide area[5,48].

Commercial Aspects

A few of the more interesting monosubstituted epoxides that were polymerized to both amorphous and crystalline high molecular weight polymers with coordination catalysts are illustrated in Table 2. The epichlorohydrin amorphous homopolymers and amorphous copolymers with ethylene oxide are especially interesting oil resistant rubbers[49,50]. These rubbers are being sold in multimillion pound quantities by Hercules Incorporated under the trademark Herclor and by B. F. Goodrich Chemical Company, our licensee, under the trademark Hydrin. The properties and advantages of these rubbers make an interesting story. Briefly, the copolymer is especially unusual - having a unique combination of properties. It has the oil resistance of nitrile rubbers combined with the good rubber properties and environmental resistance of neoprene. Indeed, in the last year or two these rubbers have begun to grow at a much greater rate, and recently Hercules announced a plant expansion to bring its capacity up to 24 million lbs./year, and more recently Goodrich has announced plans for a similar expansion. Thus, these elastomers are becoming important in the marketplace.

The copolymers of propylene oxide with small amounts of an unsaturated epoxide, such as allyl glycidyl ether, can be sulfur vulcanized to vulcanizates with excellent low temperature and

Table 2. High Polymers from Monosubstituted Epoxides

	Amorphous	Crystalline (m.p., °C)
$-CH_2-CH-O-$ 　　　$\vert$ 　　CH_2Cl		120
$-CH_2-CH-O-CH_2-CH_2O-$ 　　　$\vert$ 　　CH_2Cl	Solvent-resistant rubbers	—
$-CH_2-CH-O-CH_2-CH-O-$ 　　　$\vert$　　　　　$\vert$ 　　CH_3　　　　CH_2 　　　　　　　　　$\vert$ 　　　　　$O-CH_2-CH\!=\!CH_2$	S-Vulcanizable rubber	—
$-CH_2-CH-O-$ 　　　$\vert$ 　　$CH\!=\!CH_2$	Rubber	74

dynamic properties similar to natural rubber combined with good
ozone resistance and good heat resistance. This polyether
elastomer is available commercially through Hercules Incorporated
under the registered trademark Parel and found to be especially
useful as a specialty elastomer in applications such as automotive
engine mounts[47].

Mechanism Aspects

    Some of our mechanism studies on epoxide polymerization will
now be reviewed briefly. Much of our work in this area grew out
of our study of the polymerization of the symmetrical
disubstituted epoxides, cis and trans-2,3-epoxybutane, as shown in
Table 3.

    The trialkylaluminum-water catalyst polymerized both isomers
essentially instantaneously at dry ice temperature. This rapid
polymerization is to be contrasted with the fairly slow
polymerization of propylene oxide with this catalyst, even at room
temperature. With this catalyst, the cis-oxide gives only an
amorphous high molecular weight rubber; trans-oxide gives only a
crystalline high molecular weight plastic with a melting point of
$100°C$[51]. On the other hand, the alkylaluminum-H_2O-
acetylacetone catalyst polymerizes the cis-oxide slowly at $65°C$
and under these conditions, the product is largely a crystalline
polymer with a melting point of $162°C$[51]. This crystalline
polymer from the cis-oxide is totally different from the
crystalline polymer obtained from the trans-oxide with the
aluminum alkyl-water catalyst, having a completely different x-ray
pattern, different solubility properties, and a higher melting
point. The chelate-modified catalyst, however, caused very little

Table 3. High Polymers from Cis and Trans-2,3-Epoxybutanes[51]

$$CH_3CH-CHCH_3$$
$$\diagdown\diagup$$
$$O$$

	Catalyst	Temp. (°C)	Polymerization Rate	Polymer
cis	R_3Al-H_2O	-78	Instantaneous	Amorphous Rubber
trans	R_3Al-H_2O	-78	Instantaneous	Crystalline Polymer M.P., $100°C$
cis	R_3Al-H_2O- Acetylacetone	65	Slow	Crystalline Polymer M.P., $162°C$
trans	R_3Al-H_2O- Acetylacetone	65	Very Little	

polymerization of the trans-oxide at 65°C, indicating that the
polymerization site is too hindered to accept the chemically
similar but more hindered trans-oxide. In fact, this result alone
is our best evidence that the chelate catalyst is a coordination
catalyst involving polymerization at a hindered aluminum site.

The aluminum alkyl-H_2O catalyst, on the other hand, based on
the rapid polymerization observed at -78°C, is behaving with these
readily cationically polymerized monomers as a cationic catalyst.
Indeed, the aluminum alkyl-H_2O catalyst, as well as many other
epoxide catalysts, can polymerize by either cationic or
coordination routes. If the epoxide is easily polymerized
cationically, then the cationic process will win out. Thus, if
the trans-oxide does not polymerize with a particular epoxide
catalyst, this result provides good evidence that this catalyst is
solely a coordination catalyst, such as our chelate catalyst[5].

Our ability to polymerize the 2,3-epoxybutanes with different
catalysts to different products offered us a unique opportunity to
learn more about the mechanism of the epoxide polymerization.
Since both the main chain carbon atoms in the monomer unit of the
polymer are asymmetric, the stereochemistry of the monomer unit in
the chain would tell us whether the ring opening carbon retained
or inverted configuration during polymerization. We explored this
area by discovering a new method of cleaving polyethers with butyl
lithium to monomer, dimer and trimer diols and then establishing
the stereochemistry of these different units[5] (Table 4). The
polymers prepared by either coordination or by a cationic
mechanism were all formed by inversion of configuration of the
ring opening carbon atom. Indeed, based on our subsequent work as
well as that of Price, we have concluded that all polymerizations

Table 4.   Stereochemistry of 2,3-Epoxybutane Polymers

MONOMER		POLYMER	STEREOCHEMISTRY OF DIMER UNIT
CH_3(S) (S) H (R) (R) C——C H O CH_3 TRANS	CATIONIC → COORDINATION →	CRYST. TRACE, CRYST.	RS-RS ⎫ RS-RS ⎬ MESO-DIISOTACTIC
CH_3(R) (S) CH_3 C——C H O H CIS	CATIONIC → COORDINATION →	AMORPH. CRYST.	RR-SS DISYNDIOTACTIC RR-RR ⎫ SS-SS ⎬ RACEMIC DIISOTACTIC

of mono and disubstituted epoxides with all known anionic,
cationic, or coordination catalyst occur with complete inversion
of configuration of the ring opening carbon atom[48]. This result
appears to be a very important first principle of epoxide
polymerization[5].

Based on this result, I concluded that the mechanism of the
coordination polymerization of epoxides must involve two metal
atoms in order to make it possible to obtain a rearward attack on
the epoxide. In the mechanism proposed (Equation 1),

(1)

the epoxide is coordinated to one aluminum atom prior to its
attack by the growing chain on the other aluminum atom. The
coordination bonds in the catalyst structure are needed to move
the growing polymer chain from one metal to an adjacent one
without altering the valence of the metal. Tsurata et al have
recently determined the structure of a crystalline zinc catalytic
complex which fits in with our general proposal and indicates how
stereoselective polymerization occurs with propylene oxide[52].

The chelate catalyst also polymerizes aldehydes[53] and
oxetanes[47,54]. Presumably, both go by a coordination route.
Indeed, our work on the aldehydes, reported only by patent, may be
the first authenticated example of aldehyde polymerization via a
coordination route. Our work with trimethylene oxide is also the
first reported example of the coordination polymerization of this
monomer. This work was first reported in the patent
literature[54] and further details given in 1974[47].

The coordination copolymerization of trimethylene oxide (TMO)
with ethylene oxide (EO) with the chelate catalyst is interesting
since it gives two completely different composition copolymers
(Table 5). A copolymerization with epichlorohydrin gives similar
results[47]. This result clearly indicates that the catalyst
contains sites varying in copolymerization ability and thus
presumably varying in steric hindrance. We have seen this type of
behavior in the epoxide area also; for example, in the
copolymerization of epichlorohydrin with ethylene oxide. Even
though one takes great care to make a uniform one-to-one mole
epichlorohydrin-ethylene oxide copolymer, the product often

Table 5.   Copolymerization of TMO with Ethylene Oxide

Conditions: TMO, g.   8 g.
              EO, g.   2 g.
              Toluene, ml.   92 ml.
              Et$_3$Al-0.5 H$_2$O-0.5 AA,
                 millimoles Al   4 mm

Product (Heptane-and CH$_3$OH-insol.):

	% Conv.	η_{inh}	% TMO
water-insol.	1.4	11.0	44
water-sol.	12	6.5	8

Major product indicates EO 70 times more reactive than TMO.

contains a few percent of a copolymer which is very high in ethylene oxide[48].

    Another example where the restricted steric requirements of the polymerization site play an important role is in the copolymerization of cis-1,4-dichloro-2,3-epoxybutane with ethylene oxide[48].   This copolymerization is very unfavorable with the ethylene oxide being about 400 times more reactive than the dichloro epoxide.   This result is probably due to the much larger steric requirement of the dichloro-2,3-epoxy butane monomer as well as the decreased basicity of the oxirane with two electron withdrawing chloromethyl groups.   Therefore, this monomer would have much less tendency to coordinate at the propagation site for steric and electronic reasons and thus be less able to participate in a coordination polymerization.   In any event, the nature of the growing site in these coordination polymerizations, both steric and electronic factors, is clearly very important in determining their polymerization and their copolymerization behavior.

Other Catalyst Considerations

    In addition to the organoaluminum-H$_2$O catalyst systems discussed, we also discovered a large number of catalysts by modifying organozinc and organomagnesium compounds with polyreactive additives[5].   The performance of all these organometal derived catalysts varies a great deal with the monomer, diluent, and conditions.   The catalyst preparation must be optimized for the type of product desired (atactic, isotactic, molecular weight, etc.).   Usually it is best to have a coordinating diluent such as an ether present during the catalyst preparation.   In general, epichlorohydrin polymerizes best with the aluminum catalysts because of its reactive chlorine which apparently leads to catalyst-destroying side reactions with the more basic catalysts.

Unsymmetrical disubstituted epoxides such as isobutylene oxide (2,2-dimethyloxirane) are more difficult to polymerize with coordination catalysts but do polymerize coordinatively with certain zinc and magnesium catalysts. Isobutylene oxide polymerizes very readily with cationic catalysts and tends to give molecular weights which are too low to be useful, even at -78°C. We have reported that the R_2Mg-NH_3 catalyst[55] polymerizes isobutylene oxide at 30°C to a high molecular weight (η_{inh} 2.4), head-tail polymer which has different properties from previously reported poly(isobutylene oxides) and which is very interesting as a film and fiber former[56]. We have also found that the R_2Zn-H_2O catalysts are effective. Kamio et al have reported that amine-modified R_2Zn-H_2O catalysts are also very useful[57]. These polymerizations are evidently coordinated anionic systems. The catalyst must be carefully tuned to eliminate cationic sites. In addition, the coordination site may have to be somewhat larger to accommodate the more (or differently) hindered isobutylene oxide molecule and avoid chain-limiting side reactions. This result can apparently be more easily achieved with Zn and Mg than with Al.

The magnesium catalysts also are especially useful for making extremely high molecular weight poly(ethylene oxide)[58]. A variety of catalysts based on calcium, presumably coordination catalysts which are not based on organometallics, have been reported by Hill et al of Union Carbide for making high molecular weight poly(ethylene oxide)[59,60]. Recent modifications of these catalysts are reported to be exceptionally effective for ethylene oxide polymerization giving up to 1800 g of polymer/g of Ca[61].

We have also studied the polymerization of episulfides with coordination catalysts and presented evidence that the same principles found for epoxides apply here also[55]. Another important result of this work was the preparation of an optically active crystalline polymer from the optically inactive, meso cis-2-butene episulfide (cis-1,2-dimethyl thiirane) using a Et$_2$Zn-2.0 1-menthol coordination catalyst. This experiment proved that this crystalline polymer was racemic diisotactic, and is the first example of an unusual asymmetric polymer synthesis. Spassky, Momtaz and Sigwalt have described this as an enantiogenic process and report further details in a separate report in this book.

We do not have a very good insight into what occurs at the coordination site, only some general ideas. This conclusion is supported by the fact that epoxide coordination catalysts often

are not highly effective. One usually needs in the range of 1/2
to 1% of such catalysts to get good conversions to high polymer.
This is obviously an order of magnitude or more greater than one
might expect from the high molecular weights being made, assuming
one chain from each proposed catalyst moiety. This is to be
contrasted with the olefin polymerization area with coordination
catalysts where, in more recent years, extremely high catalyst
mileages have been obtained based on the presumably active
transition metal component - the amounts going up as high as a
million grams of polymer per gram of transition metal[62,63]. One
of the important areas for future development in epoxide
polymerization will be more efficient and perhaps even more
versatile coordination catalysts.

We need to know more about the intimate details of the
mechanism of coordination polymerizations of cyclic ethers, cyclic
sulfides, olefins and diolefins. It is quite probable that other
monomer types will be coordinately polymerizable to yield new
homopolymers and especially new, useful copolymers, perhaps even
across monomer classes, as we found with epoxides and oxetanes.
Tsuruta et al have shown that epoxides can be coordinately
copolymerized with anhydrides[64] and CO_2 [65]. Keii reports in
a separate article in this book a living propylene coordination
polymerization system and its application to block copolymer
synthesis.

Can we design polymerization sites which will accommodate
larger monomer units and facilitate copolymerizations in both the
olefin and epoxide areas? It is of interest that some of our
epoxide catalysts, particularly the Al alkyl–H_2O products, have
been found to be unique organometallic activators for transition
metal catalysts for olefin polymerization[63,66].

POLYETHERS WITH REACTIVE SIDE CHAINS

<u>General</u>

We have done extensive work on making high molecular weight
polyepoxides with reactive side chains such as esters, amines,
carboxyls and hydroxyl. Many of the catalysts that contain
reactive organometallic groups do not work well or very little in
the presence of appreciable amounts of such reactive groups. We
have reported successful polymerizations with ester groups such as
in ethyl glycidate[67] and in glycidyl methacrylate, with our Al
catalysts[68] and with <u>tert</u> amino containing epoxides such as
3-diethylamino-1,2-epoxypropane using some of our modified
organomagnesium catalysts[69]. Tucker[70] and Cantor et al[71]
have reported the preparation of high molecular weight homopolymer

and copolymers from cyanoethyl glycidyl ether using our chelate catalyst. Vogl et al[72] have reported recently preparing a variety of polyethers with ester groups using our chelate catalyst.

We have also introduced a variety of reactive groups, such as carboxyl, carboxylate, sulfoxide, amino, isothuronium, thioether, thiosulfate, azide, and phosphonate, into polyethers by the reaction of polyepichlorohydrin with various nucleophiles[73]. The latter route is somewhat limited because of the ready base cleavage of polyepichlorohydrin[74].

Polyglycidol

A hydroxyl side group as in glycidol, 3-hydroxy-1,2-epoxy-propane, is especially interesting. Some years ago we made a high polymer from glycidol by blocking the hydroxyl groups with trimethylsilyl groups, polymerizing this trimethylsilyl glycidyl ether (TMSGE) with the chelate catalyst to a very high molecular weight polymer and then converting it by methanolysis to a new polymer, high molecular weight polyglycidol, an interesting water-soluble elastomer[75] (Figure 9). This product was amorphous by x-ray, presumably atactic, with a Tg of -8°C. Subsequently, some related work on glycidol and TMSGE polymerization has appeared in the literature. Sandler and Berg

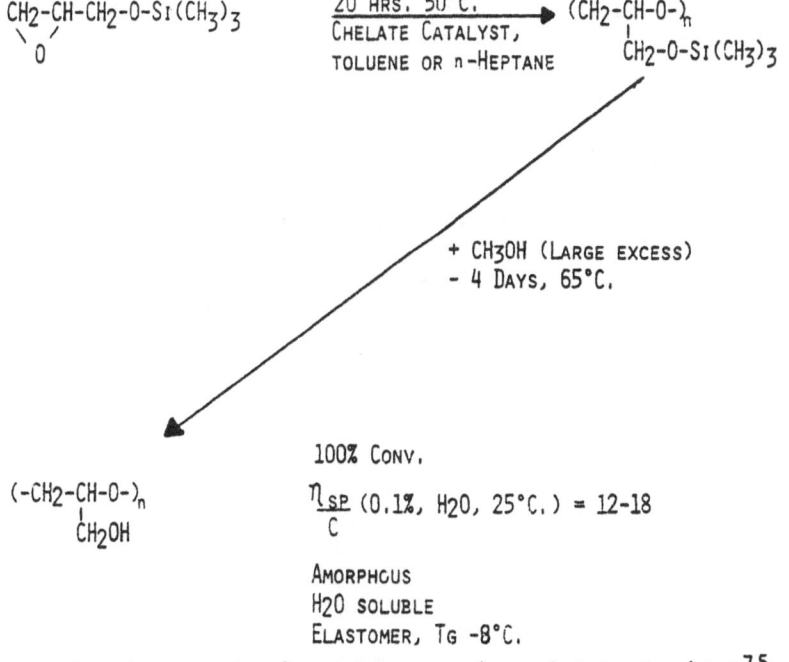

Fig. 9. Synthesis and Properties of Polyglycidol[75]

reported obtaining a liquid polyglycidol of very low molecular
weight (445) by polymerizing glycidol with various base
catalysts[76]. Tsuruta et al reported polymerizing glycidol and
TMSGE with various organometallic catalysts and then methanolyzing
the TMSGE polymer to polyglycidol[77]. With $Et_3Al-0.7\ H_2O$ and
$n-Bu_2Mg$ as catalysts, glycidol gave the low molecular weights
($[\eta]$, intrinsic viscosity, up to 0.07 in CH_3OH) reported by
Sandler, and TMSGE gave somewhat higher viscosity polyglycidol
($[\eta]$ up to 0.4). In a study of other organometallic catalysts
on TMSGE, the Et_2Zn-H_2O catalyst gave the highest molecular
weight polyglycidol with $[\eta]$ in CH_3OH of 1.8. The viscosity
of Tsuruta's polyglycidol is thus an order of magnitude lower than
we obtained with the chelate catalysts. Rates of polymerization
and conversion also appear lower. Tsuruta also reports that the
polyglycidol from glycidol polymerization has the same IR spectra
as that obtained via TMSGE.

Our polyglycidol as well as that reported in the literature
appeared to be atactic. Presumably, the isotactic form should be
a high melting crystalline polymer since its hydrocarbon analog,
isotactic poly(propylene oxide) is a crystalline polymer with a
melting point of about 70°C. Over the years we have made numerous
attempts with different catalysts to try to make this crystalline
isotactic polyglycidol without a great deal of succcess. Tsuruta
et al[77] reported on studies to initiate the asymmetric selective
polymerization of TMSGE with a Et_2Zn-1-menthol catalyst and
related catalysts, but obtained only a low degree of optical
activity in polyglycidol ($[\alpha]^{25}$ + 0.26) and a low degree of
stereoregularity. These authors did establish the absolute
configuration of the asymmetric carbon in optically active
polyglycidol in relation to its sign of rotation. Recently, we
have explored this problem further and used two approaches to the
synthesis of isotactic polyglycidol. In the first approach we
prepared the pure R enantiomer of TMSGE from the well known
R-glycidol[79] in order to utilize the method first used by
Price[3] to make isotactic poly(propylene oxide), i.e., the base
catalyzed polymerization of R-propylene oxide. The second
approach was to make isotactic poly(tert-butyl glycidyl ether) and
then convert it to polyglycidol.

In our work on polymerizing R-TMSGE, we quickly found that a
simple base catalyst such as the KOH used by Price[3] could not
be utilized because of the reactivity of the trimethyl silyl ether
group. The side reactions are complex and still under study. We
then polymerized R-TMSGE with our chelate catalyst under our usual
conditions for racemic TMSGE. The polyglycidol obtained was an
optically active polymer but it was still an amorphous elastomer,
Eq. 2,

$$
\underset{\substack{(R) \\ CH_2 - CH\text{-}CH_2\text{-}O\text{-}Si(CH_3)_3 \\ \diagdown \!\! O \!\! \diagup}}{} \quad \xrightarrow[\substack{\text{(1) Chelate} \\ \text{Catalyst} \\ \text{(2) methanolysis}}]{50\,°C} \quad \underset{\substack{(R) \\ (-CH_2\text{-}CH\text{-}O\text{-})_n \\ | \\ CH_2OH}}{} \quad (2)
$$

$[\alpha]_D^{25} = +6.8\pm1.0 \ (5\%, \ CH_2Cl_2)$ $[\alpha]_D^{25} = +5.5\pm0.5 \ (10\%, \ D_2O)$

$n_{sp}/C \ (0.1\% \ H_2O, \ 25°C) = 23$

The ^{13}C-NMR of the poly(R-glycidol) (Figure 10A) shows clearly that it is indeed isotactic polyglycidol with mainly three sharp, equal peaks at the expected locations for the three carbons of polyglycidol, based on our off-resonance decoupling studies and published ^{13}C-NMR of poly(p-chlorophenyl glycidyl ether)[80]. Contrast this with the ^{13}C-NMR of the poly(RS-glycidol) made under the same conditions where there are peaks due to other than isotactic sequences (Figure 10B). The CH group shows three peaks due to isotactic (δ 79.71), heterotactic (δ 79.59), and syndiotactic (δ 79.50) triads, and the chain CH$_2$ shows two peaks due to isotactic (δ 68.88) and syndiotactic (δ 68.60) dyads, as expected for an amorphous, atactic polymer. However, the ^{13}C-NMR of our products from racemic and R enantiomer monomer show minor peaks around δ 78.2, 70.3, 70.8, and 61.2 which, based on the work of Dworak and Jedlinski[80], is due to some head-to-head and tail-to-tail polymerization – ca. 6–10% – in addition to the usual head-to-tail polymerization. The chelate catalyst was previously found to give substantial amounts – 20–30% – of these abnormal units[5] in propylene oxide polymerization. This poly(R-glycidol) does, however, crystallize on stretching, relaxing, restretching, and then holding taut. In this way we have obtained a highly crystalline, highly-oriented x-ray pattern (Figure 11) which loses most of its crystallinity on relaxing. Atactic polyglycidol does not crystallize on stretching.

Poly(R-glycidol), nearly free of head-to-head and tail-to-tail polymerization, was prepared by polymerizing R-TMSGE at 5°C with a modified (1.5 acetyl acetone per Al) chelate catalyst (Figure 12). The specific rotation was increased about 10% as one would expect. The ^{13}C-NMR indicates that the polymer contains no more than about 1% abnormal units and is thus quite pure. However, this product was still an amorphous elastomer which crystallizes by the same stretching procedure given above, but does not give any better or more stable x-ray pattern.

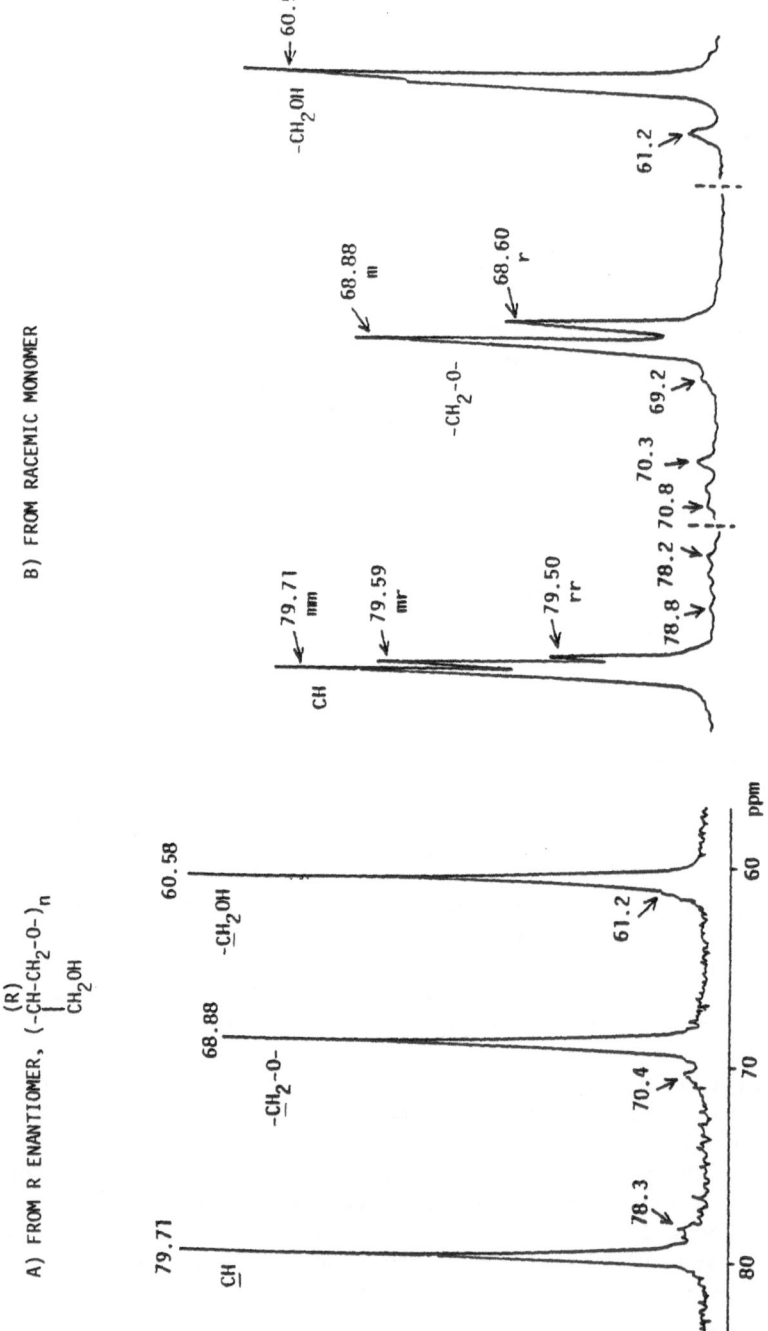

Fig. 10. ^{13}C–NMR of Polyglycidols(a) in D_2O

a) PREPARED IN TOLUENE AT 50°C. WITH CHELATE CATALYST. 62.9 MHz

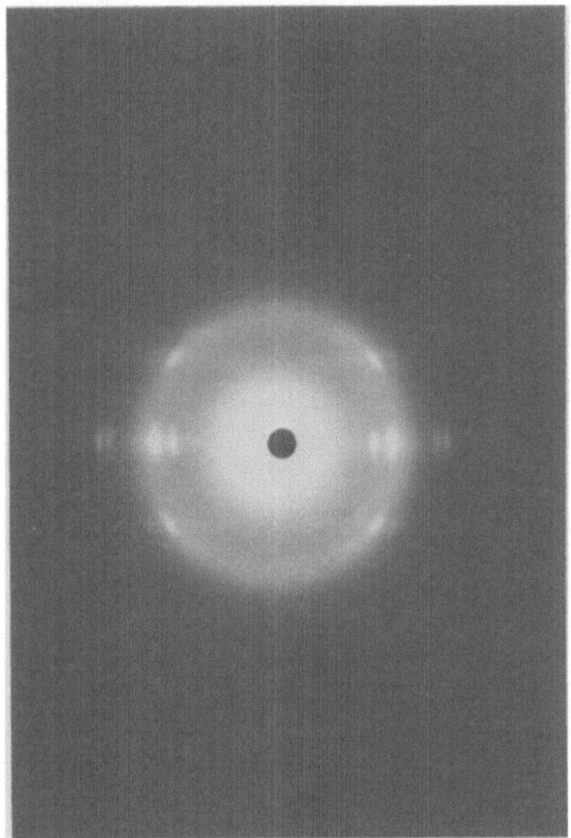

Fig. 11.   X-Ray Pattern of Oriented Isotactic Poly(R-Glycidol)

We have approached the preparation of isotactic polyglycidol
by alternate ways.   One approach was to use a catalyst which would
polymerize TMSGE directly to isotactic polymer.   For this we tried
the Teyssie-Osgan Al-Zn catalyst, specifically $(n-BuO)_2Al-O-Zn-O-Al(O\ n-Bu)_2$ kindly supplied by Professor Teyssie, which
gives isotactic polymer with propylene oxide[81].   However, the
Teyssie-Osgan catalyst does not polymerize this particular
epoxide.   Apparently, there are side reactions with the
trimethylsilyloxy group which destroy this catalyst.   Clearly
then, this is further evidence that our chelate catalyst is a
particularly preferred species for epoxide polymerization and
copolymerization.

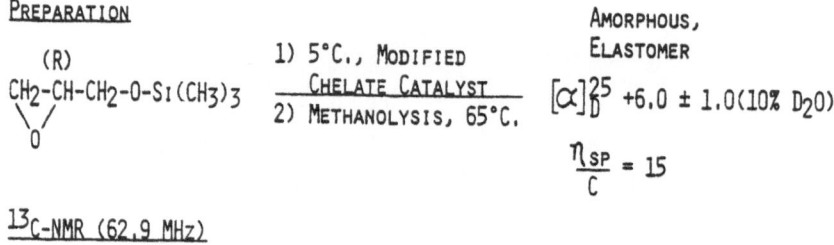

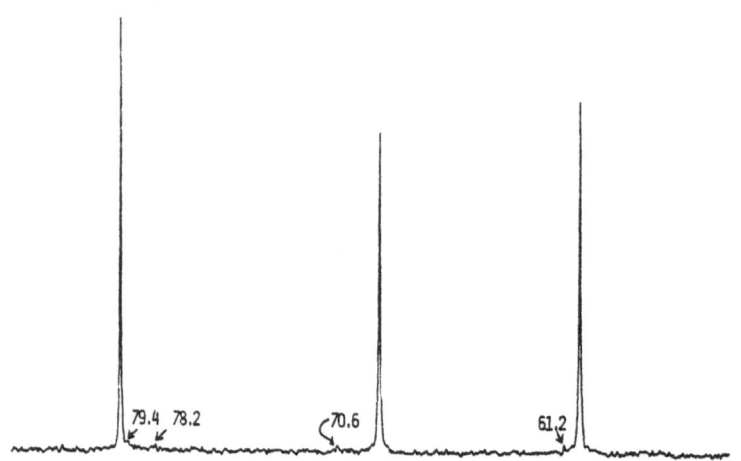

Fig. 12.   Poly(R-glycidol) Free of Head-Head Units

On the other hand, the Teyssie-Osgan catalyst readily
polymerizes the carbon analogue of TMSGE, tert-butyl glycidyl
ether, to high molecular weight crystalline polymer, presumably
isotactic poly(tert-butyl glycidyl ether) (Figure 13). After
purification by recrystallization from acetone at 5°C, the
crystalline polymer is tough, moderately crystalline and fairly
high melting (124°C). This polymer was then converted to
poly(glycidyl acetate) by treatment with toluene sulfonic acid
(5%) in 77-23 acetic anhydride-acetic acid (43 hrs. reflux) and
then saponified to polyglycidol. This product, after drying at
80°C, was an amorphous elastomer. The ^{13}C-NMR (Figure 14) is
very similar to that which we obtained from the pure enantiomer of
TMSGE, with essentially no evidence for head-to-head or
tail-to-tail units but with some minor amount of other tactic
sequences. This polymer appears to be at least 98% pure based on
the ^{13}C-NMR. We have obtained this lower molecular weight
isotactic poly(R,S-glycidol) in a crystalline form, e.g., it was
crystalline when first isolated by drying in vacuo at room
temperature. In addition, this crystalline polyglycidol (and, in

general, polyglycidol dried in vacuo at room temperature) contains 6-12% residual volatiles, presumably H_2O. This H_2O can be removed by drying in vacuo at 80°C and on cooling the polymer does not recrystallize. However, a film prepared by compression molding the dried polymer at 125°C and cooling was moderately crystalline, had a different x-ray pattern than the oriented poly(R-glycidol), melted at 53-60°C (based on birefringence disappearance) and did not recrystallize on cooling.

The crystallization differences between the isotactic poly(R-glycidol) and the isotactic poly(R,S-glycidol) are probably related, at least in part, to the large difference in molecular weight of these samples. Further work is needed to clarify this picture. However, this work does clearly indicate that isotactic polyglycidol is a low melting crystalline polymer which is reluctant to crystallize. Also, polyglycidol does pick up water from the atmosphere very readily and extreme precautions are needed to avoid this problem.

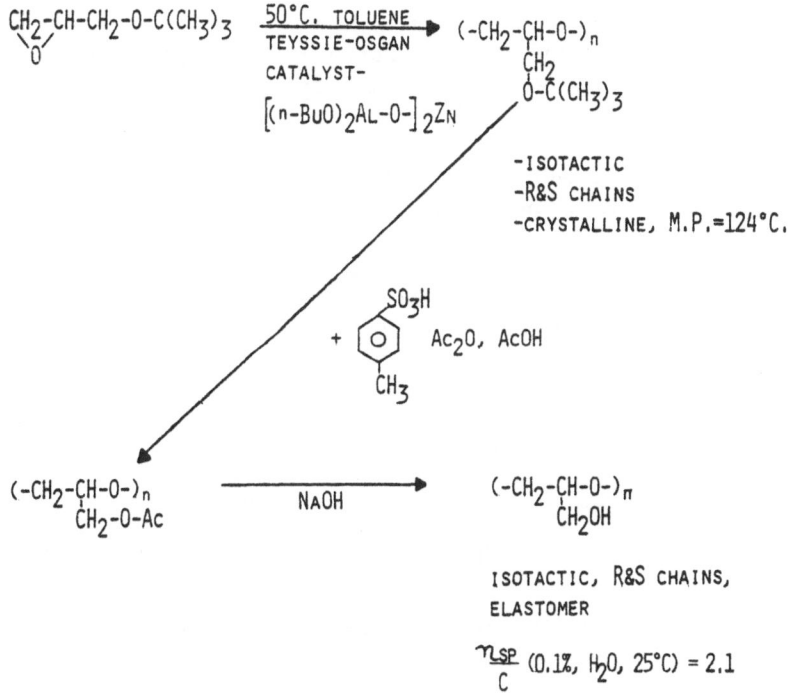

Fig. 13.   Alternate Route to Isotactic Polyglycidol

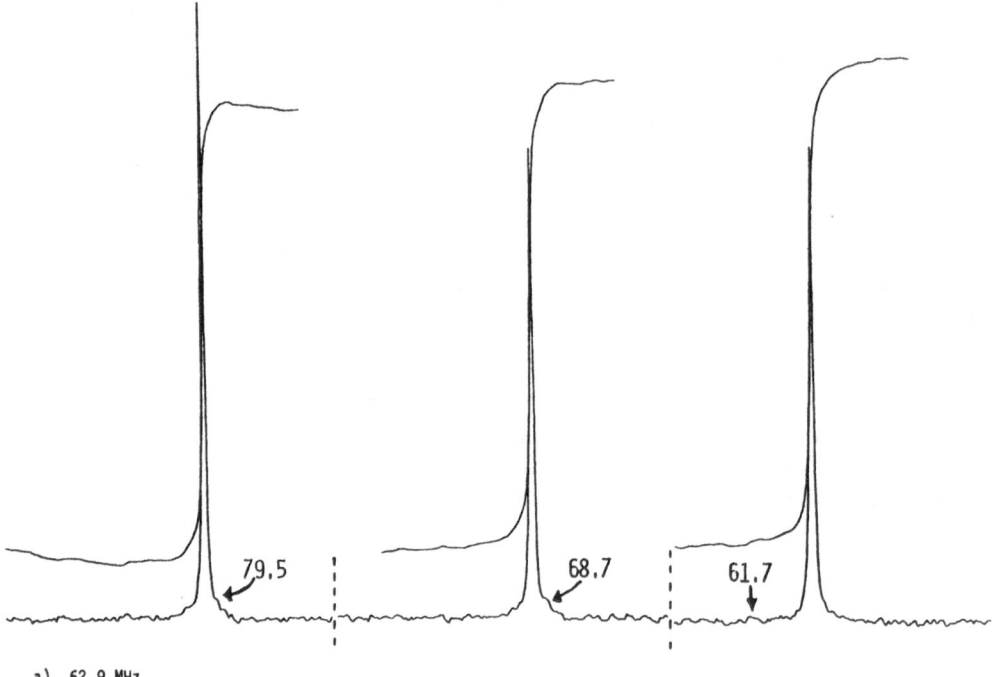

a) 62.9 MHz

Fig. 14. ^{13}C-NMR$^{(a)}$ of Isotactic Poly(R,S-glycidol) from
Isotactic Poly(R,S-tert-butyl glycidyl ether)

Small amounts of H_2O (6-12%) in polyglycidol markedly
influences its glass transition, lowering it to -30 to -40°C from
the usual -8 to -12°C in the dry state, based on DSC studies.
This effect of H_2O on Tg and the reluctance of isotactic
polyglycidol to crystallize may be due to the ready formation of
intramolecular hydrogen bonds via 5 or 6 membered rings as shown
in Figure 15.

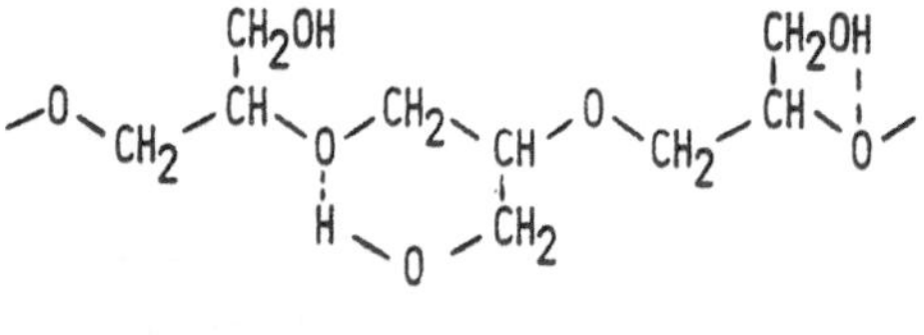

Fig. 15.   Intramolecular Hydrogen Bonding in Polyglycidol

This work on polyglycidol is obviously just a preliminary
progress report to indicate that we may expect some new, useful
products from epoxide coordination polymerization work in progress.

CONCLUSIONS

In summary, a great deal remains to be learned in the field of coordination polymerization, both of olefins and of epoxides. One can expect a great deal of important work on the mechanism aspects of these polymerizations and the ultimate development of more effective catalysts as well as the continuing preparation and development of new products and processes from this area.

ACKNOWLEDGEMENT

It is obviously impractical to cite all the individuals that have contributed to my work in Coordination Polymerization over the years. I am indebted particularly to my technician, Ms. Mary Morris, who has been very helpful in all my laboratory studies, to Mrs. M. Chris and R. Kridler for the x-ray studies on polyglycidol, and to Dr. W. Freeman and Dr. R. Crecely (University of Delaware) for the ^{13}C-NMR work on polyglycidol. I am also indebted to Professor C. C. Price for frequent stimulating discussions on the mechanism and other aspects of my work and, also, for organizing the Symposium at which this talk was presented.

REFERENCES

1. J. Boor, Jr., "Ziegler-Natta Catalysts and Polymerizations", Academic Press, Inc., New York, 1979.
2. E. J. Vandenberg and B. C. Repka, "Ziegler-Type Polymerizations", in C. E. Schildknecht, Ed., "Polymerization Processes", Wiley, 1977, Chapter 11.
3. C. C. Price and M. Osgan, J. Amer. Chem. Soc. 78, 4787 (1956).
4. M. E. Pruitt and J. M. Baggett (to Dow Chemical Co.), U.S. Patent 2,706,181, filed June 6, 1952, issued April 12, 1955.
5. E. J. Vandenberg, J. Polym. Sci., Part A1, 7, 525 (1969).
6. E. J. Vandenberg, J. Polym. Sci., Part C, 207 (1963).
7. E. J. Vandenberg and G. E. Hulse, Ind. Eng. Chem. 40, 932 (1948).
8. E. J. Vandenberg (to Hercules Incorporated), U.S. Patent 2,648,657-8, filed April 12, 1947, issued August 11, 1953.
9. E. J. Vandenberg (to Hercules Incorporated), U.S. Patent 2,648,655-6, filed November 24, 1948, issued August 11, 1953.
10. I. M. Kolthoff and A. I. Medalia, J. Polym. Sci. 5, 391 (1950).
11. C. F. Fryling, S. H. Landes, W. M. St. John and C. A. Uraneck, Ind. Eng. Chem. 41, 986 (1949).
12. W. A. Schulze, W. B. Reynolds, C. F. Fryling, L. R. Sperberg, and J. E. Troyan, India Rubber World 117, 739 (1945).

13. M. S. Kharasch, F. L. Lambert, and W. H. Urry, J. Org. Chem. 10, 298 (1945).

14. G. Natta, P. Pino, P. Corradini, F. Danusso, E. Mantica, G. Mazzanti and G. Moraglio, J. Amer. Chem. Soc. 77, 1708 (1955).

15. E. J. Vandenberg (to Hercules Incorporated), U.S. Patent 3,051,690, filed July 29, 1955, issued August 28, 1962.

16. P. Galli, "Polypropylene - A Quarter of a Century of Increasing Successful Development" presented at IUPAC Macromolecular Symposium, Florence, Italy (September, 1980). Preprints 1, 10. To be published by Pergamon Press.

17. E. J. Vandenberg (to Hercules Incorporated), U.S. Patent 3,058,963, filed April 7, 1955, issued October 16, 1962.

18. E. J. Vandenberg (to Hercules Incorporated), U.S. Patent 2,954,367, filed July 29, 1955, issued September 27, 1960.

19. E. J. Vandenberg (to Hercules Incorporated), U.S. Patent 3,108,973, filed February 16, 1961, issued October 29, 1963.

20. G. Natta, P. Corradini and G. Allegra, J. Polym. Sci. 51, 399 (1961).

21. E. J. Vandenberg (to Hercules Incorporated), U.S. Patent 3,261,821, filed December 31, 1959, issued July 19, 1966.

22. E. J. Vandenberg, "Isotactic Polypropylene" in E. L. Wittbecker, Ed., Macromolecular Syntheses, 5, 95 (1974).

23. L. W. Gamble, A. W. Langer, Jr., and A. H. Neal (Esso Research and Engineering Company), U.S. Patent 2,951,045 (August 30, 1960); D. E. Winkler and K. Nozaki (Shell Oil Company), U.S. Patent 2,971,925, (February 14, 1961); E. V. Fasce and R. J. Fritz (Esso Research and Engineering Company), U.S. Patent 2,999,086 (September 5, 1961); D. Kaufman and B. H. McMullen (to National Lead Company), U.S. Patent 3,109,822, filed May 28, 1957, issued November 5, 1963.

24. O. Ruff and F. Neuman, Z. Anorg. Chem. 128, 81 (1923).

25. E. Tornqvist, C. W. Seelbach and A. W. Langer, Jr. (to Esso Research and Engineering Co.), U.S. Patent 3,128,252, filed April 16, 1956, issued April 7, 1964.

26. Preliminary Bulletin No. 56-6, July 9, 1957, and Manual of Procedure for the Evaluation of Olefin Polymerization Catalyst, Ziegler-Natta Type, 2nd ed., Stauffer Chemical Company, 380 Madison Avenue, New York 17, New York.

27. (a) E. G. M. Tornqvist, J. T. Richardson, Z. W. Wilchinsky and R. W. Looney, J. Catal. 8, 189 (1967).
    (b) E. G. M. Tornqvist, Ann. N. Y. Acad. Sci., 155, 447 (1969).
    (c) Z. W. Wilchinski, R. W. Looney and E. G. M. Tornqvist, J. Catal. 28, 351 (1973).
    (d) E. G. M. Tornqvist, Makromol. Chem. 177, 3437 (1976).

28. J. P. Hermans and P. Henrioulle (to Solvay & Cie), U.S. Patent 4,210,738, filed March 21, 1972, issued July 1, 1980.

29. Hercules Mixer, 41: 1-10 (January, 1958).

30. K. J. Ivin, J. J. Rooney, C. D. S. Stewart, M. L. H. Green, and R. Mahtab, J. C. S. Chem. Comm., 1978, 604.

31. R. J. McKinney, J. C. S. Chem. Comm. 1980, 490.

32. G. Fink, "The Dynamic Behavior of Soluble Ziegler-Natta Catalysts", presented at IUPAC Macromolecular Symposium, Florence, Italy (September, 1980), Preprints $\underline{2}$, 56.

33. Y. I. Yermakov and V. A. Zakharov, "Coordination Polymerization", J. C. W. Chen, Ed., p. 91, Academic Press, New York, 1975.

34. C. A. Lukach and H. M. Spurlin, "Copolymers of α-Olefins" in G. E. Ham, Ed., "Copolymerization", Wiley, 1964, Chapter IVA.

35. H. Mark, B. Immergut, E. H. Immergut, L. J. Young, and K. I. Beynon, "Copolymerization Reactivity Ratios" in G. E. Ham, Ed., "Copolymerization", Wiley, 1964, Appendix A.

36. G. A. Mazzanti, A. Valvassori, G. Sartori, and G. Pajaro, Chim Ind. (Milan) $\underline{42}$, 466 (1960).

37. E. J. Vandenberg (to Hercules Incorporated), U.S. Patent 3,422,082, filed April 19, 1956 and July 30, 1956, issued January 14, 1969.

38. E. J. Vandenberg (to Hercules Incorporated), U.S. Patent 3,316,229, filed August 12, 1966, issued April 25, 1967.

39. D. S. Breslow, D. L. Christman, H. H. Espy and C. A. Lukach, J. Appl. Polym. Sci. $\underline{11}$, 73 (1967).

40. D. S. Breslow and A. Kutner, J. Polym. Sci., $\underline{9B}$, 129 (1971).

41. E. J. Vandenberg (to Hercules Incorporated), U.S. Patent 3,284,426, filed March 18, 1957, issued November 8, 1966.

42. D. L. Christman and E. J. Vandenberg (to Hercules Incorporated), U.S. Patent 3,025,282, filed September 14, 1959, issued March 13, 1962.

43. R. F. Heck and E. J. Vandenberg (to Hercules Incorporated), U.S. Patent 3,025,283, filed September 14, 1959, issued March 13, 1962.

44. E. J. Vandenberg (to Hercules Incorporated), U.S. Patent 3,159,613, filed December 27, 1960, issued December 1, 1964.

45. M. Osgan and C. C. Price, J. Polym. Sci. $\underline{34}$, 153 (1959).

46. E. J. Vandenberg (to Hercules Incorporated), U.S. Patent 3,135,705, filed May 11, 1959, issued June 2, 1964; U.S. Patent 3,219,591, issued November 23, 1965.

47. E. J. Vandenberg and A. E. Robinson, "Coordination Polymerization of Trimethylene Oxide" in "Polyethers" (edited by E. J. Vandenberg), American Chemical Society, Washington, D.C., 101 (1975).

48. E. J. Vandenberg, Pure and Applied Chem., $\underline{48}$, 295 (1976).

49. E. J. Vandenberg (to Hercules Incorporated), U.S. Patent 3,158,580, filed March 11, 1960, issued November 24, 1964.

50. E. J. Vandenberg (to Hercules Incorporated), U.S. Patent 3,158,580, filed March 11, 1960, issued November 24, 1964.

51. E. J. Vandenberg (to Hercules Incorporated), U.S. Patent 3,065,187, filed July 26, 1960, issued November 20, 1962.

52. T. Hagiwara, M. Ishimori, and T. Tsuruta, Makromol. Chem.
    182, 501 (1981).
53. E. J. Vandenberg (to Hercules Incorporated), U.S. Patent
    3,208,975, filed December 22, 1961, issued September 28, 1965.
54. E. J. Vandenberg (to Hercules Incorporated), U.S. Patent
    3,205,183, filed December 13, 1960, issued September 17, 1965.
55. E. J. Vandenberg, J. Polym. Sci., Part A1, 10, 329 (1972).
56. E. J. Vandenberg (to Hercules Incorporated), U.S. Patent
    3,354,097, filed June 29, 1961, issued November 21, 1967.
57. K. Kamio, M. Kuwana and S. Nakada (to Nippon Carbide Kogyo
    K.K.), U.S. Patent 3,509,074 (April 28, 1970).
58. E. J. Vandenberg (to Hercules Incorporated), U.S. Patent
    3,415,761, filed March 31, 1960, issued December 10, 1968.
59. F. N. Hill, F. E. Bailey, Jr. and J. T. Fitzpatrick, Ind.
    Eng. chem. 50, 5 (1958).
60. F. N. Hill, F. E. Bailey, Jr. and J. F. Fitzpatrick (to Union
    Carbide Corp.), U.S. Patent 2,969,402, filed December 29,
    1958, issued January 24, 1961). Also, U.S. Patent 3,062,755
    (1962), and U.S. Patent 3,167,519 (1965).
61. G. L. Goeke and F. J. Karol (to Union Carbide Corp.), U.S.
    Patent 4,193,892 (1980).
62. A. Andresen, H. G. Cordes, J. Herwig, W. Kaminsky, A. Merck,
    R. Mattweiler, J. Pein, H. Sinn, H. Vollmer., Angew. Chem.
    Int. Ed. Engl. 15, 630 (1976).
63. W. Kaminsky, H. Sinn and R. Woldt, "Polymerization and
    Copolymerization of α-Olefins with Soluble Ziegler-Natta
    Catalysts of Extremely High Activity", presented at IUPAC
    Macromolecular Symposium, Florence, Italy (September, 1980).
    Preprints 2, 59.
64. S. Inoue, K. Kitamura and T. Tsuruta, Makromol. Chem. 126,
    250 (1969).
65. M. Kobayashi, S. Inoue, and T. Tsuruta, J. Polym. Sci., Chem.
    Ed. 11, 2383 (1973).
66. W. P. Long (to Hercules Incorporated), U.S. Patent 3,152,105,
    filed August 16, 1960, issued October 6, 1964.
67. E. J. Vandenberg (to Hercules Incorporated), U.S. Patent
    3,106,549, filed August 24, 1959, issued October 8, 1963.
68. E. J. Vandenberg (to Hercules Incorporated), U.S. Patent
    3,285,870, filed September 17, 1964, issued November 15, 1966.
69. E. J. Vandenberg (to Hercules Incorporated), U.S. Patent
    3,403,114, filed May 21, 1964, issued September 24, 1968.
70. H. A. Tucker (to B. F. Goodrich Company), U.S. Patent
    3,410,810, filed May 3, 1966, issued November 12, 1968.
71. S. E. Cantor, G. D. Brindell and T. J. Brett, Jr., J.
    Macromol. Sci.-Chem., A7, 1483 (1973).
72. O. Vogl, J. Muggee and D. Bansleben, Polymer J. 12, 677
    (1980).
73. E. J. Vandenberg in Kirk-Othmer: Encyl. Chem. Tech. 8, Third
    Ed. 568 (1979).
74. E. J. Vandenberg, J. Polym. Sci. Chem. Ed., 10, 2903 (1972).

75. E. J. Vandenberg (to Hercules Incorporated), U.S. Patent
     3,446,757, filed September 26, 1966, issued May 27, 1969.
76. S. R. Sandler and F. R. Berg, J. Polym. Sci., Part Al, $\underline{4}$,
     1253 (1966).
77. T. Tsuruta, S. Inoue and H. Koenuma, Makromol. Chem. $\underline{112}$, 58
     (1968).
78. P. H. Khanh, H. Koinuma, S. Inoue and T. Tsuruta, Makromol.
     Chem. $\underline{134}$, 253 (1970).
79. J. C. Sowden and H. O. L Fischer, J. Am. Chem. Soc., $\underline{64}$, 1291
     (1942); J. J. Baldwin, K. Mensler and G. S. Ponticello, J.
     Org. Chem. $\underline{43}$, 4878 (1978).
80. A. Dworak and Z. Jedlinski, Polymer 21, 93 (1980).
81. M. Osgan and Ph. Teyssie, J. Polym. Sci., Part B, $\underline{5}$, 789
     (1967).

MOLECULAR LEVEL ELUCIDATION OF STEREOSPECIFIC POLYMERIZATION OF

METHYLOXIRANE USING A WELL-DEFINED ORGANOZINC COMPLEX

Teiji Tsuruta[*], Tokio Hagiwara and Michihiro Ishimori

Department of Synthetic Chemistry
Faculty of Engineering, The University of Tokyo
Bunkyo-ku, Tokyo, Japan

Since Professor Giulio Natta postulated the enantiomorphic structure for active sites of Ziegler-Natta catalyst in the propylene polymerization[1], it has widely been believed that the steric regulation in the isotactic propagation stage is effectuated by the chiral structure around the catalyst sites. A number of proposals concerning possible structures of the catalyst site have been presented. Most of the proposed mechanism for the steric course of polymerization were based on the chiral structure around titanium atoms which were located at the surface crystal lattice of α- or γ-type of $TiCl_3$[2]. The lattice structure of the $TiCl_3$ crystal has well been elucidated in the field of crystallography[3,4,5], but isolation and characterization of a catalyst species in a well-defined form has not yet been successful in any of studies on the stereospecific polymerization of propylene in which the formation of isotactic enchainments of the monomeric units is brought about. Owing to "the third generation catalyst", polypropylene industries are achieving dramatic innovation in terms of productivity and of energy conservation. Mechanistic elucidation of the role of the catalyst components, such as $MgCl_2$ or ethyl benzoate, in determining reactivity and stereospecificity of "the third generation catalyst" will be one of the most important unsolved problems both in science and technology of the olefin polymerization.

We reported previously that the reaction mechanism for the stereospecific polymerization of methyloxirane catalyzed by zinc dimethoxide was satisfactorily explained in terms of the model of enantiomorphic catalyst sites[6]. We assumed the symmetrical dis-

---

* Present Address : Department of Industrial Chemistry,
  Faculty of Engineering, Science University of Tokyo, Kagurazaka,
  Shinjuku-ku, Tokyo, 162 Japan

tribution of d^*- and l^*- catalyst sites in the active zinc dimethoxide. The highly isotactic poly(methyloxirane) molecules are
formed at the catalyst sites having the high degree of stereospecificity. On the other hand, catalyst sites having lower degree of
stereospecificity can produce polymers having lower isotacticity.
    Mechanistic principle for steric control operative in the
stereospecific polymerization of methyloxirane can be considered
to fall in the same category as that of propylene polymerization
initiated with Ziegler-Natta catalyst. It should be noted, however,
that methyloxirane,in contrast with propylene, is a chiral monomer
having an asymmetric carbon atom at each molecule. We can carry
out copolymerization reactions between D- and L-methyloxirane by
using zinc dimethoxide as catalyst, and results of the copolymerization will be useful for mechanistic considerations on the steric
course in the stereospecific polymerization of methyloxirane. The
simple and important copolymerization formula,

$$d[D]/d[L] = [D]/[L]$$

was obtained from a series of the D- and L-copolymerization experiments, in conformity with the concept of the enantiomorphic catalyst sites model.
    As the writer discussed previously, there are, in principle,
two modes of stereoregulation in polymerization reactions: (i)
growing chain control mechanism,and (ii) enantiomorphic catalyst
sites control mechanism. In contrast with the growing chain control
mechanism, there are rather fewer examples for polymerizations in
which the enantiomorphic catalyst sites mechanism is operative,
the only example other than the above-mentioned ones being the
stereospecific polymerization[7] of methyl vinyl ether initiated by
a heterogeneous catalyst system consisting of $Al_2(SO_4)_3$ and H_2SO_4.
    The present paper is concerned with a series of studies which
were undertaken to elucidate "the enantiomorphic catalyst sites
control mechanism" in terms of the moleculer level considerations.

AN ORGANOZINC COMPLEX, $[Zn(OCH_3)_2 \cdot (C_2H_5ZnOCH_3)_6]$, AS A MODEL FOR THE ENANTIOMORPHIC CATALYST SITES

    An organozinc complex, $[Zn(OCH_3)_2 \cdot (C_2H_5ZnOCH_3)_6]$ (I), was
prepared by the addition of 16 mmol methanol to 14 mmol diethylzinc
in 10 ml of heptane under an argon atmosphere. Complex (I) has
some significant advantages as model compound for the enantiomorphic catalyst sites.
    (i) Complex (I) could be isolated as a single crystal, the
structure of which was determined by x-ray diffraction analysis[8].
Complex (I) consists of the two enantiomorphic distorted cubes
which share a corner as shown in Fig. 1.
    (ii) Complex (I) is soluble in most of organic solvents. The
molecular structure of Complex (I) in a single crystal was proved
to be retained even in solution. Cryoscopic measurement with
Complex (I) in benzene showed strict agreement with the formula
$[Zn(OCH_3)_2 \cdot (C_2H_5ZnOCH_3)_6]$. 13C-PFT NMR of Complex (I) in C_6D_6

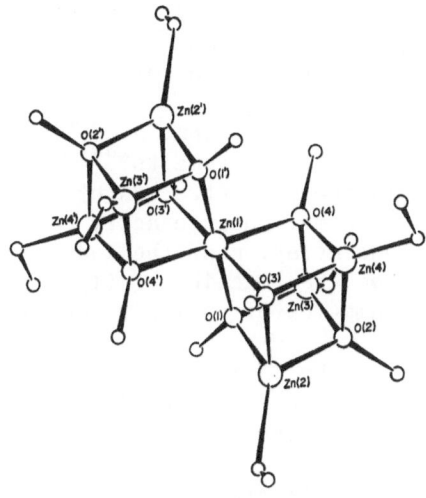

Fig. 1
Molecular structure of
$[Zn(OCH_3)_2 \cdot (C_2H_5ZnOCH_3)_6]$
(Complex (I))

[from Bull.Chem.Soc.
Jpn.49,1165(1976)]

showed no sign of dissociation or association in the temperature
range from 5 to 80°C.

(iii) Complex (I) in benzene did not polymerize methyloxirane
at 30°C, whereas the complex exhibited catalytic activity at higher
temperatures. Figure 2 shows that the reactivity of Complex (I)
increases suddenly at reaction temperatures between 60°C and 80°C.

(iv) Observed distribution of triad tacticities of poly(methyloxirane) prepared by Complex (I) as catalyst was found to obey
the statistics of the enantiomorphic control model :

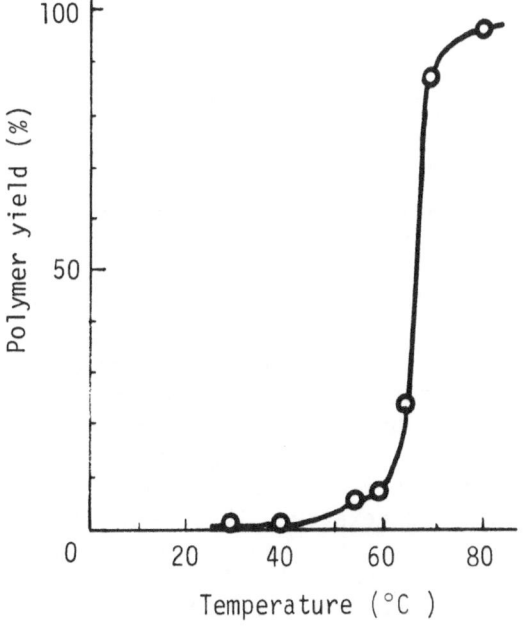

Fig. 2
Temperature dependence
of the catalytic activ-
ity of Complex (I) in
the polymerization of
methyloxirane.
[Complex (I)] 0.07 mol/l,
[MO] 4.8 mol/l,in benzene
for 3050 min.

$I = 1-3\sigma_2(1-\sigma_2)$, $S = \sigma_2(1-\sigma_2)$,   and $H = 2\sigma_2(1-\sigma_2)$, where σ_2 refers to the probability of D-configuration persistence at a D-preferring site or the probability of persistence for L at an L-site[9].

The reaction of Complex (I) with methyloxirane at 80°C was studied by means of quantitative ^{13}C NMR measurements[10,11]. Result obtained showed that the increase in the concentration of the terminal methoxy group of the growing polymer chain exactly corresponded with the decrease in the concentration of the inner methoxy group of Complex (I), which indicated the reaction of (I) with methyloxirane to take place by consumption of an inner methoxy group of Complex (I).

The enhancement in reactivity of the inner methoxy groups, but not of the outer ones, at higher temperatures could reasonably be understood in the light of results obtained from a series of measurements of the spin-lattice relaxation time, T_1, at increasing temperatures. Results of the measurements are shown in Fig. 3.

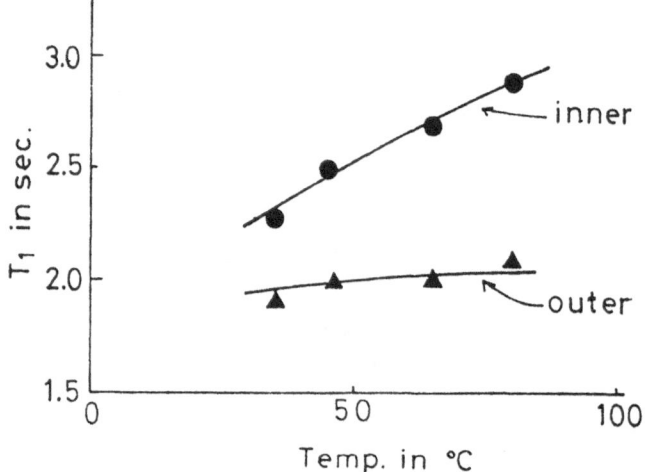

Fig. 3.   Plots of spin-lattice relaxation times of methoxy carbons of [Complex (I)] in variation of temperature.

Although the spin-lattice relaxation time normally becomes longer with the elevation of temperature, Fig. 3 demonstrates that there is a marked difference between inner and outer methoxy groups in terms of the temperature dependency of the relaxation time. The T_1 value of inner methoxy group increases more steeply than that of outer methoxy group with increasing temperature. This means the mobility of inner methoxy group increases more than that of the outer one. Based on these results, it was suggested[10,11] that the stereospecific polymerization of methyloxirane initiated with Complex (I) proceeded according to the mechanism given in Scheme 1.

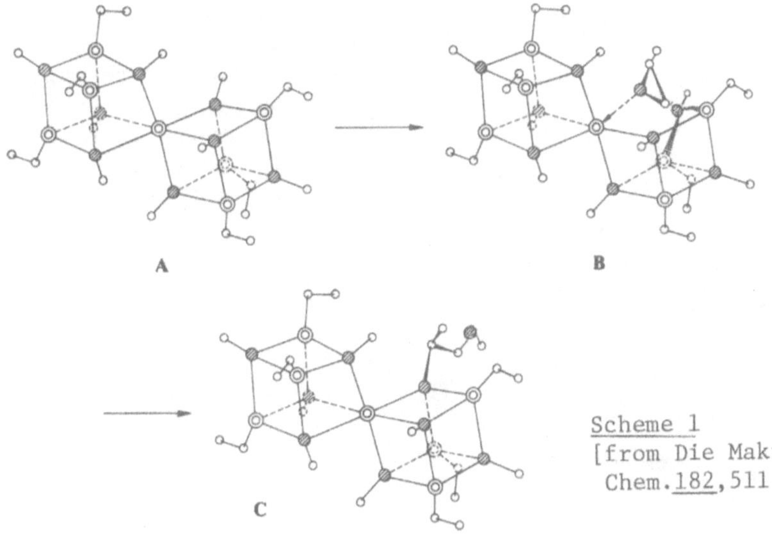

A

B

C

Scheme 1
[from Die Makromol.
Chem.182,511(1981)]

As stated before, Complex (I) consists of two enantiomorphic distorted cubes, each of which can be regarded as d- and l-cubes, respectively. At 80°C, one of the longest bonds (in d-cube, for instance) to the central zinc atom will be loosened.

The ring opening reaction of the coordinated methyloxirane takes place by a nucleophilic attack of the inner methoxy group to give $Zn-OCH(CH_3)CH_2OCH_3$ unit in C. The chiral structure around the central zinc atom in B prefers the coordination of (L)-methyloxirane before the (D)-isomer. This may be the origin of the l-catalyst site.

When the reaction takes place by loosening of the Zn-O bond in l-cube, a catalyst site having d-chiral nature should be generated. Since the probability of the bond loosening in the d-cube is exactly the same as that in the l-cube, an equal number of d-sites and l-sites will be generated, which leads to the formation of poly(oxypropylene) consisting of an equal number of D-selective polymer molecules and L-selective polymer molecules.

KINETIC BEHAVIOR OF THE INITIATION REACTION OF METHYLOXIRANE WITH COMPLEX (I)

The kinetic behavior of the initiation reaction of methyloxirane with Complex (I) was studied in the presence of excess methyloxirane. As the reaction proceeded, the concentration of Complex (I) decreased according to the first order kinetics:

$$-d[\text{Complex (I)}]/dt = k'[\text{Complex (I)}] \qquad (1)$$

where k' is a pseudo first order rate constant. Figure 4 shows three examples of the kinetic measurement.

Each of the three kinetic curves obeys Equation (1). It is surprising to find the smaller k' value for a reaction which was

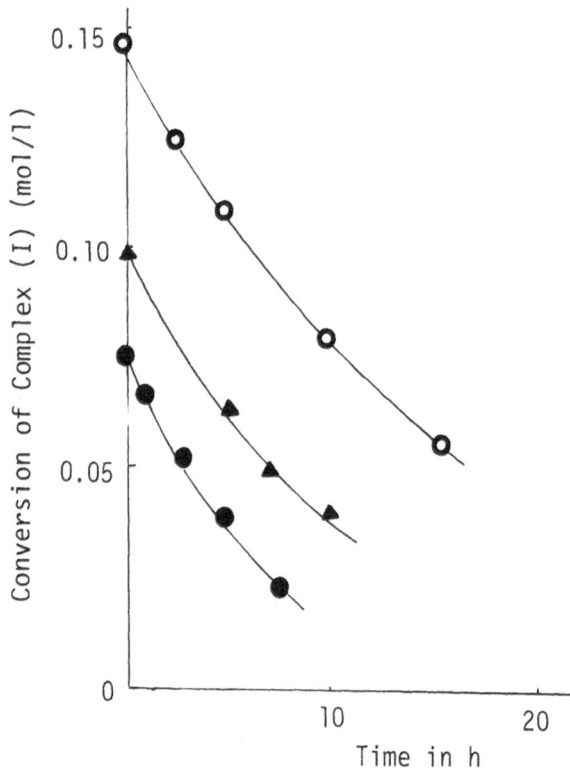

Fig. 4
Time-conversion
curves for reactions
between Complex (I)
and methyloxirane.
The initial concen-
tration of methyl-
oxirane: 2.93 mol/l.

Table I Pseudo first-order rate constants for
reactions between Complex(I) and methyloxirane

Complex(I)*(mol·l^{-1})	k'(h^{-1})
0.15	0.06
0.10	0.09
0.08	0.13

Monomer concentration: 2.93 mol/l.
* Initial concentration.

run with the larger initial concentration of Complex (I), as shown
in TABLE I.   This is also obvious at a glance at the time-conversion
curves in Fig. 4.   Under these reaction conditions, in which methyloxirane was present in a large excess in the reaction system and
percent conversion of methyloxirane was about 10% at largest, the
"monomer concentration" could be considered to remain essentially
unchanged in harmony with the observed "first order dependence"
of the overall reaction rate upon the concentration of Complex (I).
     It would not be probable to assume for Complex (I) to change
its reactivity toward methyloxirane in response to the concentration of Complex (I), because neither dissociation nor association
of Complex (I) molecules was proven to take place on this concen-

tration range.

   We interpret the anomalous correlation between k' value and the concentration of Complex (I) in terms of "the effective concentration" of methyloxirane in the reaction system with Complex (I). "The effective concentration" of methyloxirane is defined as the concentration of methyloxirane molecules which are not "captured" by any of the four-coordinated zinc atoms in Complex (I). The monomer "capturing" may be understood as the same concept as that for "solvation" of the four-coordinated zinc atoms by methyloxirane molecules.

   If we assume that the monomer molecules captured by the four-coordinated zinc atoms have no ability to enter the polymerization reactions, the effective concentration of methyloxirane may become smaller in reaction systems in which larger quantity of Complex (I) is present. The effective concentration of methyloxirane, $[MO]_{eff}$, is approximated as Eq. (2):

$$[MO]_{eff} = [MO] - n[Complex\ (I)]_0 \qquad (2)$$

where [MO] and $[Complex\ (I)]_0$ are the concentration of methyloxirane and the initial concentration of Complex (I), respectively, in the reaction system. n is the number of captured monomer molecules. As stated above, [MO] can be approximated to be $[MO]_0$, the initial concentration of the monomer, so that the rate of the initiation reaction should be given by Eq. (3):

$$-d[Complex(I)]/dt = k([MO]_0 - n[Complex(I)]_0)[Complex(I)] \qquad (3)$$

where k is a rate constant.

   Eq. (4) is derived by comparing Eq. (1) with Eq. (3).

$$k' = k([MO]_0 - n[Complex\ (I)_0]) \qquad (4)$$

   From the values for k' given in Table I, n is estimated to be 12 - 18, which means that each of the four-coordinated zinc atoms is solvated by two or three molecules of methyloxirane.

Table II ^{13}C Spin-lattice relaxation times (T_1) for systems of Complex (I), Complex (I)/methyloxirane(MO), and Complex (I)/tetrahydrofuran(THF)*

	Zn–O–$\underline{C}H_3$		Zn–$\underline{C}H_2$–C	Zn–C–$\underline{C}H_3$
System	outer	inner		
Complex (I)	1.9	2.3	4.6	11.4
Complex (I)/MO	3.8	4.0	3.8	4.6
Complex (I)/THF	3.7	3.8	3.7	4.1

Note: T_1 in sec.

*   Measured at 35°C, immediately after the samples were prepared. Solvent C_6D_6, concentration of Complex (I) ca.0.06 mol·l^{-1}, Complex (I)/[cyclic ether]≈0.1(mol/mol).

Table III ^{13}C Spin-lattice relaxation time $(T_1{}^{a)})$ of organozinc compounds$^{b)}$

System	Zn-O-$\underline{C}$H$_3$	Zn-C-$\underline{C}$H$_3$		Zn-$\underline{C}$H$_2$-C		
	δ in ppm	δ in ppm		δ in ppm		
	55.79	13.21	10.03	6.50	3.78	-4.64
(C$_2$H$_5$ZnOCH$_3$)$_4$	3.4	11.0	-	-	-	5.5
(C$_2$H$_5$ZnOCH$_3$)$_4$/MO	4.2	6.8	-	-	-	5.3
(C$_2$H$_5$)$_2$Zn	-	-	11.2	10.5	-	-
(C$_2$H$_5$)$_2$Zn/MO	-	-	8.3	-	8.3	-

a) in second.
b) Measured at 35°C, solvent: C$_6$D$_6$, concentration of organozinc compound: ca. 0.06 mol/l, [Org-Zn cpd]/[MO]≈0.1 (mol/mol).

        The solvation of the four-coordinated zinc atoms by a cyclic
ether such as methyloxirane was also assumed to take place from
results of the T_1 measurements with Complex (I) in the presence of
cyclic ether. As shown in Table II, the values of T_1 of the ethyl
carbons of Complex (I) were observed to become lower in the presence
of cyclic ether. The change in the T_1 values for the ethyl carbons
suggests that "solvation" by the cyclic ether molecules takes place
at the four-coordinated zinc atoms in Complex (I), because the free
internal motions of the ethyl group are hindered by the solvated
cyclic ether. The higher values of T_1 of methoxy carbons observed
in the systems with cyclic ether may imply "loosening" of Zn-OCH$_3$
bonds which was caused by the solvation of the cyclic ether at the
four-coordinated zinc atoms of Complex (I).
        These considerations are also supported by another series of
T_1-measurements with (C$_2$H$_5$)$_2$Zn and (C$_2$H$_5$ZnOCH$_3$)$_4$. T_1 data in
Table III exhibit a similar tendency to those in Table II.
        It is to be noted that either (C$_2$H$_5$)$_2$Zn or (C$_2$H$_5$ZnOCH$_3$)$_4$ exhib-
ited no ability to initiate the polymerization of methyloxirane, in
spite of their ability to capture a number of monomer molecules at
their solvation sites. The chemical circumstances around the four-
coordinated zinc atoms in Complex (I) are very similar to those
around zinc atoms in (C$_2$H$_5$ZnOCH$_3$)$_4$.
        When a kinetic study of the reaction of Complex(I) with methyl-
oxirane was carried out in the presence of a comparable amount of
(C$_2$H$_5$ZnOCH$_3$)$_4$, a marked decrease in the reaction rate was observed
which resulted from the lowering of effective concentration of
methyloxirane in the reaction system.

CONCLUSION

Using a well-defined organozinc complex, $[Zn(OCH_3)_2(C_2H_5ZnOCH_3)_6]$ Complex (I), as a model for the enantiomorphic catalyst sites, the mechanism of the stereospecific polymerization of methyloxirane was studied by means of quantitative[13]C NMR measurements. The origin of D-(or L-)selective catalyst sites was discussed at molecular level based on the information of structure and reactivity of Complex (I). Kinetic study of the initiation reaction revealed that the reaction proceeded according to the first order rate law with respect to the concentration of Complex (I) and the observed rate constant k' decreased with increasing concentration of Complex (I). The anomalous behavior of k' was explained in terms of "the effective" concentration of methyloxirane in the reaction system. The effective concentration of methyloxirane was defined as the concentration of methyloxirane molecules which were not "captured" by any of the four coordinated zinc atoms in Complex (I). The results on the spin-lattice relaxation time, T_1, for C_2H_5- and CH_3O-groups in Complex (I) seemed to support the concept of the effective concentration of methyloxirane in the reaction system with Complex (I).

REFERENCES

1. G. Natta, J. Inorg. Nuclear Chem.,8,589(1958); J. Polymer Sci., 34,21;109(1959).
2. G. Allegra, Makromol. Chem.,145,235(1971).
3. P. Cossee, J. Catalysis,3,80(1964).
4. E. J. Arlman, J. Catalysis,3,89(1964).
5. E. J. Arlman and P. Cossee, J. Catalysis,3,99(1964).
6. T. Tsuruta, J. Polymer Sci.,D6,179(1972).
7. T. Higashimura, Y. Ohsumi, K. Kuroda and S. Okamura, J. Polymer Sci.,[A-1]4,863(1967).
8. M. Ishimori, T. Hagiwara, T. Tsuruta, Y. Kai, N. Yasuoka and N. Kasai, Bull. Chem. Soc. Japan,49,1165(1976).
9. M. Ishimori, T. Hagiwara and T. Tsuruta, Makromol. Chem.,179, 2337(1978).
10. T. Tsuruta, J. Polymer Sci.,Polymer Symposium,67,73(1980).
11. T. Hagiwara, M. Ishimori and T. Tsuruta, Makromol. Chem.,182, 501(1981).

ACKNOWLEDGMENT
The authors are grateful to Professor Tominaga Keii for his valuable discussion on the results of our kinetic measurements.

POLYMERIZATION OF EPOXIDES CATALYZED BY CONDENSATES OF ORGANOTINS OR FRIEDEL-CRAFTS TYPE METAL HALIDES WITH ALKYL PHOSPHATES

Tetsuya Nakata

Research Laboratories, Osaka Soda Co., Ltd.
9 Otakasu-cho, Amagasaki City
660, Japan

SYNOPSIS

Some Friedel-Crafts catalyst type metal halides and organotins condense with alkyl phosphates on heating to give condensates which are catalytically reactive for the polymerization of epoxides to give high molecular weight polyethers. Condensates of organotins with alkyl phosphates are highly reactive catalysts and give highly crystalline polymers of epoxides. The condensates obtained from Friedel-Crafts type metal halides give less crystalline polymers.

The initiation of epoxide polymerization by the organic tin condensate catalyst appears to be due to phosphate anion and there appears to be appreciable transfer reaction to polymer except for the case of ethylene oxide polymerization. For propagation reactions, it is presumed that polymeric phosphate structure of the catalyst plays an important role.

The organotin condensates change their acid strength depending on the degree of condensation. Also, the ^{119}Sn in the catalyst exhibits NMR signals at extremely high field. Thes results suggest that highly coordinated and/or monoalkylated tin atoms are important to the catalytic activity observed.

INTRODUCTION

Extensive studies on epoxide polymerization were reported from 1950 - 70 and several catalyst systems which give high molecular weight polymers of epoxides were disclosed, such as condensates of aluminum alkyls[1] or zinc alkyls[2] with limited quantities of water.

55

The mechanism for the polymerization of epoxides by these catalysts to a high molecular weight polymer has been well established as "coordination polymerization"[3]. The catalyst nature for the polymerization, that is, the necessity of having a polynuclear character at the active site of the catalyst[4,5] and the existence of metal-oxygen (hetero atom)-metal groups in their structure[1], has been confirmed as well.

This presentation is concerned with new catalysts i.e., condensates of an alkyl phosphate with some Friedel-Crafts type metal halides and especially with organotin compounds, which are different in both their skeletal structures and metal atoms from the above described catalysts.

EXPERIMENTAL

Materials

Organotins were obtained from Nitto Kasei Co., Ltd. Freshly distilled or recrystallized organotins were used except for dialkyltin oxides which were used as received. Metal halides were commercial reagents which were used as received or used after pulverizing if necessary. Alkyl phosphates from Daihachi Chemical Industry Co., Ltd., were distilled and then stored over a molecular sieve.

Epichlorohydrin was the product of Osaka Soda Co., Ltd. and propylene oxide was a commercial reagent. These monomers were distilled and usually diluted with hexane and stored over a molecular sieve. Ethylene oxide was the product of Nippon Soda Co., Ltd., and was distilled before use. Other epoxides were commercial reagents which were distilled and stored over a molecular sieve. A commercial hexane stored over a molecular sieve was used.

General Procedure of Catalyst Preparation

Organotin Condensate. In a three-necked glass flask equipped with a thermometer, a mechanical stirrer and a distilling head connected through a condenser with a receiver and a gas holder for storing evolved gases, a mixture of an organotin and an alkyl phosphate is stirred and heated at 240 to 250°C under nitrogen atmosphere. The reaction starts usually at around 220°C and above 240°C is accompanied by a violent evolution of gases. Viscosity of the mixture increases gradually and finally a porous solid is obtained. Overheating caused decomposition to a colored, undesirable material. The reaction is stopped by cooling at any desirable point of condensation. The degree of condensation can be estimated approximately by measuring an amount of distillates. The obtained condensate is pulverized and stored.

An organotin – alkyl phosphate condensate catalyst is quite stable but adsorbs moisture when stored in air. Such water adsorption causes an induction period in the polymerizations and reduces the molecular weight of the polymer. The air-exposed catalyst exhibits absorption bands in its infrared spectra at $3450 cm^{-1}$ and $1630 cm^{-1}$ due to adsorbed water. These bands disappear completely after drying the catalyst under a high vacuum and the catalytic reactivity of the catalyst is restored by this treatment.

Condensates from Friedel-Crafts Type Metal Halides. The condensates are obtained by an analogous procedure to that of the organotin condensates but at 140 to 170°C. It is necessary to avoid atmospheric humidity.

Polymerization Procedures

Polymerization of Epoxides other than Ethylene Oxide. Polymerization were carried out generally in sealed 50ml. glass tubes shaken in a constant temperature water bath. Solvents and monomers were dried to under 10 ppm of water content over a molecular sieve. The pulverized solid catalyst was dried out in the polymerization tube by heating thirty minutes at 160°C under a flow of nitrogen. After cooling, solvent and then monomer was injected into the tube. and the tube was sealed.

Polymers, which were generally obtained as solid precipitates or slurries, were usually dissolved in hot benzene containing an antioxidant and washed with dilute hydrochloric acid, sodium bicarbonate solution, water and dried in vacuum.

Polymerization of Ethylene Oxide. The catalyst is weighed and placed into a three necked flask in a constant temperature water bath equipped with a mechanical stirrer, an inlet, an outlet through a reflux condenser and a thermometer, and is dried by heating at 160°C under a flow of nitrogen. After cooling, dried hexane was introduced into the flask. Ethylene oxide was bubbled into the hexane with stirring at 20°C. After polymerization an antioxidant was added to the polymerization vessel. The obtained polymers were treated similarly to the other polymers except washing treatments were omitted.

Polymerization of Gaseous Ethylene Oxide. The polymerization of gaseous ethylene oxide was carried out in a 200ml. glass apparatus which was connected with a vacuum line, a nitrogen line, an ethylene oxide line and a pressure gauge, and in which was fitted a heated glass plate whose temperature was taken with a thermocouple.

After 100mg of pulverized catalyst was placed on the plate the apparatus was evacuated to 10^{-5} mmHg. Then gaseous ethylene oxide was introduced to a particular pressure and immediately the changes of the pressure recorded.

Identification of the product was made by a direct measurement of the infrared spectra of the product on the plate, and a growth of the absorption band at 1345cm^{-1} was noted.

Physical Measurements

Molecular weights of the polymers of ethylene oxide, propylene oxide and epichlorohydrin were determined viscometricaly using the following equations:

Ethylene oxide[6] $[\eta] = 6.40 \times 10^{-5} (\overline{Mv})^{0.81}$
Propylene oxide[7] $[\eta] = 11.20 \times 10^{-5} (\overline{Mw})^{0.77}$
Epichlorohydrin[8] $[\eta] = 8.93 \times 10^{-5} (\overline{Mw})^{0.73}$

Determination of the molecular weight of catalysts was made in a toluene solution using a HITACHI PERKIN-ELMER 115 vapor pressure osmometer.

For Measurements of infrared spectra a JAS. Co.112A-2 was used.

The values of ^{119}Sn chemical shifts were obtained using a Jeol FX-100 spectrometer in 30% chloroform solution. The δ values are relative to external tetramethyl tin.

RESULTS AND DISCUSSION

Catalyst

Condensation Reaction of Organotins with Alkyl Phosphate. A rather simple condensation reaction of alkyltin halides with an alkyl phosphate to give alkyltin phosphates has been described in a patent application[9] (Equation 1).

$$3R_3SnX + R'_3PO_4 \xrightarrow{-R'X} (R_3Sn)R_2'PO_4$$

$$\xrightarrow{-R'X} (R_3Sn)_2R'PO_4 \xrightarrow{-R'X} (R_3Sn)_3PO_4 \qquad (1)$$

Because the catalyst was prepared by the reaction of organotin halides with an excess of an alkyl phosphate, the catalyst forma-tion reactions seemed to consist of the condensation of alkyl phos-phates themselves and substitution or elimination reactions of alkyls attached to tin atoms by alkyl phosphates in addition to the reaction of the type of Equation (1).

Fig. 1. indicates the carbon contents and the apparent molecu-lar weights of the condensates at various stages obtained by the reaction of one mole of tributyl tin chloride and two moles of tri-butyl phosphate on heating at 240 to 250°C. In the condensation reaction butane, butenes, butyl chloride, butyl alcohol and dibutyl

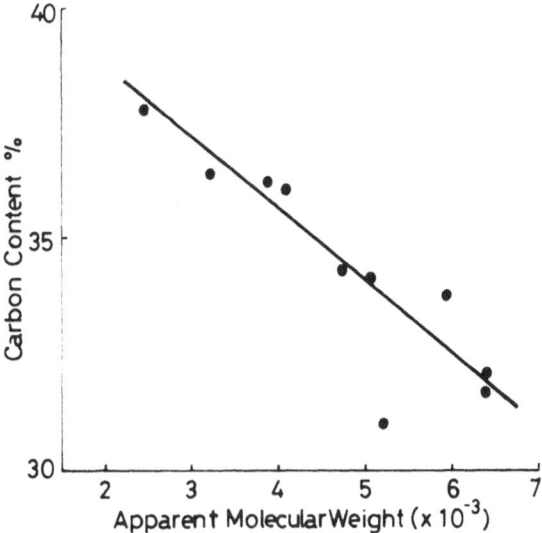

Fig. 1. Relation between carbon content and apparent molecular weight for the condensate of tributyl tin chloride with tributyl phosphate (1:2).

ether were obtained as distillates. The condensates were soluble at the initial stage of the reaction in a variety of organic solvents but with an increase of the degree of condensation, it changed to insoluble because of network formation. In this case, molecular weight of the insoluble material was estimated to be around eight thousand.

In the case of dibutyl tin dichloride, insolubilization occurs above twelve thousand of apparent molecular weight. The condensation reaction of dibutyltin oxide with an alkyl phosphate also gave an analogous condensate and distillates except for absence of butyl chloride, to those from dibutyltin dichloride or tributyltin chloride. It seems, however, that the nature of the reaction is more complicated than with the halides.

The infrared spectra of the condensates from tributyltin chloride (Fig. 2), dibutyltin dichloride and dibutyltin oxide are undistinguishable from each other even at the initial stage of condensation. In fact, catalytic reactivities do not vary appreciably with the nature of the starting organotins.

Catalysts Obtained from Metal Halides. In the work on condensates of Friedel-Crafts type metal halides with alkyl phosphates, the condensates from zirconium chloride, hafnium chloride and stannous chloride with an alkyl phosphate induced polymerization of epoxides to high molecular weight.

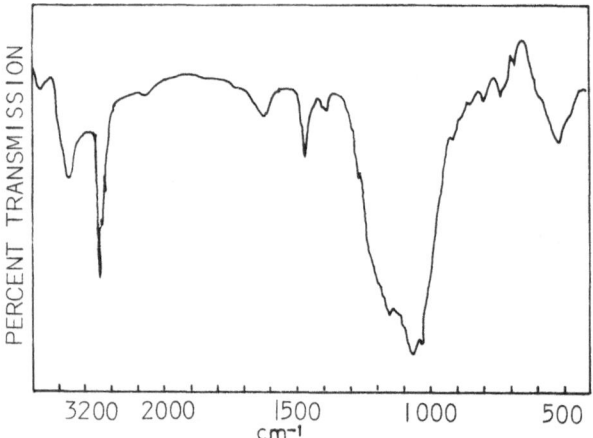

Fig. 2.   Infrared spectra of the condensate of tributyl tin
          chloride with tributyl phosphate.

     Stannic chloride, aluminum chloride and ferric chloride have
failed to give condensates which are catalytically active for the
high molecular weight polymerization. Aluminum chloride, however,
in combination with silicon chloride gave a condensate with an alkyl phosphate, which polymerizes epichlorohydrin to an amorphous
high molecular weight polymer, though it is not clear yet whether
water participates in the formation of the catalyst or not.

     General Reactivity of the Catalysts. Several examples of poly-
merization by the condensate catalysts are given in Tables 1 and 2.
The catalysts obtained from Friedel-Crafts metal halides give considerably less crystalline polymers of epichlorohydrin compared
with the polymers formed by an organotin condensate catalyst.

     The catalysts obtained from organotins have failed to polymer-
ize cyclic ethers other than monosubstituted ethylene oxides.

Characteristics of Organotin – Alkyl Phosphate Condensate Catalysts

     Initiation Reaction. It is well known that organotins are
distinctly different in chemical properties from aluminum alkyls or
zinc alkyls, e.g. Lewis-type acid strength of organotins is very weak
and nucleophilicity of carbon-tin linkage has scarcely been known.
No initiation reaction of an epoxide polymerization by an organotin
has ever been disclosed though a few nucleophilic reactions based
on heteroatom-tin linkages are known[10].

     As shown in Fig. 3, the infrared spectra of a soluble propylene
oxide polymer in hexane which was obtained by a low conversion poly-

Table 1

Epichlorohydrin Polymers derived from Various Metal Chloride/Butyl Phosphate Catalysts

	$AlCl_3+SiCl_4$	$ZrCl_4$	$SnCl_4$	Bu_3SnCl
Conversion %	95	90	60	90
$\eta sp/c$ 0.1g/100ml (80°C, Chlorobenzene)	1.8	1.0	0.8	1.8
Relative Crystallinity(a) (DSC)	amorphous	0.12	0.26	1.0
M.P. °C		106	111	115
Polymerization Temperature °C	45	45	45	25
Polymerization Time hr.	48	48	48	48
Catalyst : ($\frac{\text{Metal Halide}}{2Bu_3PO_4}$) ; wt%	0.15	0.15	0.15	0.08
Polymerization Solvent	none	none	none	none

(a) Compared to the product from Bu_3SnCl Catalyst as 1.0

Table 2

Polymerization with Organotin Condensates (*)

	Solvent	Conversion	$\eta sp/c.$
Ethylene Oxide*	hexane	ca.100	
Propylene Oxide	hexane	ca.100	5.5**
Epichlorohydrin	none	>90	2.0***
Styrene Oxide	none	50	
Phenyl Glycidyl Ether	none	80	

Polymerization Temperature 25°C, Polymerization Time 48hr., Catalyst 0.1 wt%

* 10g Monomer   ** 0.1g/100cc Benzene, 50°C

  25g Hexane

  0.05g Catalyst   *** 0.1g/100cc Chlorobenzene, 80°C

(*) All polymers were crystalline by X ray.

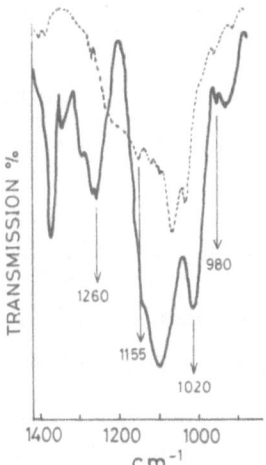

Fig. 3.   Infrared spectra of low molecular weight polypropylene
           oxide obtained with an insoluble organotin condensate
           catalyst from dibutyltin oxide and tributyl phosphate
           (—) and original catalyst (····).

merization of propylene oxide catalyzed by an insoluble organotin
condensate catalyst gave some absorption bands due to an ester
phosphate[12], that is, at $1260cm^{-1}$(P=O), $1020cm^{-1}$(P-O-C) and to a
pyrophosphate[12] at $980cm^{-1}$(P-O-P). The last two absorptions are
less definitive since the absorption at $1020cm^{-1}$ is generally found
in the infrared spectra of polypropylene oxide and the absorption at
$980cm^{-1}$ is found in the infrared spectrum of the catalyst. The absorption at $1260cm^{-1}$ indicates clearly the formation of an ester
phosphate linkage suggesting that the initiation of polymerization
has been caused by a phosphoxy anion. Also low conversion crystalline polymer from epichlorohydrin gave the same absorption bands in
its infrared spectra. This fact excludes the possibility that the
ester phosphate initiation is limited to amorphous polymer formation.

     Effect of Apparent Molecular Weight of Catalyst. Table 3
shows the effect of catalyst molecular weight on epichlorohydrin
polymerization. The highest molecular catalyst was substantially
more effective.

     Apparent catalyst molecular weight also seems to affect the
crystallinity of an epoxide polymer. Table 4 indicates the results
of an examination of crystallinity of the epichlorohydrin polymers
obtained with catalysts of different carbon contents. A carbon
content of 33.5 percent corresponds to an apparent molecular weight
of about four thousand. The lower carbon content catalyst is insoluble and its molecular weight, although probably higher, is not

Table 3. Effect of Apparent Molecular Weight of Catalyst on
          the Polymerization of Epichlorohydrin.

Molecular Weight of Catalyst	Conversion at 3 hr. (%)	
	$30^\circ C$	$50^\circ C$
4000	13	35
10000	35	76

Catalyst :   Condensate of $\dfrac{(C_4H_9)_2 \; SnCl_2}{2(C_4H_9)_3PO_4}$

                0.105mg atom of Sn

Monomer :   216mmole / 10cc Hexane

known. This result indicates that the catalyst less crowded by
alkyls gives a more crystalline polymer.

Also a considerable effect of the catalyst molecular weight on
the composition of epichlorohydrin-ethylene oxide copolymer was
observed (Fig. 4). The points on these curves were determined from
the monomer concentration analysis. Because of nonhomogeneity of
the polymerization system, the plotted values of the polymer composition are not precise. However, the trends appear reasonable.
The reactivity ratios for these same catalysts are given in Table 5
and show a large effect of molecular weight.

Saegusa[12] indicated previously a correlation of the reactivities of various catalysts for ring opening polymerizations with
their acid strength using Hammet's indicators. According to this
method, rough examination of the acid strengths of two organotin -
alkyl phosphate condensates different in apparent molecular weight
revealed that the acid strength of the catalyst increases with an
increase of the apparent molecular weight of the condensate (Table
6). The catalyst of apparent molecular weight of ten thousand indi-

Table 4[(*)]. Effect of Carbon Content of Catalyst on
              Crystallinity of Epichlorohydrin Polymer.

Carbon Content %	Acetone Insoluble %
33.5	63
24.8	82

Catalyst : $\dfrac{(C_4H_9)_2 \; SnCl_2}{2(C_4H_9)_3PO_4}$

(*) Crystalline portions obtained by cooling to $10^\circ C$ 24hr.
    after dissolved in boiling acetone.

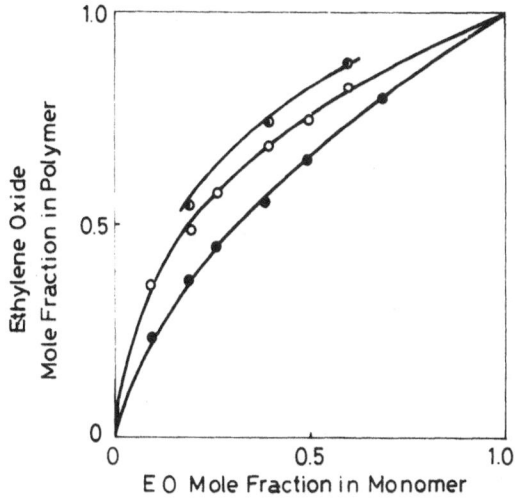

Fig. 4. Effect of apparent molecular weight of catalyst on copolymer composition.

Polymerization Temperature   15°C

Monomer Concentration   25% in Hexane

Catalyst : $\dfrac{Bu_2SnCl_2}{2Bu_3PO_4}$

Catalyst Concentration   0.32 wt% of Monomer

◑   Molecular Weight   4000

○   "   6600

●   "   10000

Table 5. Effect of Apparent Molecular Weight of Catalyst on Reactivity Ratio of Copolymerization.

Molecular Weight	r_1(EO)	r_2(ECH)
4000	4.97	0.23
6600	2.55	0.15
10000	1.22	0.29

Catalyst :   Condensate of $Bu_2SnCl_2/2Bu_3PO_4$ 0.120mg atom of Sn.

Monomer Concentration :   30 weight %

Solvent   :   Hexane

Total Charge   :   30g

Table 6 (*)

Acid Strength of Condensate Catalysts Different in Apparent Molecular Weight.

	pKa	Zn Et$_2$/H$_2$O	Condensate-Catalyst		Organotin-Dialkyl Phosphate*
			MW 10000	MW 4000	MW 650
Phenylazonaphthylamine	+4.0	A	A	A	B
Butter Yellow	+3.3	A	A	B	B
Phenylazodiphenylamine	+1.5	A	A	B	B
Dicinnamalacetone	-3.0	A-B	B	B	B
Benzalacetophenone	-5.0	B	B	B	B

* Dibutyltin-bis-dibutyl phosphate

(*) "A" denotes an acidic, and "B" denotes a basic state of indicators.

cates acidity in the pKa region of phenyl-azo-diphenylamine, while
the other of molecular weight of four thousand does not show acidity
until the pKa range of phenyl-azo-naphthyl amine. In this table the
acid strength of diethyl zinc - water (1:1) catalyst for which value
was as indicated by Saegusa[12], and that of a low molecular weight
organotin alkyl phosphate are shown for comparison. It is interesting that the acid strength of the diethyl zinc - water catalyst
is in same region as that of the more active condensate catalyst.

    Table 7 gives the results of preparations of several organotin
alkyl phosphates and pyrophosphates and of the examination of their
catalytic reactivities and their acid strengths. It was found in
these studies that the dibutyl tin dibutyl pyrophosphate (II) alone
induced the polymerization of epichlorohydrin and its acid strength
was between the two kinds of the condensate catalyst in Table 6.
(II) is the only polymeric structure in the compounds studied
in Table 7.

    Though dibutyltin-bis-dibutyl phosphate (I), triphenyltin-
dibutylphosphate (III) and di-triphenyltin-dibutyl-pyrophosphate (IV)
themselves are catalytically inactive, a condensate of (I) itself
and condensates of (III) and (IV) with an alkyl phosphate showed
catalytic reactivities similar to that of a condensate of an organotin halide with an alkyl phosphate. It is presumed that a polymeric
structure containing pyro- or poly-phosphate linkages is necessary
to obtain catalytic reactivity.

^{119}Sn NMR spectroscopy revealed interesting results in connection with catalytic reactivity. It was found that two broad absorptions occured in the NMR spectrum of every catalytically active
compound examined (Table 8). The higher $\delta(^{119}Sn)$ value of each is
660 $\pm$ 40 ppm upfield relative to tetra methyltin. This result
probably suggests the existence of monoalkylated and/or highly coordinated tin atoms in the organotin-condensate catalysts.

    Polymerization of Gaseous Ethylene Oxide. The organotin
catalyst polymerizes gaseous ethylene oxide smoothly even at low
temperature, as shown by the decrease of the pressure of ethylene
oxide at 0°C and 50°C (Fig. 5). At 0°C there is recognized a
distinct induction time but the final rate of polymerization is
similar to the rate as 50°C. The results suggest the propagation
is due to a process with little separation of charge. At higher
temperatures a process of different mechanism would probably apply.

    Transfer and Termination Reactions. Fig. 6 and Fig. 7 show
respectively how the number of polymer chains vary with conversion
for the polymerization of ethylene oxide and for the polymerizations
of propylene oxide and epichlorohydrin by an organotin - alkyl
phosphate condensate catalyst. These results indicate that termination and transfer reactions are very small in the polymerization

Table 7 (*)

Examination of Catalytic Reactivities for Epichlorohydrin Polymerization of Organotin – Alkyl Phosphates or Organotin – Alkyl Pyrophosphates and Their Acid Strengths.

	(I)	(II)	(III)	(IV)
	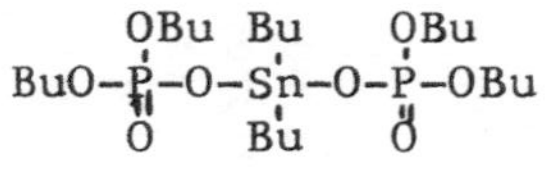 $\begin{array}{c}\text{OBu}\quad\text{Bu}\quad\text{OBu}\\ \text{BuO–P–O–Sn–O–P–OBu}\\ \quad\text{O}\quad\text{Bu}\quad\text{O}\end{array}$	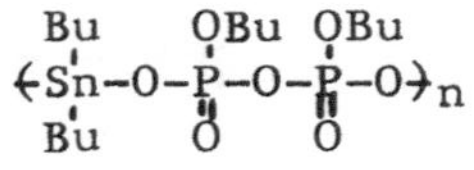 $\begin{array}{c}\text{Bu}\quad\text{OBu}\ \text{OBu}\\ (\text{Sn–O–P–O–P–O})_n\\ \text{Bu}\quad\text{O}\quad\text{O}\end{array}$	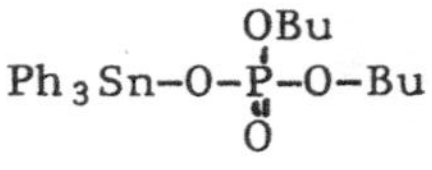 $\begin{array}{c}\text{OBu}\\ \text{Ph}_3\text{Sn–O–P–O–Bu}\\ \text{O}\end{array}$	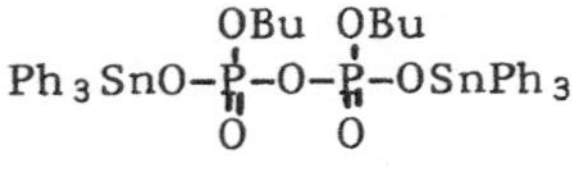 $\begin{array}{c}\text{OBu}\ \text{OBu}\\ \text{Ph}_3\text{SnO–P–O–P–OSnPh}_3\\ \text{O}\quad\text{O}\end{array}$
Conversion %	none	11*	none	none
pKa				
+ 4.0	B	A	B	B
+ 3.3	B	A	B	B
+ 1.5	B	B	B	B
− 3.0	B	B	B	B
− 5.0	B	B	B	B

Melting Temperatures

(I)	252°C
(II)	not determined
(III)	91 – 93°C
(IV)	not determined

Polymerization Conditions

Monomer	10g
Hexane	10g
Catalyst	0.1g
Reaction Temperature	50°C
Reaction Time	48hrs.

* at 8hr.

(*) "A" and "B" are as indicated in Table 6.

Table 8. ^{119}Sn Chemical Shift (Data in CHCl$_3$)

	Compound	Activity	$\delta(^{119}Sn)$ppm
(I)	BuO Bu OBu BuOPOSnOPOBu O Bu O	none	−466
(II)	Condensate of (I)	active	−699, −264
(III)	Bu Bu Bu O O (SnOPOPO)$_n$ Bu O O	active	−644, −257
(IV)	Further Condensate of MW 4000 Catalyst	active	−626, −263

of ethylene oxide, while, to the contrary, considerable transfer
reactions occur in propylene oxide and epichlorohydrin polymerizations. The large difference in transfer reaction between ethylene
oxide and the other monomers is not understandable excepting to
attribute it to a difference in the polymerization rates.

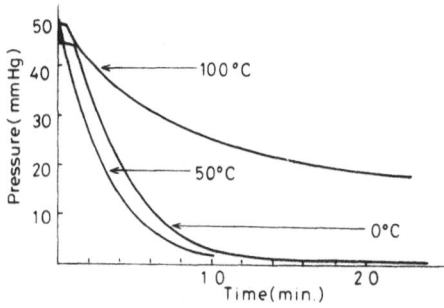

Fig. 5. Polymerization of gaseous ethylene oxide.

Catalyst : $\dfrac{Bu_2SnCl_2}{2Bu_3PO_4}$ 100mg

Reactor Volume   200ml

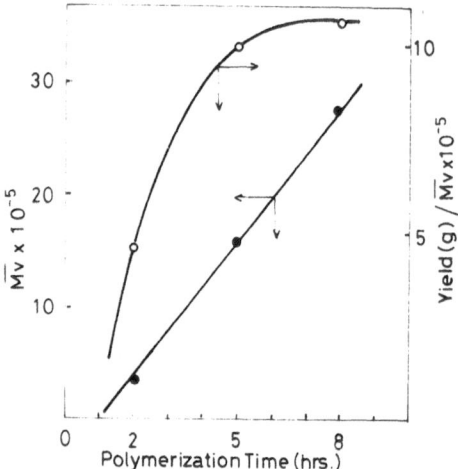

Fig. 6. Molecular weight and comparative data on number of
polymer molecules in ethylene oxide polymerization as
function of polymerization time.

Catalyst : $\dfrac{Bu_3SnCl}{2Bu_3PO_4}$, 0.5g/1.4ℓ Hexane

Ethylene Oxide Feed, 1g/min.
Polymerization Temperature 20°C

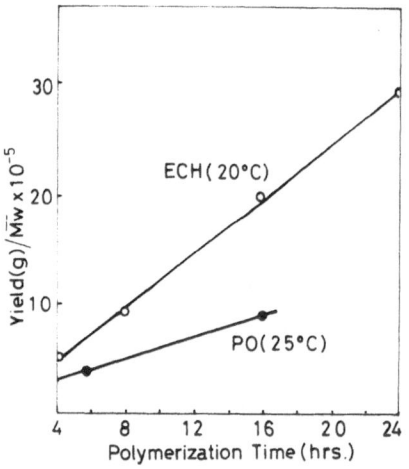

Fig. 7. Variation in number of polymer molecules in propylene
oxide (PO) and epichlorohydrin (ECH) polymerizations
as function of polymerization time.

Catalyst : $\dfrac{Bu_3SnCl}{2Bu_3PO_4}$ Solvent : Hexane

Alcohols behave as transfer agents in the polymerization of epichlorohydrin using an organotin - alkyl phosphate condensate as catalyst. Plotting of the reciprocal of molecular weight of the obtained polymers vs. the mole ratio of added methanol to epichloro-hydrin gave a good straight line (Fig. 8). The intercept of the ordinate suggests existence of transfer reactions to the catalyst itself, to the produced polymer and/or to the monomer, ignoring a transfer to other impurities. It seems, however, that the transfer reaction to the polymer are most important in view of the result of polymerization indicated in Fig. 7.

An organotin - alkyl phosphate condensate catalyzes the addi-tion reaction of an alcohol to an epoxide as does an ordinary strong acid (Equation 2). In this sense, the catalyst is cationic.

$$ROH + C\overset{}{\underset{O}{\diagup\diagdown}}C\text{-}R' \qquad \overset{}{\underset{OR\ OH}{C\text{—}C\text{—}R'}} + \overset{}{\underset{OH\ OR}{C\text{—}C\text{—}R'}} \qquad (2)$$

As compared with this addition reaction, it seems reasonable to assume that the transfer reaction to an alcohol or to a polymer is based on an activated proton of the alcohol or an activated alkyl

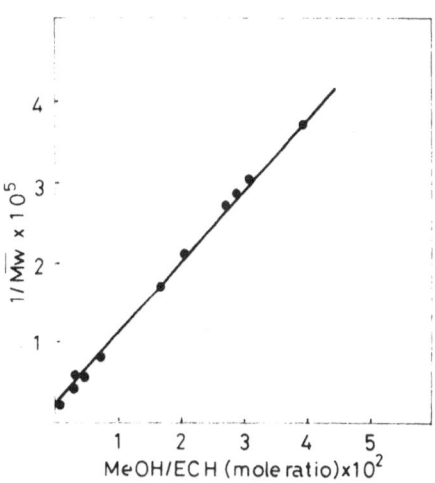

Fig. 8.   Effect of mole ratio of added methanol to epichloro-hydrin on molecular weight.

Catalyst   : $\dfrac{Bu_3SnCl}{2Bu_3PO_4}$

Bulk Polymerization

Polymerization Temperature $25°C$

of the polymer as shown in Fig. 9. However, direct confirmation
of the end groups of the polymers has never been made but it has
been observed that addition of a small quantity of alcohol to polymerization system increases the polymerization rate. This may be
due to the participation of an alkoxy anion left on a tin atom.

Conclusions

The acidity of the organotin condensate catalysts has not been
explained in relation to ^{119}Sn NMR chemical shift data. It is reasonable, however, to propose a participation of a specific poly-
pyrophosphate structure in the catalytic site in which nearby
"acidic" tin atoms participate by inter- or intramolecular coordination. The catalytic site structures of Fig. 10 are probable for
the propagation steps in view of the small charge separation
(Fig. 11).

The structure of Fig. 10-I is known as that of an organotin
phosphinate[14] and the delocalized structure of Fig. 10-II was assigned from an NMR study[15].

Fig. 9. Transfer reactions in polymerizations catalyzed by
organotin condensate catalysts.

Rearward attack by a phosphoryl group on a coordinated epoxide
at the initiation step is possible as shown in Fig. 12. In view of
the possibility of a delocalized structure, this six-membered ring
transition state is equivalent to the structure for the initiation
step shown in Fig. 11. It is not necessary to limit catalytic site
structure for propagation as in Fig. 11. It appears, however, that
such structures would be effective for steric control of propagation.

(I)   (II)

Fig. 10.   Intra or inter molecular cyclic structure of the
           catalyst.

The author is indebted to Dr. J. Otera of Okayama University
of Science for his NMR works and discussions of the results and to
Dr. T. Nakata of Hokkaido University for her IR and other physical
studies.

The author is also indebted to Mr. K. Aoki of this company for
his collaboration in conducting the experiments.

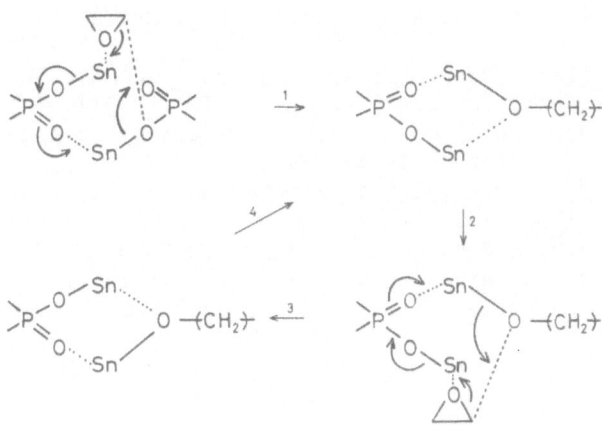

Fig. 11.   Assumed scheme of propagation.

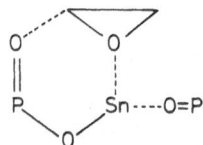

Fig. 12.   Rearward attack by phosphoryl group

References

1. E.J. Vandenberg, J.Polymer Sci. 47 486 (1960).
2. J. Furukawa, T. Tsuruta, R. Sakata, and T. Saegusa, Macromol. Chem. 32 90 (1959).
3. C.C. Price, and M. Osgan, J.Am.Chem.Soc. 78 690 (1956).
4. E.J. Vandenberg, J.Am.Chem.Soc. 83 3538 (1961).
5. C.C. Price, and R. Spector, J.Am.Chem.Sec. 88 4171 (1966).
6. G. Allen, C. Booth, and M.N. Jones, Polymer 5 195 (1964).
7. E.J. Vandenberg, "Macromolecular Syntheses" ed. by W.J. Bailey, John Wiley & Sons, Inc., New York (1972), p.49.
8. H. Coats, and P.A.T.Hoye, B.P. 1040297 (Aug. 2, 1963).
9. S. Sakai, K. Itoh, and Y. Ishii, Org. Synthetic Chem. 28 1109 (1970).
10. L.J. Bellamy, "The Infrared Spectra of Complex Molecules" 2nd. ed. Methuen & Co. Ltd., London (1958), P.311.
11. T. Saegusa, "Ring Opening Polymerization" (Japanese) Kagakudojin, Tokyo (1974), P.58.
12. J. Otera, T. Hinoishi, Y. Kawabe, and R. Okawara, Chemistry Letters 1981 237.
13. R. Okawara, and M. Ohara, "Organotin Compounds" vol 2, ed. by A.K. Sawyer, Marcel Dekker, Inc., New York (1962), P.253.
14. R.E. Ridenour, and E.E. Flagg, J.Organometal,Chem. 16 393 (1969).

NOVEL CATALYST SYSTEMS, BASED ON ORGANOZINC COMPOUNDS AND

GROUP IVB ORGANOMETALLIC COMPOUNDS, FOR THE

STEREOSPECIFIC POLYMERIZATION OF OXIRANES*

H. C. W. M. Buys, H. G. J. Overmars
and J. G. Noltes

Institute for Organic Chemistry TNO
P.O. Box 5009, 3502 JA Utrecht
The Netherlands

SYNOPSIS

Highly active catalyst systems suitable for the polymerization of methyloxirane have been prepared by reacting diethylzinc with a variety of group IVB organometallic compounds.

Triphenyl group-IVB-metal hydroxides, Ph_3MOH (M = Sn, Pb), as well as group IVB organometallic oxides, sulphides and selenides of the type R_2MX or R_3MXMR_3 (X = O, S, Se) yield very active catalysts upon 1/1 interaction with Et_2Zn. The reaction products of 2-hydroxyethyl(triphenyl group-IVB-metal)thiolates, $Ph_3MSCH_2CH_2OH$ (M = Sn, Pb), with Et_2Zn represent a third group of very effective catalysts. The latter catalysts are particularly useful for obtaining high-molecular weight, highly crystalline poly(methyloxirane).

Attention is given to the characterization of these novel catalyst systems and to the elucidation of the nature of the catalytically active species. Physical and mechanical properties of some of the polymeric products obtained are reported.

---

*Based on lectures presented at the Bad Nauheim Polymer Symposium of the Verein Deutscher Chemiker, Bad Nauheim, April 21-22, 1980 (H.C.W.M.B.) and at the Research Center of Hercules, Incorporated, Wilmington, Del., July 30, 1981 (J.G.N.)

INTRODUCTION

It is well-known that the polymerization of oxiranes using
organozinc compounds as basic components of the catalyst system is
strongly affected by variation of the cocatalyst. Depending on
the nature of the cocatalyst the various catalyst systems show
different activities and the resulting poly(oxiranes) vary
considerably in molecular weight and crystallinity. Diorganozinc
compounds R_2Zn (R = alkyl or aryl) exert catalytic activity in
the polymerization of oxiranes in the presence of cocatalysts such
as water, alcohols and oxygen [1,2], sulphur [5], ketones [3,4] and
amines [6,7]. The interaction of such agents with diorganozinc
compounds gives rise to the formation of Zn-X bonds (X = N, O or
S). The presence of such bonds is a prerequisite for catalytic
activity [8].

In our work Zn-O bonds have been generated by reacting
diorganozinc compounds with group IVB organometallic hydroxides
R_3MOH (R = alkyl or aryl; M = Si, Ge, Sn or Pb). The product of
the reaction between Me_2Zn and Me_3SiOH has already been
described as tetrameric methyl(trimethylsiloxy)zinc $[MeZnOSiMe_3]_4$
which has a cubic structure, as confirmed by x-ray structure
analysis [10]. The reactions between Ph_2Zn or Et_2Zn and Ph_3SiOH
or Et_3SiOH which have been examined as well [14,15] lead to
triorganosiloxy zinc derivatives of the type $[RZnOSiR'_3]_n$. Only
one report on the polymerization of oxiranes with such systems has
been published. In 1965 Saegusa et al.[9] described the attempted
polymerization of monochloromethyloxirane catalyzed by Et_2Zn and
Et_3SiOH which were present in a 1:1 and 1:2 molar ratio. However,
no activity was found.

Well-defined alkylzinc monoalkoxide compounds $[RZnOR']_n$ were
obtained from the reaction of tert-carbinols (M = C) and diorganozinc compounds [12]. These were found to be active in the polymerization of methyloxirane[13].

At Utrecht we have carried out an extensive study of the
possible suitability of representatives of compounds R_3MOH (M =
Si, Ge, Sn, Pb) as cocatalyst components in the polymerization of
oxiranes with organozinc compounds. Such compounds were available
at our Institute from a research program directed towards the
exploration of group IVB organometallic chemistry.

The successful results obtained prompted us to test combinations of diorganozinc compounds and various other types of
group IVB organometallic compounds for catalytic activity in the
polymerization of oxiranes. In the extensive series of
experiments in which combinations of organozinc compounds with a

variety of group IVB organometallic compounds were screened, the 2-hydroxyethyl(triorgano group-IVB-metal)thiolates emerged as very effective cocatalysts. Moreover, the poly(methyloxirane), obtained with these cocatalysts, was a highly crystalline polymer of high molecular weight.

In the following we will summarize some of our work on these organozinc-based catalyst systems and on the stereospecific polymerization of methyloxirane using these catalysts.

RESULTS AND DISCUSSION

In our research, catalyst systems containing Zn–O bonds have been generated by reacting diorganozinc compounds R_2Zn with triphenyl group-IVB-metal hydroxides, Ph_3MOH (M = Si, Ge, Sn or Pb). Methyloxirane was the principal monomer used in these studies.

The catalysts were prepared in situ by mixing the triphenyl group-IVB-metal hydroxide and the diorganozinc compound in a 1:1 molar ratio in benzene and stirring for half an hour or one hour at a definite temperature in a nitrogen atmosphere. The monomer was then added and polymerization was allowed to take place at room temperature. The results of a series of polymerization experiments are shown in Table 1.

Table 1. Polymerization of methyloxirane with
the systems Et_2Zn/Ph_3MOH[a]

Cocatalyst Ph_3MOH M	Conditions of Preparation of the Catalyst Systems		Solvent	Polymer Yield wt. %
Si	25°C	1 h	benzene	0
Si	80°C	1 h	benzene	0
Ge	80°C	1 h	benzene	0
Ge	130–140°C	1/2 h	m-xylene	0
Ge	170–200°C	1/2 h	no solvent	0
Sn	25°C	1 h	benzene	0
Sn	80°C	1 h	benzene	0
Sn	110°C	1/2 h	toluene	100
Sn	130–140°C	1/2 h	m-xylene	100
Pb	25°C	1 h	benzene	0
Pb	80°C	1 h	benzene	60
Pb	110°C	1/2 h	toluene	100

[a]Polymerization conditions are: monomer, methyloxirane, 10.5 ml; solvent, 10 ml; Zn/M IVB group compound molar ratio, 1 mol/mol; organozinc compound/monomer molar ratio, 0.01 mol/mol; polymerization time, 20 h at room temperature.

Under the conditions indicated, systems containing triphenyl-
silanol, Ph₃SiOH, and triphenylgermanium hydroxide, Ph₃GeOH, do
not display any activity. On the other hand the cocatalysts
triphenyltin hydroxide, Ph₃SnOH, and triphenyllead hydroxide,
Ph₃PbOH, give rise to excellent activity, if prior to the addition
of the monomer the catalyst system is heated at about 110°C in
toluene for half an hour. Evidently a chemical reaction must take
place in order to get a catalytically active species.

The equimolar reaction of diethylzinc with triphenyl group-
IVB-metal hydroxides was followed by measuring the rate of
ethane-formation in this reaction. In view of the high reactivity
of diethylzinc towards proton-active compounds the following
reaction may be assumed to take place:

$$Et_2Zn + Ph_3M^{IV}OH \longrightarrow EtZnOM^{IV}Ph_3 + EtH.$$

In benzene at 25°C this reaction is so fast that the relative
rates for the metals M are difficult to establish. When the
reaction is slowed down by using the complexing solvent
1,2-dimethoxyethane the graphs shown in Fig. 1 are obtained.

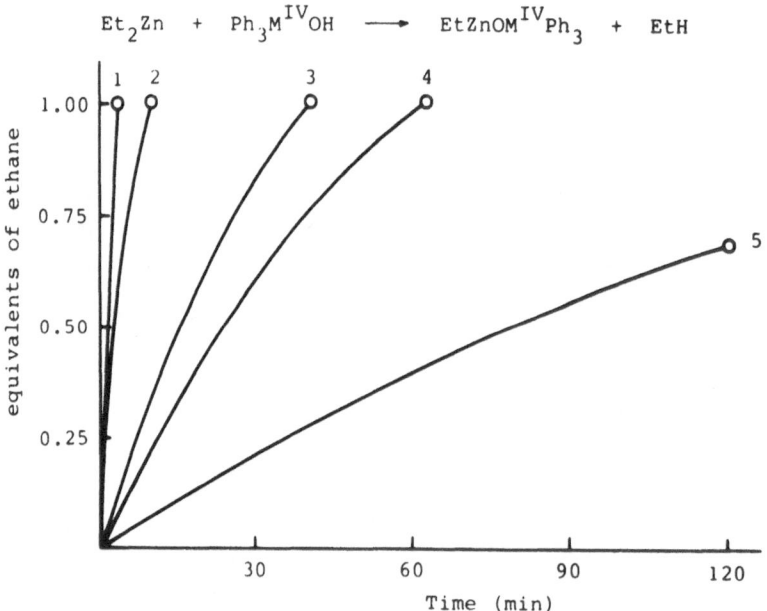

Fig. 1.   Reaction of diethylzinc with triphenyl group-IVB-metal
hydroxides, Ph₃MOH, for M = Si (1), Ge (2), C (3), Sn (4)
and Pb (5).   Equimolar reaction in 1,2-dimethoxy- ethane at
25°C.   Initial concentration of reactants: 0.75 M.

In the 1:1 molar reaction of Ph_3SiOH and Et_2Zn the formation
of one equivalent of ethane is very fast. Triphenylgermanium
hydroxide reacts less rapidly and a much slower evolution of
ethane is observed with the triphenyl tin and -lead hydroxides.
When the reaction is performed in benzene solution at 80°C the
formation of one equivalent of ethane is complete within two
hours. The difference in reactivity of diethylzinc towards the
triphenyl group-IV-metal hydroxides arises from the different
acidities of the hydroxyl groups in these compounds.

The reaction of Et_2Zn with Ph_3SiOH and Ph_3GeOH gives rise to
well-defined products which have been isolated and fully
characterized (see Table 2). The reaction products isolated are
indeed the expected acidolysis products which in benzene are
dimeric in the case of Si and trimeric in the case of Ge. These
reactions have already been described in the literature, but a
degree of association of 1.75 has been reported for the germanium
compound [11].

A ^{1}H=NMR study revealed a different picture for the interaction of Et_2Zn with Ph_3SnOH or with Ph_3PbOH. The ^{1}H-NMR
spectrum of the reaction product of Et_2Zn and Ph_3SnOH is more
complicated than would be anticipated if only the formation of
(triphenylstannoxy)ethylzinc and ethane had taken place. In
Fig. 2 only the resonance signals at high field are shown.

Spectrum A represents the situation after stirring for 1 hr at
room temperature. Two quartets are observed at 0.19 and 0.56 ppm
which must be assigned to the methylene group in ethylzinc units.
Apparently two different types of ethylzinc units are present.
The corresponding resonances of the methyl group of the ethylzinc
units appear at 0.96 and 1.11 ppm. The resonance at 0.80 ppm
belongs to ethane. Additional peaks, appearing around 1.30 ppm,
are assigned to the Et group in Ph_3SnEt from the system A by
extraction with pentane.

Table 2. Melting Points and Analytical Data for
compounds $EtZnOMPh_3$ (M = Si and Ge)

Compound	m.p. (°C)	Analyses, Found (calcd.) %		Molecular Association
		Zn	M	
$EtZnOSiPh_3$	212	17.20 (17.67)	7.43 (7.59)	2.2
$EtZnOGePh_3$	192	15.79 (15.78)	17.53 (17.52)	3.0

H. C. W. M. BUYS ET AL.

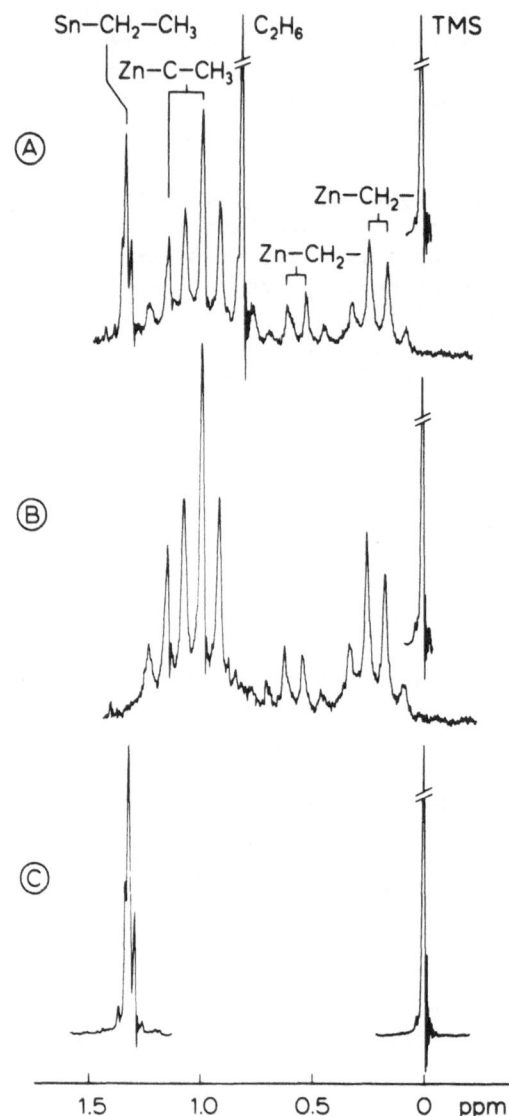

Fig. 2. High-field part of the ^{1}H-NMR spectrum of the mixture obtained upon reacting Et_2Zn with Ph_3SnOH in benzene solution.

Spectrum B is the spectrum obtained after removal of Ph_3SnEt from the system A by extraction with pentane.

Spectrum C is the spectrum of Ph_3SnEt isolated from the pentane extract.

The NMR results clearly prove the formation of triphenyl-ethyltin, Ph_3SnEt, upon interaction of Et_2Zn with Ph_3SnOH. Formation of Ph_3SnEt occurs to the extent of 20-30%. In the process, more than one ethylzinc species is formed. The formation of Ph_3SnEt must be accompanied by the formation of products containing more than one -ZnO- unit. It is plausible to assume that a primary reaction takes place in which ethyl(triphenyl-stannoxy)zinc is formed as an intermediate species. This compound is then believed to eliminate Ph_3SnEt with formation of benzene-soluble species in which n is larger than 1.

$$Et_2Zn \ + \ Ph_3SnOH \ \longrightarrow \ [EtZnOSnPh_3] \ + \ EtH$$

$$n \ EtZnOSnPh_3 \ \longrightarrow \ (n-1) \ Ph_3SnEt \ + \ Et(ZnO)_n SnPh_3$$

$$\text{(20-30%)} \qquad \text{(benzene soluble)}$$

The diethylzinc/triphenyltin hydroxide system shows catalytic activity only after heating the components at elevated temperatures (110-140°C). At these temperatures a clear, pale yellow solution was obtained from which Ph_4Sn crystallizes upon cooling (about 10-20 mol.%). Besides Ph_3SnEt and Ph_4Sn, also Ph_2SnEt_2 was isolated by repeated extraction of the catalyst system with n-pentane. These results are rationalized as presented in the following equations:

$$Ph_3SnOH \ + \ Et_2Zn \ \longrightarrow \ Ph_3SnOZnEt$$
$$\text{(I)}$$

$$\text{(I)} \ \xrightarrow[\text{(20-30 mol.%)}]{-Ph_3SnEt} \ Ph_3Sn(OZn)_n Et$$
$$\text{(II)}$$

$$(II) \xrightarrow[\text{(10-20 mol.\%)}]{-Ph_4Sn} Ph_2Sn[(OZn)_pEt]_2$$
$$(III)$$

$$(III) \xrightarrow[\text{(}\sim\text{10 mol.\%)}]{-Ph_2SnEt_2} Ph_2Sn[(OZn)_qEt]_2$$
$$(IV)$$

The primary intermediate (I) is susceptible to alkylation of the Sn-O bond by the EtZn group, resulting in the formation of Ph_3SnEt and the intermediate (II). Upon further heating (II) may undergo a redistribution reaction resulting in the formation of Ph_4Sn and the intermediate species of type (III). The latter compounds will be alkylated to diphenyldiethyltin by ethylzinc units still present in the catalyst system, giving rise to the formation of Ph_2SnEt_2 and species of type (IV).

The removal of Ph_3SnEt, Ph_4Sn and Ph_2SnEt_2 (all inactive in the polymerization of methyloxirane) from the system Et_2Zn/Ph_3SnOH results in a light yellow-brown catalyst powder which dissolves in benzene to a yellow solution.

The formation of Ph_4Sn during catalyst preparation was observed more or less accidentally when poly(methyloxirane) films cast from a $CHCl_3$ solution were studied under the microscope. A picture of the Ph_4Sn crystallized in PPO is shown in Figure 3.

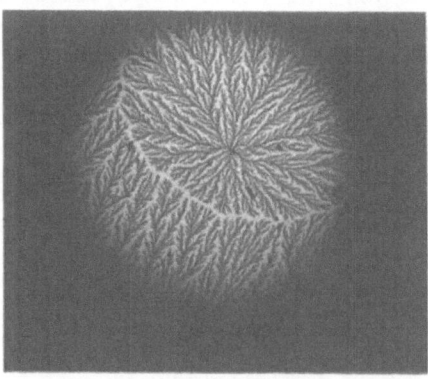

Fig. 3.   Crystallization of Ph_4Sn in film of poly(methyloxirane) cast from chloroform solution.

For the system Et_2Zn/Ph_3PbOH, analogous phenomena have been observed. Unlike the Sn-containing system the formation of a catalytically active species in the system Et_2Zn/Ph_3PbOH takes

place at somewhat lower temperatures (80-110°C) and is accompanied
by the formation of a white, catalytically active precipitate.
The chemistry of the formation of the Pb-Zn system is the same as
that for the Sn-Zn system.

$$Et_2Zn + Ph_3PbOH \longrightarrow [EtZnOPbPh_3] + EtH$$

$$n\ EtZnOPhPh_3 \longrightarrow (n-1)\ Ph_3PbEt + Et(ZnO)_nPbPh_3$$
$$\sim 70\%$$
benzene insoluble
catalytically active

$$\text{---}ZnEt + Ph_3PbO\text{---} \longrightarrow Ph_3PbEt + \text{---}ZnO\text{---}$$

We believe that the high catalytic activity of these systems
depends on the presence of a network of ZnO units, modified by the
presence of Ph_3M and EtZn groups.

The poly(methyloxirane) polymers prepared with these systems
are obtained as white, somewhat rubbery materials with high
molecular weights.

Poly(methyloxirane), prepared with the system Et_2Zn/Ph_3SnOH
(1/1), has been separated into two fractions by fractional
precipitation from an isopropanol solution of the polymer. A more
amorphous and a more crystalline fraction were obtained in this
way. The latter is about 20% by weight.

Extensive information is available in the literature about the
microstructural characterization of poly(methyloxiranes) by
^{13}C-NMR spectroscopy [16]. Fig. 4 shows the ^{13}C-NMR spectra in
the methylene and methine carbon region of crude poly(methyloxirane) and its amorphous and crystalline fractions. Without
going into detail it is clear that the ^{13}C-NMR spectrum confirms
the stereoregular character of the crystalline fraction.
Resonances due to structural irregularities, such as head-to-head
or tail-to-tail structures, are not observed.

It may thus be concluded that the catalyst system
Et_2Zn/Ph_3SnOH (1/1) provides initiating species, from which a
regular head-to-tail poly(methyloxirane) chain propagates. The
occurrence of stereospecific propagation in this polymerization
follows from the ^{13}C-NMR spectra of the polymer.

The catalytic activity of the Et_2Zn/Ph_3MOH systems is
connected with the further conversion of the primary species
$EtZnOMPh_3$ into species of the type $Et(ZnO)_nMPh_3$. The latter
conversion occurs for M = Sn or Pb, but not for M = Si or Ge. The
intermediate formation of similar species has been realized in the

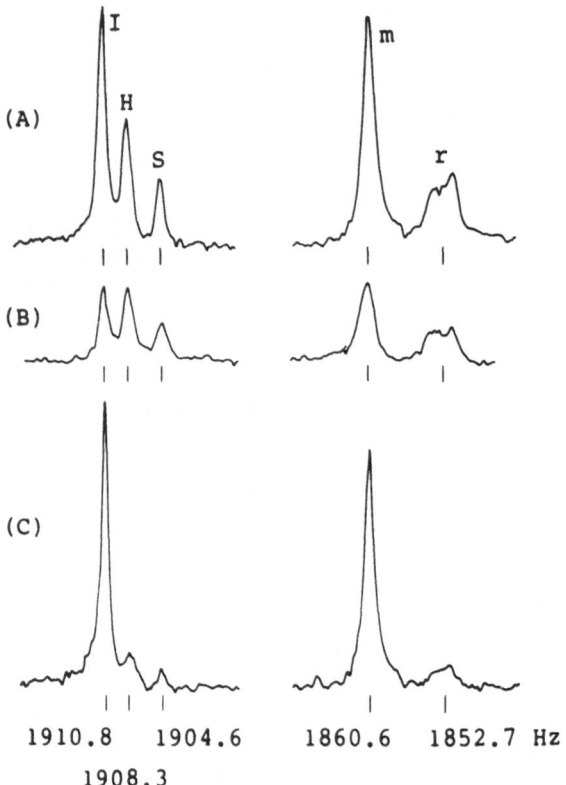

Fig. 4   Proton noise decoupled 25.1 MHz ^{13}C-NMR spectra in the
          methylene and methine carbon region of (A) crude
          poly(methyloxirane) prepared with the catalyst system
          Et_2Zn/Ph_3SnOH (1/1) and (B) the amorphous and (C) the
          crystalline fractions.

reaction of diorganozinc compounds with various group IVB organometallic oxides, sulfides and selenides, viz. by an insertion-type
reaction (1) or a ligand-exchange reaction (2). In both cases
intermediate species of the type $RZnXMR'_3$ are formed which may
then further react with elimination of a tetraorganometal compound
to give a catalytically active species containing a network of ZnX
units.

$$R'_2MX \ + \ R_2Zn \ \longrightarrow \ RZnXMR'_2R \qquad\qquad (1)$$
$$X = 0, \ S \ or \ Se$$

$$(R'_3M)_2X \; + \; R_2Zn \longrightarrow R'_3RM \; + \; RZnXMR'_3 \quad (2)$$
$$X = O, \; S \; or \; Se$$

$$n \; RZn-X-MR_3 \longrightarrow n-1 \; R_4M \; + \; R-(ZnX)_n-MR_3$$
$$X = O, \; S \; or \; Se$$

Combinations of diorganozinc compounds and these various group
IVB organometallic compounds have been tested for catalytic
activity in the polymerization of methyloxirane. Some results
have been collected in Tables 3 and 4.

The combination Ph_2Zn/Ph_2SnO represents a particularly
active combination (see Table 3). Other combinations give an
active catalyst only at elevated temperature. The organo-Ge and
-Pb systems are generally inactive.

The replacement of oxygen by sulphur or selenium in these
oxides results in strongly enhanced catalytic activity. In most
cases polymer yields are quantitative within 1-2 hours at room
temperature. High molecular weight poly(methyloxirane) is
obtained. The tacticity of the polymers obtained is very similar
to that for the Et_2Zn/Ph_3SnOH system, i.e. the crystalline
fraction amounts to 20%. Surprisingly, the system $Et_2Zn/$
$(Ph_3Sn)_2S$ has no catalytic activity at all, whereas the
corresponding system containing Ph_2Zn is highly active.

Table 3.   Polymerization of methyloxirane with Ph_2Zn or
Et_2Zn in combination with R_2MO or $(R_3M)_2O$ in benzene

Catalyst Components		Polymer Yield (wt.%)
Ph_2PbO	Ph_2Zn	22
$(Ph_3Pb)_2O$	Ph_2Zn	0
Ph_2SnO	Ph_2Zn	100 (100)[a]
Ph_2SnO	Et_2Zn	18 (18)[a]
Me_2SnO	Ph_2Zn	0 (0)[a]
$(Ph_3Sn)_2O$	Ph_2Zn	24 (100)[a]
$(Ph_3Sn)_2O$	Et_2Zn	0 (100)[a]
$(Ph_3Ge)_2O$	Ph_2Zn	0 (0)[a]
$(Ph_3Ge)_2O$	Et_2Zn	0 (0)[a]

[a]Catalyst preparation at elevated temperatures (110-140°C)

Table 4.   Polymerization of methyloxirane with Ph_2Zn or
           Et_2Zn in combination with R_2MX or $(R_3M)_2X$
           (X = S or Se) in benzene

Catalyst Components		Polymer Yield (wt.%)
$(Ph_3Pb)_2Sn$	Et_2Zn	100
Ph_2SnS	Et_2Zn	100
Ph_2SnS	Ph_2Zn	100
Bu_2SnS	Et_2Zn	100
Bu_2SnS	Ph_2Zn	100
$(Ph_3Sn)_2S$	Et_2Zn	0[a]
$(Ph_3Sn)_2S$	Ph_2Zn	100
$(MePh_2Sn)_2S$	Ph_2Zn	100
$(PrPh_2Sn)_2S$	Ph_2Zn	100
$(BuPh_2Sn)_2S$	Ph_2Zn	100
$(HexPh_2Sn)_2S$	Ph_2Zn	100
$(Et_3Sn)_2S$	Ph_2Zn	100
$(Bu_3Sn)_2S$	Ph_2Zn	100
$(Ph_3Ge)_2S$	Ph_2Zn	100
Ph_2SnSe	Ph_2Zn	100
$(Ph_3Sn)_2Se$	Ph_2Zn	100

[a]Catalyst preparation at elevated temperatures (110-140°C).

    In view of the high catalytic activity of these systems the
reaction of diethylzinc with diphenyltin sulphide was investigated
in greater detail in particular for the purpose of identifying the
nature of the actual cataytic species.

    The following sequences of reactions have been proved to occur
by NMR spectroscopy, isolation of Ph_2SnEt_2 and elemental
analysis of the catalyst:

    Ph_2SnS + Et_2Zn $\longrightarrow$ $EtZnSSnPh_2Et$

    n $EtZnSSnPh_2Et$ $\longrightarrow$ (n-1) Ph_2SnEt_2 + $Et-(ZnS)_n-SnPh_2Et$

                                                                (n = 6)

Elimination of Ph_2SnEt_2 results in the formation of $(ZnS)_n$
oligomers with Et and Ph_2EtSn end groups. It is the build-up of
this inorganic $(ZnS)_n$ oligomer which is believed to be connected
with the generation of catalytic activity.

Table 5.   Effect of the type of solvent on the polymerization
            rate of methyloxirane with the catalyst system
            Et_2Zn/Ph_2SnS (1/1)[a]

Solvent	Yield (%)	
	1 hour	20 hours
benzene	80	100
n-pentane	71	100
diethylether	58	100
dioxane	24	100
tetrahydrofuran	23	100
pyridine	1	1
tetramethylethylenediamine	4	4

[a]Polymerization conditions: monomer, methyloxirane, 5.25 ml;
solvent, 10.5 ml; organozinc compound: monomer molar ratio, 0.01
mol/mol, 30°C.

     Table 5 shows the effect of the type of solvent on the polymer-
ization rate of methyloxirane using the Et_2Zn/Ph_2SnS (1/1) catalyst
system.   In the presence of complexing solvents such as dioxane,
tetrahydrofuran and pyridine a remarkable retardation occurs in
the polymerization rate.   This is likely to be due to the
coordination of solvent to catalytically active zinc atoms.   With
dioxane and tetrahydrofuran as a solvating medium labile complexes
will be formed, which exchange rapidly with methyloxirane as a
ligand.   The stronger coordinating power of pyridine prevents
coordination of methyloxirane to zinc.   It is, therefore,
concluded that the polymerization reaction proceeds through
coordination of methyloxirane to coordinatively unsaturated zinc
atoms present in the $(ZnS)_n$ oligomer.

     Fig. 5 shows an x-ray diagram of the powdery catalyst
precipitate obtained from the Et_2Zn/Ph_2SnS (1/1) system.   This
x-ray diagram points to an amorphous structure for the $(ZnS)_n$
oligomer.   The sharp diffraction lines, observed in the diagram of
crystalline zinc sulphide, have disappeared; only two broad lines
are observed.   Crystalline ZnS is catalytically inactive, the
$Et(ZnS)_n$ oligomer is extremely active.

     In the x-ray diagram of the precipitate, resulting from the
system $Et_2Zn/(Ph_3Sn)_2S$ prepared in benzene, (Fig. 6), the
reflections are concentrated at the positions of the sharp
reflection in the diagram of crystalline zinc sulphide.   However,
the lines are much more diffuse.   It is concluded that this system
exhibits some crystallinity.   No activity has been found for this
system in the polymerization of methyloxirane.

X-ray powder diagram of the catalyst precipitate
from the system Et$_2$Zn/Ph$_2$SnS (1/1)

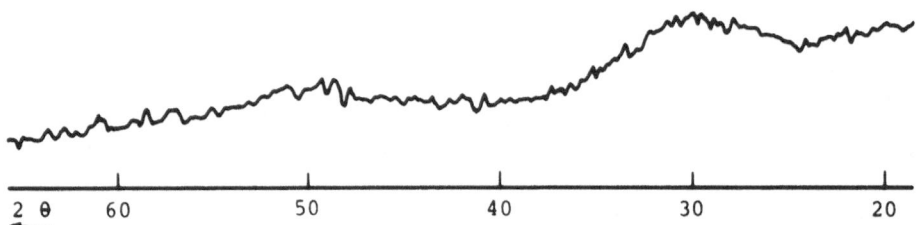

X-ray powder diagram of crystalline ZnS

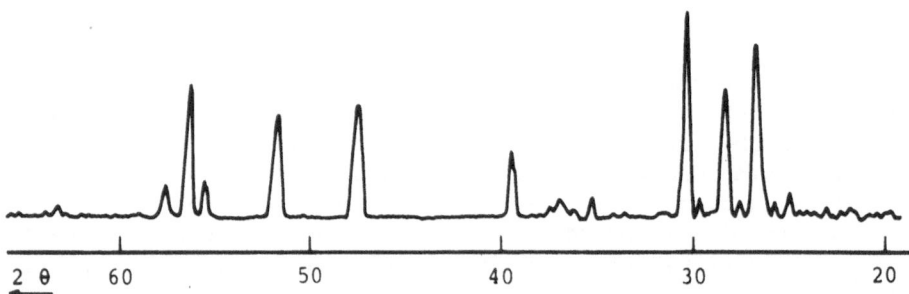

Fig. 5.   X-ray powder diagram of Et$_2$Zn/Ph$_2$SnS (1/1) catalyst
and of crystalline zinc sulphide.

However, when the catalyst is prepared in the presence of
dioxane the x-ray diagram indicates a decrease in crystallinity of
the precipitated catalyst. For this more amorphous precipitate a
good catalytic activity has been found in the polymerization of
methyloxirane. It appears then that a certain degree of disorder
in the ZnS catalyst seems to be essential for catalytic activity
in these systems.

Results for a third group of very effective catalysts for the
polymerization of oxiranes are given in Table 6.

X-ray powder diagram of the precipitate from the system
$Et_2Zn/(Ph_3Sn)_2S$ (1/1), prepared in benzene

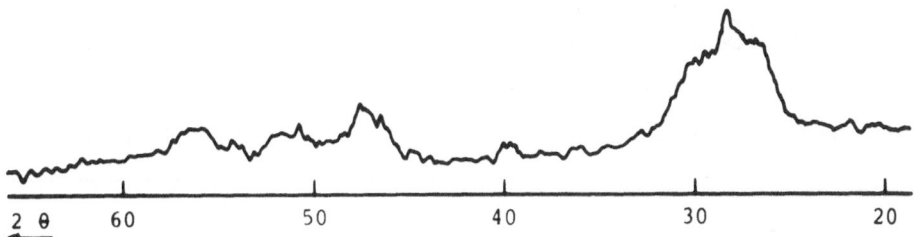

X-ray powder diagram of the precipitate from the system
$Et_2Zn/(Ph_3Sn)_2S$ (1/1), prepared in benzene/dioxane

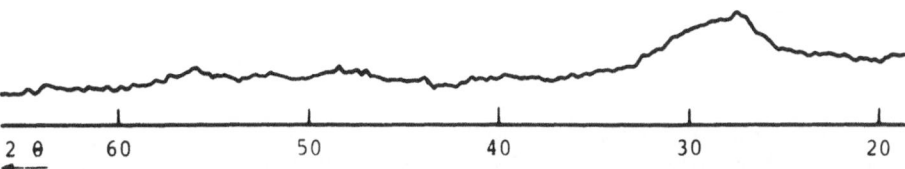

Fig. 6.   X-ray powder diagram of $Et_2Zn/(Ph_3Sn)_2S$ (1/1) catalyst
prepared in benzene and benzene/dioxane (1/1) mixture,
respectively.

The 2-hydroxyethyl(triorgano group-IVB-metal)thiolates emerged as
particularly effective cocatalysts. Moreover, the poly-
(methyloxirane) obtained with these cocatalysts is a highly
crystalline polymer of high molecular weight. With the Pb and Sn
compound extremely active systems are obtained. Quantitative
polymerization to high molecular weight product occurs using catalyst
concentrations in the order of 0.1 mole %. In the case of the Ge
compound no activity has been found. The building-up of the
catalytically active species (II) from the catalyst components is
presented for the zinc/lead system:

$$Et_2Zn + Ph_3PbSCH_2CH_2OH \longrightarrow [EtZnOCH_2CH_2SPbPh_3] + EtH$$

$$n\ EtZnOCH_2CH_2SPbPh_3 \longrightarrow (n-1)\ Ph_3PbEt + Et(ZnOCH_2CH_2S)_nPbPh_3$$

$$80-90\% \qquad\qquad (II)$$

$$\bar{n} = 8-9$$

Table 6. Polymerization of methyloxirane using catalyst
         combinations of R_2Zn (R = Et or Ph) and
         $Ph_3MSCH_2CH_2OH$ (M = Ph, Sn or Ge) or

         $Ph_2\overline{MSCH_2CH_2S}$ (M = Pb or Sn)

Catalyst Components		Polymer Yield (Wt.%)
Ph_2Zn	$Ph_3PbSCH_2CH_2OH$	100
Et_2Zn	$Ph_3PbSCH_2CH_2OH$	100
Ph_2Zn	$Ph_3SnSCH_2CH_2OH$	100
Et_2Zn	$Ph_3SnSCH_2CH_2OH$	80
Et_2Zn	$Ph_3GeSCH_2CH_2OH$	0
Ph_2Zn	$Ph_2\overline{PbSCH_2CH_2S}$	36
Ph_2Zn	$Ph_2\overline{SnSCH_2CH_2S}$	40
Et_2Zn	$Ph_2\overline{SnSCH_2CH_2S}$	0

Again, one equivalent of EtH is evolved in the first step
which is then followed by elimination of Ph_3PbEt to give the
catalytically active zinc species. The actual catalyst consists
of $[-ZnOCH_2CH_2S-]$ oligomers with n = 8-9. These will be
interconnected via Zn-O and Zn-S coordinate bonds to give a
polymeric structure end-capped by Et and Ph_3Pb groups.

The poly(methyloxirane) obtained with this catalyst differs
from the polymers obtained with the Et_2Zn/Ph_3SnOh system in
that both the molecular weight and the degree of crystallinity are
considerably higher (90-100% isotactic).

Already in 1956 Price and Osgan have obtained optically active
poly(methyloxirane) starting from one enantiomer using the anionic
catalyst KOH [17]. A low molecular weight, solid isotactic poly-
(methyloxirane) resulted. Low molecular weight polymer in the
form of a viscous liquid was obtained starting from racemic
monomer and the same KOH catalyst. Anionic polymerization with
KOH results in CH_2-O ring-opening with retention of the
configuration of the CH carbon atom. Optically active monomer
will give an optically active polymer.

We have used the $Et_2Zn/Ph_3PbSCH_2CH_2OH$ (1/1) catalyst to
polymerize one of the enantiomers of methyloxirane. Starting from
R-methyloxirane our catalyst likewise produced an optically active
polymer. However, the molecular weight was much higher.
According to ^{13}C-NMR the polymer was isotactic. Upon polymer-

ization of racemic methyloxirane a polymer with identical melting
point, tacticity and molecular weight was obtained. In this case
the polymer is optically inactive and it is assumed that the
polymer consists of a 1/1 mixture of fully isotactic chains of
poly-R- and poly-S-methyloxirane. The high degree of tacticity of
the poly(methyloxirane) is also illustrated by the ready formation
of spherulites.

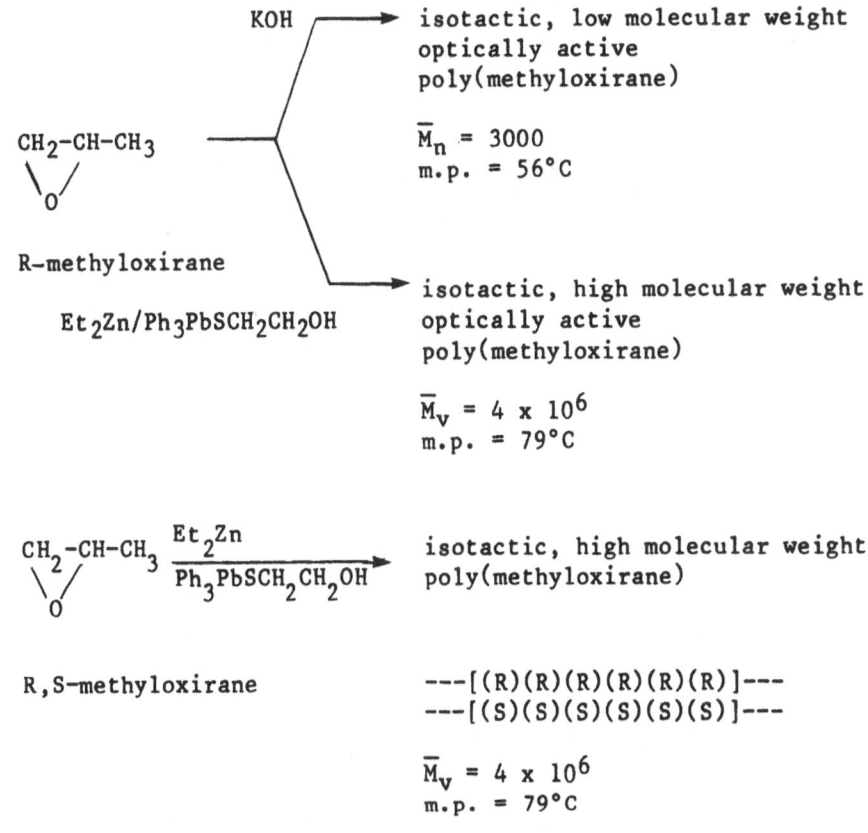

KOH → isotactic, low molecular weight
optically active
poly(methyloxirane)

$\overline{M}_n$ = 3000
m.p. = 56°C

CH$_2$-CH-CH$_3$
\O/

R-methyloxirane

Et$_2$Zn/Ph$_3$PbSCH$_2$CH$_2$OH → isotactic, high molecular weight
optically active
poly(methyloxirane)

$\overline{M}_v$ = 4 x 10^6
m.p. = 79°C

CH$_2$-CH-CH$_3$ $\xrightarrow{\text{Et}_2\text{Zn} / \text{Ph}_3\text{PbSCH}_2\text{CH}_2\text{OH}}$ isotactic, high molecular weight
\O/   poly(methyloxirane)

R,S-methyloxirane

---[(R)(R)(R)(R)(R)(R)]---
---[(S)(S)(S)(S)(S)(S)]---

$\overline{M}_v$ = 4 x 10^6
m.p. = 79°C

Extension of the chain length of the triphenyllead thiolate
results in a strongly reduced tacticity of the polymer (see Table
7).

For catalysts derived from triphenyllead thiolates containing
an n-propylene or n-butylene instead of an ethylene group the
percentage of isotactic polymer formed is the same as that in the
Et$_2$Zn/Ph$_3$SnOH system. Perhaps intramolecular Zn-S coordination
in the glycolate system plays a part. It is known in zinc-complex
chemistry that coordination resulting in a 5-membered chelate ring
is strongly favored over formation of a 6- or 7-membered chelate
ring.

Table 7.   Influence of chain length of thiolate on
tacticity of polymer

Catalyst System	Poly(methyloxirane)
$Et_2Zn/Ph_3PbSCH_2CH_2OH$	90-100% isotactic
$Et_2Zn/Ph_3PbSCH_2CH_2CH_2OH$	20% isotactic
$Et_2Zn/Ph_3PbSCH_2CH_2CH_2CH_2OH$	20% isotactic
Et_2Zn/Ph_3SnOH	20% isotactic

Table 8 presents some mechanical properties of the poly-
(methyloxirane) polymers prepared with our zinc catalysts. Data
for some other commercial polymers have been added for comparison.

Table 8.   Mechanical properties of different poly(methyloxirane) samples, compared with some other products

Product	Tensile Strength at Break $(dyn/cm^2 \times 10^{-8})$	Elongation at Break (%)	Tensile Modulus $(dyn/cm^2 \times 10^{-8})$
Low crystalline (cast) poly(methyloxirane)[a]	0.33	319	0.47
Highly crystalline (cast) poly(methyloxirane), (melt-pressed) prepared in benzene[b]	3.56 2.55	372 +400	3.75 5.18
Highly crystalline (cast) poly(methyloxirane), melt-pressed) prep. in petr. ether 60-80°C[b]	2.72* 2.94*	226 395	21.37 9.41
Hycar[R] 2679	0.76	443	0.32
Cariflex[R] TR 3200	1.25	917	1.73
P.U. from Corfam[R]	4.51	434	1.51
Lycra[R]	5.36	689	0.87

*Probably higher.   [a]Et_2Zn/H_2O catalyst.   [b]$Et_2Zn/Ph_3PbSCH_2CH_2OH$
catalyst.

The high value of the elongation at break for the highly
crystalline, melt-pressed polymer and the high values for the
tensile modulus deserve attention. A melt-pressed strip of the
highly crystalline poly(methyloxirane) can be stretched. Cold-
drawing of this strip is possible. Sections of drawn and undrawn
polymer can clearly be distinguished. Stretching up to 400% is
possible. If the stretched film is heated to a temperature of
50-60°C it reverts to its original dimensions. The films can be
stretched several times. These properties are believed to be
connected with the high amount of micro-crystalline domains in the
polymer. The microcrystalline regions act as physical cross-links
rendering the polymer a macromolecular three-dimensional network
structure.

REFERENCES

1.   J. Furukawa, T. Tsuruta, R. Sakata, T. Saegusa, and
     A. Kawasaki, Makromol. Chem., 32, 90 (1959).
2.   R. Sakata, T. Tsuruta, T. Saegusa, and J. Furukawa,
     Makromol. Chem., 40, 64 (1960).
3.   K. T. Garty, T. B. Gibb, Jr., and A. Clendinning,
     J. Polymer Sci., A1, 1, 85 (1963).
4.   T. Tsuruta and R. Fujio, Makromol. Chem., 75, 208 (1964).
5.   J. Lal, J. Polymer Sci., A1, 4, 1163 (1966).
6.   E. J. Vandenberg, J. Polymer Sci., 47, 486 (1960).
7.   H. Tani, T. Araki, N. Oguni, and N. Ueyama, J. Amer.
     Chem. Soc., 89, 173 (1967).
8.   J. Furukawa, and T. Saegusa, Polymerization of aldehydes
     and oxides, Polymer Reviews (H. F. Mark and E. H.
     Immergut, Eds.), Interscience Publishers, Wiley and
     Sons, New York, 1963, Volume 3, p. 155.
9.   T. Saegusa, T. Ueshima, T. Nakajima, and J. Furukawa,
     Kogyo Kagaku Zasshi, 68 (12), 2514 (1965); C.A., 65,
     12290 (1965).
10. F. Schindler, H. Schmidbauer, and U. Krüger, Angew.
     Chem., 77 (19) 865 (1965).
11. Yu. N. Krasnov, R. F. Galiullina, G. G. Petukhov, and
     M. D. Bogoyavlenskaya, Tr.Khim.Khim.Tekhnol., 2, 16
     (1973) ; C.A., 80, 83161 (1974).
12. G. E. Coates and D. Ridley, J. Chem. Soc., A, 1064
     (1966).
13. M. Nakaniwa, I. Kameoka, R. Hirai, and J. Furukawa,
     Makromol. Chem., 155, 197 (1972).
14. H. Schmidbaur, Advances in Organometallic Chemistry (F.
     G. A. Stone and R. West, Eds.), Academic Press, New
     York, 1970, Volume 9, p. 286.

15.   G. G. Petukhov, R. F. Galiullina, Yu. N. Krasnov, and
      A. D. Chernova, J. Gen. Chem. (USSR), $\underline{42}$, 1037 (1972).
16.   N. Oguni, K. Lee, and H. Tani, Macromolecules, $\underline{5}$, 819
      (1972).
17.   C. C. Price and M. Osgan, J. Am. Chem. Sco., $\underline{78}$, 4787
      (1956).

POLYMERS AND COPOLYMERS OF METHYL ω-EPOXYALKANOATES[*]

O. Vogl, P. Loeffler, D. Bansleben and J. Muggee

Polymer Science & Engineering Department
University of Massachusetts
Amherst, Massachusetts   01003

SYNOPSIS

Polymers and copolymers of methyl ω-epoxyalkanoates with 0 to 8 methylene groups separating the epoxy- and carbomethoxy groups have been prepared with the aluminumalkyl/water/acetyl-acetone (1/0.5/1) initiator system. The polymers are of high molecular weight with a M_n of 250,000 and a polydispersity slightly more than the most probable molecular weight distribution of 2. Homopolymers and copolymers with cyclic ethers have been characterized by spectroscopic analysis and gel permeation chromatography but also by the determination of glass transition temperatures and where applicable by x-ray identification. Selected polymers have been hydrolyzed to the ionomers and transformed into the acid. Methyl ω-epoxyalkanoates were prepared by epoxidation of methyl alkenoates with m-chloroperbenzoic acid. The rate of epoxidation and of subsequent polymerization depended on the length of the spacer length of methylene groups. The rates were low when the spacer length was 0 or 1 and reached a constant value when the spacer group had more than 3 methylene groups.

---

[*]Part VI on Poly(alkylene oxide) Ionomers.
Part V: O. Vogl, J. Muggee and D. Bansleben, Polymer J.
(Japan), 12(9), 677 (1980).

INTRODUCTION

    The discoveries that aluminum alkyls, particularly triethyl-
aluminum and zinc alkyls, particularly diethylzinc (1,2), could be
modified with water, alcohols or other compounds (3-5), have led
to the use of these reaction mixtures as the most important general initiator systems for the polymerization of epoxides (6), cyclic
ethers (2,7) and lactones to polymers of high molecular weight.
Promoters of cyclic ether polymerization are also used in combination with the modified organometallic initiator systems; prominent
among those rank epichlorohydrin (8), acetylacetone or
other carbonyl compounds, esters, ethers, and selected amines,
particularly aromatic amines (1,6,9).

    Initially these initiator systems were developed for the poly-
merization of ethylene oxide (1-7) and propylene oxide. Ethylene
oxide could be polymerized to very high molecular weight. The
polymerization of propylene oxide with these aluminum- or zinc-
alkyl/water modified systems gave rise to many important contributions in the area of stereoselective and stereoelective polymerizations (10-13). It was shown that propylene oxide could be polymerized stereospecifically to isotactic polymers if the proper combination of organometallic compounds, water and modifier was
used. Optically active modifiers, like phenylalanine (13) or born-
eol (11) in these organometallic initiating systems resulted in the
formation of polymers with optical activity.

    Generally, the polymerization of propylene oxide is not com-
pletely stereoelective and only a 10% to 50% electivity (and optical
activity) of the maximum value could be obtained (10-13). Such
stereoelective polymerization reactions have made the separation
of optical antipodes of cyclic ethers possible. By selective polymerization of one antipode the remaining monomeric antipode of
opposite sign could be isolated, either in pure or enriched form
(14).

    Details of the polymerization mechanism of cyclic ethers with
modified aluminumalkyls remain an unsolved problem (15), even
though a number of mechanisms have been suggested. Most of
these proposals account for the stereospecific propagation through
coordination with one metal atom; others require several metal
atoms for the effectiveness of the initiating system (16).

    In cases where aluminumalkyls were modified with aniline or
with methyl benzoate, molecular structures of some purified reaction products, believed to be responsible for initiation, have been
determined by x-ray analysis (17); x-ray structure analyses of
some modified zincalkyl systems have also been reported. Stereoelective chain propagation and the optical activity of the resulting
polymers were explained by describing a chemical compound con-

taining several zinc atoms which have a chirality and by virtue of this an optically active compound results.

In all cases of modified aluminumalkyl and zincalkyl initiating systems, large amounts, sometimes in the range of 5 to 10 mole %, of these initiators had to be used. Interestingly, even at such high initiator concentrations polymers of very high molecular weights were obtained. This result makes it quite clear that only a very small percentage of the initiator system is truly active and utilized for the polymerization. Although it was often claimed that a proton was responsible for initiation in such systems, the question whether a proton is necessary or is the actual initiator in the definition of Kennedy (18) remains open. One research group claims that protons are the actual initiators and the complicated aluminum compounds serve only as effective and stable non-nucleophilic counterions (16). Yet the stereospecificity and stereoelectivity of these systems are strong evidence that coordination and polymerization actually do occur on specific sites of the initiating systems with aluminum or zinc as the coordinating metal. The controversies which have existed are still not resolved and the actual initiating species are unknown. Many attempts have been made to isolate and identify them, but most of these efforts have been unsuccessful. This problem is similar to the attempted identification of certain coordination polymerization initiating systems involving transition metal compounds. The halides of titanium, vanadium, hafnium, chromium, tungsten, molybdenum and others, when reacted with aluminum alkyls or other reducing agents give highly effective initiators for the polymerization of ethylene and α-olefins (19). As in aluminumalkyl/water or alcohol systems for cyclic ether polymerizations, only small amounts, perhaps 1 to 3 mole % of the compound mixture, are actually active centers in such coordination initiators for polymerization of α-olefins (19). Many proposals for initiation and propagation of polymerization have been developed also for these systems, but the true nature of the actual initiator complex and mechanism remains unknown (19).

The only specific compounds which are active in the polymerization of four-membered lactones and of some cyclic ethers and which have been identified, have the following structures: $R_2Al-OZn(\text{or } Mg)-O-AlR_2$ (20,21). Even in these cases it is not clear whether these compounds themselves or their reaction products with monomers are the actual polymerization initiators.

Modified aluminum- or zincalkyls have been used for initiating polymerizations of epoxides, oxetanes and lactones. It is well known that all cyclic ethers also undergo polymerizations with cationic initiators and that epoxides can also be polymerized with anionic initiators. The capability of these compounds to polymerize depends upon the ring strain and the electron density

at the heteroatom in the ring. For example, ethylene oxide and
propylene oxide can be readily polymerized with cationic and anionic initiators (2,7) as well as with modified organometallic initiators;
tetrafluoroethylene oxide (22) and hexafluoropropylene oxide (22)
can be polymerized only with anionic initiators. Modified organometallic initiators were used successfully for the polymerization of
terminal epoxides with aliphatic side chains for internal epoxides
such as cyclohexene oxide and 2-butene oxide (1,9) or for epoxides with a chloromethyl side chain like epichlorohydrin (1,6), or
with an ether side chain like phenyl glycidyl ethers (1,9). In
some cases the stereospecificity of the polymerization has been
studied. A particularly elegant work described the stereochemistry of the addition for the polymerization of oxirane. Either inversion or retention of configuration could be observed in the polymerization of 2,3-epoxybutane depending on the initiating system
and polymerization conditions (9). Several oxetanes have also
been polymerized with these initiator systems. Most noteworthy,
other than the polymerization of oxetane itself, is the polymerization of 3,3-bischloromethyl oxetane (23).

Aluminumalkyl initiator systems were also used for the polymerization of ethyl glycidate (EG); a polymer was obtained which
was partially crystalline (24). In subsequent years, the polymerization of EG as the comonomer in the copolymerization with trioxane was studied (25-32). It was found that trioxane (TO) copolymerized with EG in the gas phase or in solution with BF_3 as initiator; it also copolymerized with glycidonitrile. The EG/TO copolymers could be converted to polyoxymethylene (POM) based ionomers (25-32). These copolymers were of relatively low molecular
weight; high molecular weight polymers were obtained when 1,3-
dioxolane was used in amounts of 5 to 10 mole % as the termonomer
and suggests that EG probably acted as a chain transfer agent
(25-32). The incorporation of EG into POM was poor, only one
tenth of the comonomer feed amount was incorporated into the POM
copolymer. Other workers studied the use of classical cationic initiators such as triflate esters for the homo- and copolymerization
of EG. Only low molecular weight homopolymers were obtained;
while copolymerization of EG with tetrahydrofuran (THF) gave only low molecular weight copolymers containing the components in a
ratio of 1:2 (16).

Copolymerizations of EG with propylene oxide (PO), 1-butene
oxide (BO), and oxetane (Ox) were also carried out with triethylaluminum alkyl/water/acetylacetone systems (1/0.3/0.5). As in
the case of TO copolymerizations with BF_3, with this initiator system too, only one tenth of the comonomer feed ratio of EG was incorporated into the copolymer (33). It was further found that
with increasing amounts of EG in the comonomer feed, the molecular weight of the copolymer was much lower. While copolymers
with an inherent viscosity of 3 to 4 dL/g could be obtained with

PO at comonomer ratios of 20:1, the inherent viscosity of the copolymer decreased to about 0.5 dL/g if the feed ratio was 3:1, yet
the incorporation of EG was still only 10% (3 mole % comonomer content).

Copolymerization of EG with a number of substituted epoxides
or Ox and with aluminumalkyl/water/acetylacetone initiator systems, (1.0/0.3/0.5) gave copolymers of high molecular weight. Incorporation of EG, however, was only between 1 and 4 mole %, although the feed ratio of the monomers in these copolymerizations
ranged from 10 to 30 mole %. EG could not be incorporated into
copolymers with EO at all; only homopolymers of EO were isolated
(33).

Studies of the effectiveness of modified aluminumalkyl/water
systems as the initiator for the polymerization of cyclic ethers and
functionally substituted oxiranes continue in our laboratory.

In a previous paper we described preliminary results of our
work dealing with the syntheses of methyl ω-alkenoates and methyl
ω-epoxyalkanoates with methylene groups as spacers between the
ester group and the oxirane ring (34). We have also described
the initial results of the homopolymerizations of several functionally substituted oxiranes and their copolymerization with PO, BO
and other epoxides.

In the present paper we discuss the continuation of our work
with the objective to investigate: (a) The effect of spacer length
between the olefinic double bond and the carboxylic ester group
on the rate of epoxidation of methyl ω-alkenoates. (b) The poly-
merizability of methyl ω-epoxyalkanoates. (c) The effect of the
ester functionality on the reactivity of the methyl ω-epoxyalkanoates with the triethylaluminum/water/acetylacetone system. (d)
The properties of the resulting homo- and copolymers.

EXPERIMENTAL

Measurements

The transition temperatures for the polymers were determined
on a Perkin-Elmer DSC-2 differential scanning calorimeter at a
heating rate of 20°C/minute in a nitrogen atmosphere; the mid-
point of the ΔCp was taken as the glass transition temperature.

Infrared spectra were recorded (either as liquids between
salt plates or as thin films cast from solution) on an Infracord or
a Perkin-Elmer Model 283 infrared spectrometer.

Materials

The materials purchased for this work and the purification of starting materials and solvents are described elsewhere (34).

Procedures

Epoxidation with m-chloroperbenzoic acid followed the earlier procedures (35,36). Preparation of the initiator system and typical homo- and copolymerizations were described in detail in our previous paper (34). Reaction conditions for the polymerizations discussed in this article are summarized in Table 1.

TABLE 1

Copolymerization of Oxiranes and Oxetane with MEA
Comonomer Feed: 30 mole % MEA, Initiating System:
$AlEt_3/H_2O/AcAc$: 1/0.5/1(a)

	MEP			MEO			MEU		
Type of Comonomer	Yield, in %	η_{inh}(b) in dL/g	Comonomer(c) Content in Polymer	Yield, in %	η_{inh}(b) in dL/g	Comonomer(c) Content in Copolymer	Yield in %	η_{inh}(b) in dL/g	Comonomer(c) Content in Copolymer
None	13	--	100	88	1.0	100	85	2.0	100
EO	--	--	--	--	--	--	57	4.8	15
PO	63	--	27	64	1.2	30	90	1.0	20
BO	51	1.3	25	75	1.8	30	90	2.7	25
HO	22	1.1	25	93	1.8	30	83	0.5	25
PhGE	45	0.8	25	71	0.8	30	55	--	25
ECH	20	0.9	30	89	0.8	30	60	2.6	17
TCCBO	--	--	--	--	--	--	35	0.2	40
Ox	11	0.9	17	--	--	--	--	--	--

MEA = methyl ω-epoxyalkanoate MEO = methyl-7,8-octanoate

MEP = methyl-4,5-epoxypentanoate MEU = methyl-10,11-epoxyundecanoate

(a) Initiator Concentration: 5 mole % EO = ethylene oxide PhGE = phenyl glycidyl ether

(b) 0.5% in Benzene Solution at 30°C PO = propylene oxide ECH = epichlorohydrin

(c) Reaction Time: 2-4 weeks BO = 1-butene oxide TCCBO = 4,4,4-trichlorobutylene oxide

HO = 1-hexene oxide Ox = oxetane

RESULTS AND DISCUSSION

Epoxides which have a carboxylate group at the other end of the molecule and are separated from it by 0 to 8 methylene spacer groups, have been synthesized and characterized. The epoxides were prepared from the methyl ω-alkenoates and m-chloroperbenzoic acid, an excellent and safe peroxidizing agent.

Methyl ω-alkenoates in turn were synthesized by various methods, all of which have been described earlier (34). They included pyrolysis of cyclic lactones, oxidative decarboxylation of methyl half esters of dicarboxylic acids with lead tetraacetate and copper salts or other special methods. Where needed, the esterification of the acid was carried out with methanol and sulfuric acid; no shift of the terminal double bond was observed during esterification.

The epoxidation of the terminal double bond of the methyl 1-alkenoates with m-chloroperbenzoic acid in methylene chloride proceeded rapidly at room temperature for the higher homologues in in the series and gave very high yields (Eqn. 1). For the lower

$$n = 0\text{—}8$$

$$(1)$$

members in the series, with $n < 3$, the reaction does not proceed that smoothly. When we attempted to epoxidize methyl acrylate (n=0) with m-chloroperbenzoic acid at room temperature, only a very low yield of methyl glycidate (MG) was obtained. Only under forcing conditions, namely by heating to reflux for 14 hours in boiling 1,2-dichloromethane (83°C), a 31% yield of MG was obtained. It had been reported previously (36) that epoxidation of ethyl acrylate has to be carried out with CF_3CO_3H and H_2O_2 (90%), which is a much more aggressive epoxidation agent.

We suspected that the proximity of the carboxylate group to the olefinic double bond might retard the rate of epoxidation and studied the reaction in more detail. Under identical reaction conditions, at room temperature and with dichloromethane as the solvent, only 10% of methyl acrylate (n=0) was epoxidized in one day to MG, while 15% of methyl 3-butenoate (n=1) was epoxidized in about 1 hour to methyl 3,4-epoxybutanoate (MEB). A 60% yield of MEB was obtained if the reaction was allowed to proceed

for 24 hours at room temperature and additional 4 hours at 40°C.
Methyl 4-pentenoate (n=2) is not as reactive as the higher homologues, yet in 6 hours at room temperature 90% were converted to
methyl 4,5-epoxypentanoate (MEP) (Figure 1).

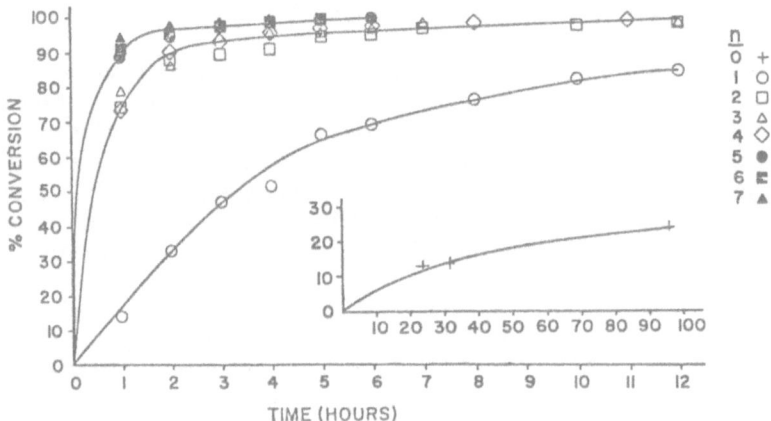

FIGURE 1:   Rate of Epoxidation of Methyl ω-Alkenoates

        Previous attempts to polymerize methyl 1-epoxyalkanoates
(MEA's) with classical cationic initiators such as $BF_3 \cdot O(C_2H_5)_2$
resulted only in the formation of low molecular weight, oligomeric
polymers.   The use of an aluminumalkyl/water system (1/0.5) also
resulted primarily in low molecular weight polymers, although some
higher molecular weight solids were also observed.   A modified
triethylaluminum/water/acetylacetone system proved to be a much
better initiator for homo- and copolymerizations of the MEA's and
was carefully investigated with the aim of finding the optimum parameters which would lead to high yields and high molecular
weights (34).   A ratio of 1/0.5/1 for the components triethylalu-
minum/water/acetylacetone was found to give the best results,
however, not only the ratio of the components but also the sequence of mixing was found to be critical.   It was important that
first acetylacetone was added to the solution of triethylaluminum
in ether and allowed to react at low temperature.   Only after the
noticeable reaction had ceased, the appropriate amount of water
was added.   With this system we successfully homopolymerized
the series of ω-epoxyalkanoates with 0 to 8 methylene spacer
groups (n=0-8) between the terminal epoxide and carboxylic ester
groups.

        The polymerization of MEB (n=1) proceeded slowly and gave
only a low yield.   Homopolymerization of methyl 4,5-epoxypentan-
oate (MEP) (n=2) and methyl 5,6-epoxyhexanoate (MEH) (n=3)
was also slower than the homopolymerization of MEA's with n>3.

For example, the yield of poly-MEP was only 13%, while the polymerization of methyl 10,11-epoxyundecanoate (MEU) (n=8) proceeded smoothly and gave polymers in 50-80% yield with an inherent viscosity of up to 1.5 dL/g.

Copolymerization of the methyl ω-epoxyalkanoates was carried out with EO, PO, BO, 1-hexene oxide (HO), epichlorohydrin (ECH), phenylglycidyl ether (PhGE) and Ox as comonomers (Table 1). The copolymerization of MEP (n=2) with aluminumalkyl/water/acetylacetone initiator systems (1/0.5/1) was investigated systematically. Copolymers with PO were obtained in 63% yield, with BO in 51% yield, with HO in 22% yield and with ECH in 20% yield. With PhGE the yield of copolymer was 45% and with Ox only 11% (Eqn. 2). The attempted copolymerization of MEP with THF gave only a homopolymer of MEP in low yield. The inherent viscosities of all copolymers ranged from 0.8 to 1.3 dL/g. Incorporation of the individual cyclic ethers into the MEP copolymers was good; with a feed ratio of 30 mole %, 25 to 30% of MEP were incorporated. In the case of the Ox copolymerization however, incorporation was lower; only 17% of MEP units were found in the copolymer.

$$R = -H, -CH_3, -C_2H_5, -CH_2Cl,$$
$$-CH_2-O-C_6H_5, etc.$$

(2)

MEA's with more than three methylene units as spacers, as in methyl 7,8-epoxyoctanoate (MEO), (n=5), are more reactive and with our triethylaluminum/water/acetylacetone initiator system an 88% yield of homopolymer with an inherent viscosity of 1.0 dL/g was obtained. Copolymers of MEO with PO, BO, HO, ECH and PhGE were obtained in 4-93% yield and had inherent viscosities ranging from 0.8 to 1.8 dL/g. Incorporation of MEO into the copolymers was excellent in all cases; 30% MEO units were incorporated with a monomer feed ratio of 30% MEO. Attempted copolymerization of MEO with THF was not successful, but resulted in the conversion of 43% of MEO into its homopolymer with an inherent viscosity of 1.0 dL/g.

Methyl 10,11-epoxyundecanoate (MEU) (n=8) was copolymerized with a number of cyclic ethers with a feed ratio of 30 mole %
of MEU. With EO a 57% yield of a copolymer was obtained which
had an inherent viscosity of 4.8 dL/g and 15 mole % of MEU comonomer incorporated. PO and   BO copolymers of MEU were obtained in 90% yield and had inherent viscosities of 1.0 and 2.7
dL/g respectively. A copolymer of MEU with   HO was obtained
in 83% yield, but it had a somewhat lower inherent viscosity of 0.5
dL/g (Eqn. 2). 20 to 25 mole % of MEU were incorporated into
these copolymers. A 25 mole % incorporation of MEU was also
found with PhGE as the comonomer. In this case, however, the
material was inhomogeneous because the polymer mixture also contained a crystalline fraction. This indicates that some stereoregular polymerization of the PhGE had occurred.

ECH gave a copolymer with MEU in 60% yield and with an inherent viscosity of 2.6 dL/g but only 17% of comonomer was incorporated. Copolymerization of 4,4,4-trichlorobutylene oxide
(TClBO) with MEU gave a 30% yield of a copolymer with an inherent
viscosity of 0.2 dL/g; in this case MEU was incorporated to 40
mole % indicating that MEU was the more reactive comonomer, the
only case where this phenomenon was observed.

In addition to these systematic qualitative and quantitative
polymerization studies we also made a comparative rate study of the
homopolymerizations of MEU and of 1-dodecene oxide (DO). Both
of these molecules have a chain of ten carbon atoms attached to
the oxirane ring. We found that polymerization of MEU levelled
off after 6 days, with about 40% of the monomer converted to polymer. This result is compared to 60% conversion of DO to poly-DO
in the same time period. The initial rate of polymerization of DO
was faster than that of MEU as 45% of it polymerized within 2 hrs.
while only 17% of MEU polymerized in that time period. After 4
hours nearly 55% of DO was converted to polymer but only 22% of
MEU (Fig. 2). This slightly higher rate of polymerization of a
nonfunctionally substituted epoxide monomer with about the same
side chain length as an ω-epoxyalkanoate is not surprising. In
both cases a linear side chain consisting of 10 carbon atoms is
attached to the oxirane ring, but in MEU the last two members of
the chain are part of an ester function which, although separated
from the epoxide functionality by eight methylene groups, still
interferes to some degree with the propagation step and/or the
initiating site. This might be due to association of carboxylate
groups with each other or with the initiator system. The homopolymer of MEU had the most probable molecular weight distribution. From the GPC curve, a polydispersity of 2.3 was observed
with a M_n of 250,000.

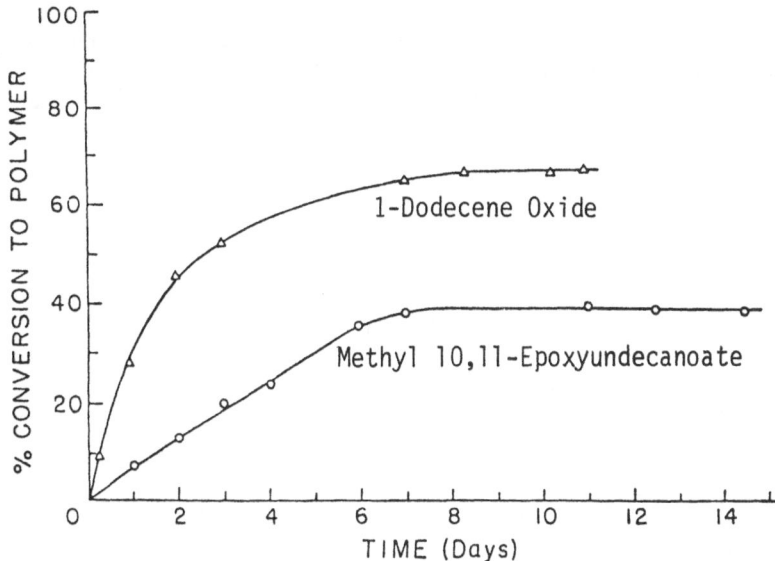

FIGURE 2: Comparative Rate of Polymerization of Epoxides
with Long Side Chains, with and without
Carboxylate Functional Groups.

The thermal transition behavior of the copolymers of MEU
with cyclic ethers was also studied. While the homopolymer
showed a glass transition temperature of -33°C, the copolymers
with comonomer contents of 15-25% showed glass transition temperatures ranging from -58°C (BO) to -68°C (HO). The ECH copolymer with 17 mole % of MEU had a T_g of -45°C, while the copolymer with TClBO had a T_g of -35°C. In this polymer a small
amount of crystallinity was observed with a T_m of 41°C. A T_m of
43°C was also noted for the copolymer of EO containing 15 mole %
of MEU, indicating moderate crystallinity. We have not determined whether the T_m of 43°C of the poly(MEU-co-EO)
is caused by side chain crystallinity (paraffin type) or by backbone crystallinity (poly-EO type).

Homopolymers of MEA's, particularly poly-MEU (n=8), were
hydrolyzed with 4 N aqueous sodium hydroxide solution in dioxane to the polymeric salt which in turn could be transformed by
treatment with aqueous acetic acid to the polyacid (Eqn. 3).

$$\left(\!\!\!\begin{array}{c} CH_2\!-\!CH\!-\!O\!-\!(CH_2)_n\!-\!O \\ | \\ (CH_2)_n \\ | \\ COOCH_3 \end{array}\!\!\!\right) \xrightarrow{NaOH} \left(\!\!\!\begin{array}{c} CH_2\!-\!CH\!-\!O\!-\!(CH_2)_n\!-\!O \\ | \\ (CH_2)_n \\ | \\ COO^-\cdot Na^+ \end{array}\!\!\!\right) \tag{3}$$

Copolymers of MEU and EO were also hydrolyzed with 4 N
aqueous sodium hydroxide solution in dioxane and gave the poly-
EO ionomer containing about 15 mole % of carboxylate groups.
This ionomer was then treated with aqueous acetic acid to give
the free acid.

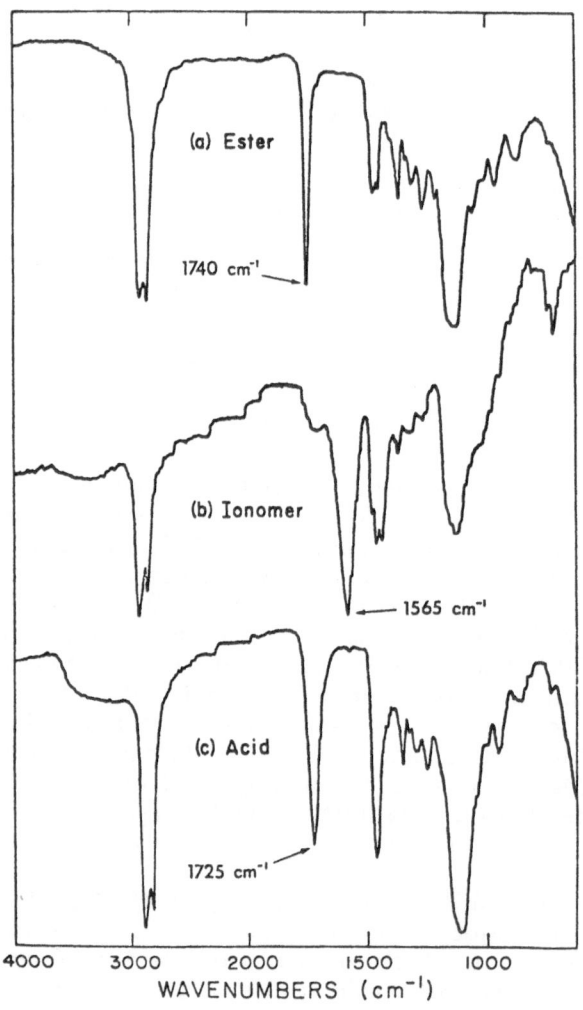

(a) Ester

1740 cm^{-1}

(b) Ionomer

1565 cm^{-1}

(c) Acid

1725 cm^{-1}

4000   3000   2000   1500   1000

WAVENUMBERS (cm^{-1})

FIGURE 3: Infrared Spectra (Films) of

(A) EO/MEU Copolymer

(B) EO/MEU Copolymer Salt

(C) EO/MEU Copolymer Acid

The transformation of the ester groups to the polymeric anion (ionomer) and of the polymeric anion to the free carboxylic
acid form can be demonstrated by infrared spectroscopy.
Poly(MEU-co-BO) showed a characteristic ester carbonyl absorption band at 1740 cm^{-1}, the ionomer absorption band at 1565 cm^{-1},
and the polymeric acid carbonyl band at 1725 cm^{-1} (Fig. 3).

Other ionomers were obtained by a rather different method.
A copolymer of MEU and trioxane (TO) and particularly its ter-
polymer with 1,3-dioxolane was treated at 150°–160°C and under
pressure of about 300 lbs of nitrogen, with methanol. This procedure is known to degrade the unstable hemiacetal endgroups to
the thermally stable ethylene glycol endgroups (25-27). After
precipitation, thermally   stable MEU modified POM was obtained in
80% yield. When the treatment was done in the presence of sodium
hydroxide, further hydrolysis of the ester groups took place.
Not only were the unstable hemiacetal endgroups of POM degraded
but also the carbomethoxy groups were hydrolyzed to the polymeric sodium carboxylate. By sodium analysis the copolymers
were found to contain about 2 mole % of carboxylate groups.

To summarize, we have demonstrated that high molecular
weight polyether elastomers with pendant ester groups can be
prepared by coordinative anionic polymerization with aluminum-
alkyl/water/acetylacetone initiator systems. Good levels of incorporation of polar group containing epoxy monomers were achieved
in copolymerization reactions with cyclic ethers. The pendant
ester groups can be converted readily to novel ion containing
polymers.

ACKNOWLEDGEMENTS

        This work was supported in part by a grant from the Mater-
ials Division of the National Science Foundation and by a grant
from the National Science Foundation to the Materials Research
Laboratory of the University of Massachusetts.

REFERENCES

1.   E.J. Vandenberg, J. Polym. Sci., 47, 486 (1960).
2.   J. Furukawa and T. Saegusa, Polymerization of Aldehydes
      and Epoxides, Wiley Interscience, New York, 1963.
3.   J. Furukawa, T. Tsuruta, R. Sakata, T. Saegusa, A. Kawa-
      saki, Makromol. Chem. 32, 90 (1959).
4.   J. Furukawa, T. Tsuruta, T. Saegusa, R. Sakata, G. Kako-
      gawa, A. Kawasaki, and I. Harada, J. Chem. Soc., Ind.
      Chem. Sec. 62, 1269 (1959).
5.   R. Sakata, T. Tsuruta, T. Saegusa and J. Furukawa,
      Makromol. Chem. 40, 64 (1960).
6.   E.J. Vandenberg, U.S. 3,158,580 (1964).

7.   N.G. Gaylord, Polyethers, Part I: Polyalkylene Oxides and Other Polyethers, Wiley Interscience, New York, 1963.

8.   T. Saegusa, H. Imai, and J. Furukawa, Makromol. Chem. 65, 70 (1963).

9.   E.J. Vandenberg, J. Polym. Sci., A1, 7, 525 (1969).

10. T. Tsuruta, S. Inoue, N. Yoshida and Y. Yokota, Makromol. Chem. 81, 191 (1965).

11. S. Inoue, T. Tsuruta and J. Furukawa, Makromol. Chem. 53, 215 (1962).

12. S. Inoue, T. Tsuruta and N. Yoshida, Makromol. Chem. 79, 34 (1964).

13. T. Tsuruta, J. Polym. Sci., D, Symposia, 180 (1972).

14. J. Furukawa, T. Saegusa, S. Yasui and S. Akutsu, Makromol. Chem. 94, 74 (1966).

15. C.C. Price, Polyethers, in Polyethers, E.J. Vandenberg, Ed., 1975, p. 1.

16. H. Tani, Adv. in Polym. Sci. 11, 40 (1973).

17. Y. Kai, N. Kasai, M. Kakudo, H. Yasuda and H. Tani, Chem. Comm. 1968, 1332.

18. J.P. Kennedy, Cationic Polymerization of Olefins: A Critical Inventory, Wiley Interscience, New York, 1975.

19. J. Boor, Jr., The Nature of the Active Site in the Ziegler Type Catalyst, in Macromolecular Reviews, A. Peterlin, M. Goodman, S. Okamura, B.H. Zimm, and H.F. Mark, Eds., Wiley Interscience, New York, 1967.

20. P. Teyssie, T. Ouhadi and J.P. Bioul, Int. Rev. Sci., Phys. Chem., Ser. Two, 8, 191 (1975).

21. P. Teyssie, T. Ouhadi, J.P. Bioul and L. Hocks, Chemtech, 7(3), 192 (1977).

22. H.S. Eleuterio, J. Macromol. Sci.-Chem., A6(6), 1027 (1972).

23. A.C. Farthing and R.J.W. Reynolds, J. Polym. Sci. 12, 503 (1954).

24. E.J. Vandenberg, U.S. Pat. 3,106,549 (1963).

25. K.V. Martin and O.F.L. Vogl, U.S. Pat. 3,284,411 (1966).

26. K.V. Martin and O.F.L. Vogl, Brit. Pat. 1,005,761 (1964).

27. K.V. Martin and O.F.L. Vogl, Fr. Pat. 1,401,637 (1965).

28. L. DeMejo, W.J. MacKnight and O. Vogl, Polymer (London) 19(8), 856 (1978).

29. T. Saegusa, T. Kobayashi, S. Kobayashi, S. Lund-Couchman and O. Vogl, Polymer J. (Japan), 11(6), 463 (1979).

30. L. DeMejo, W.J. MacKnight and O. Vogl, Polymer J. (Japan) 11(1), 15 (1979).

31. L. DeMejo, W.J. MacKnight and O. Vogl, Abstracts, SPE ANTEC, 1978, p. 5.

32. L. DeMejo, W.J. MacKnight and O. Vogl, Acta Polymerica, 31, 617 (1980).

33. D. Tirrell, O. Vogl, T. Saegusa, S. Kobayashi and T. Kobayashi, Macromolecules 13, 1041 (1980).

34. O. Vogl, J. Muggee and D. Bansleben, Polymer J. 12(9), 677 (1980).

35. N.N. Schwartz and J.H. Blumberg, J. Org. Chem. $\underline{29}$, 1976 (1964).
36. H.Y. Aboul-Enem, Synth. Commun. $\underline{4}$, 255 (1974).
37. P. Dreyfuss and J.P. Kennedy, Annal. Chem. $\underline{47}$, 771 (1975).
38. J. Muggee, Ph.D. Dissertation, University of Massachusetts, 1982.

STEREOCHEMICAL ASPECTS OF THE POLYMERIZATION OF

CIS AND TRANS THIIRANES USING CHIRAL INITIATORS

Nicolas Spassky, Ardéchir Momtaz and Pierre Sigwalt

Laboratoire de Chimie Macromoléculaire, LA 24
Université Pierre et Marie Curie
4, Place Jussieu   75230 Paris Cédex 05, France

SYNOPSIS

   Stereochemical rules and kinetic behavior of the resolution
reaction were established in the caso of ring-opening polymerization of monosubstituted cyclic monomers using chiral initiators.

   The use of the same initiators is now applied to disubstituted monomers.

   Trans dimethylthiirane, a racemic monomer, mixture of two
stereoisomers bearing two asymmetric carbons of identical configuration, give by polymerization optically inactive composed of -RS-
units. This confirms the complete inversion of the attached assyme-
tric carbon during ring-opening as established previously by
Vandenberg. Polymers obtained are crystalline and of high molecular
weight. Unusual kinetics were observed for the resolution reaction
in which the sign of the unreacted monomer changed at a definite
conversion.

   The polymerization of cis dimethylthiirane, and achiral monomer of meso configuration, produces optically active polymers. This
result shows that the initiator is able to recognize the donfigura-
tion of an asymmetric carbon in a molecule and to direct an attack

on a preferential side in agreement with a homochirality rule. The enantiomeric distribution in the polymer chain, i.e. the proportion of -RR- and -SS- type configurational units, was established by NMR and from cationic degradation studies. The formation of disulfide linkages in substantial amounts was observed with some of the initiator systems used. However, the polymers are generally crystalline and their melting points are different from those of optically inactive polymers prepared with a chiral initiators.

INTRODUCTION

The first work on the stereochemistry of ring-opening polymerization was reported by Osgan and Price[1] in 1956 who polymerized the S-enantiomer of propylene oxide. This crystalline optically active polymer obtained by anionic initiation is usually given as the example of isotactic structure for polymers prepared by ring-opening of monosubstituted heterocycles.

The polymerization of disubstituted oxiranes bearing two vicinal asymmetric carbons of identical configuration (trans monomer) or of opposite configuration (cis monomer) has been extensively studied by Vandenberg[2] who demonstrated that ring-opening proceeds with complete inversion of the asymmetric carbon. Diisotactic structures were proposed for the crystalline polymers obtained on the basis of polymer cleavage studies with organometallic compounds.

Similar studies have been made in the field of thiiranes. We have prepared optically active polymers of propylene sulfide and other monosubstituted thiiranes by polymerization of one enantiomer using anionic, stereospecific and cationic initiators[3,4] . In the latter case inversion of the asymmetric carbon occurred to some extent.

The mechanistic aspects of the polymerization of disubstituted thiiranes and oxiranes using the same type of initiators were compared by Vandenberg[5] . The properties of polyoxiranes and polythiiranes with proposed identical structures, e.g. mesodiisotactic, were quite different (table 1).

Table 1. Compared properties of polymers obtained from cis- and trans- dimethyloxiranes and dimethylthiiranes using various initiators

Work of E.J. Vandenberg (2,5)

	Initiator	Oxirane Cryst.	Oxirane m.p. (°C)	Oxirane Struct.	Thiirane Cryst.	Thiirane m.p. (°C)	Thiirane Struct.
Trans	$ZnEt_2$–1.0 H_2O	high	91		mod. (a)	192	RS–RS
	cationic	high	100	RS–RS	none	–	
	Al chelated	no	polymer		none (a)	–	
Cis	$ZnEt_2$–1.0 H_2O	none	–		high	74/99	
	cationic	none	–		high	105/158	
	Al chelated	mod.	162	RR–RR SS–SS	mod. (a)	156	RR–RR SS–SS
	$ZnEt_2$–ℓ–menthol				$[\alpha_p]$ + 5.8	136	

(a) Yield of only 1-2 % obtained

The synthesis of optically active polymers can also be realized by polymerizing a racemic monomer in the presence of optically
active initiators. Such a process, usually called "stereoelective"
was described[6] for the first time by Inoue, Tsuruta and Furukawa[6]
who polymerized racemic propylene oxide with a diethylzinc-d-borneol
initiator system. The polymerization of propylene sulfide using the
same type of initiator system was then reported simultaneously by
Furukawa[7] and by ourselves[8]. In the course of our studies we were
able to apply successfully the stereoelective process to a series
of monosubstituted oxiranes, thiiranes and more recently to
β-propiolactones. High stereoelectivities could be obtained in several cases. Many of these results have been already reviewed[9,10,11].

A few years later we began to examine the aspects of the polymerization of disubstituted monomers, namely cis[12] and trans[13] di-
methylthiiranes in the presence of chiral initiators.

The aim of this paper is to report recent results obtained in
this field, to discuss the structure of the polymers and to examine
the stereochemistry of the process in the light of existing chiral
relationships between the configurations of the monomer, of the initiator and of the resulting products.

In a preliminary chapter it is necessary to discuss the main
aspects of the polymerization of monosubstituted cyclic monomers.

STEREOCHEMICAL RELATIONSHIPS IN THE RING-OPENING POLYMERIZATION OF
MONOSUBSTITUTED CYCLIC MONOMERS USING CHIRAL INITIATORS

The stereochemical rules governing the ring-opening polymerization of cyclic monomers were recently discussed[14]. They seem to
be quite general and were verified with a series of oxiranes, thiiranes and β-propiolactones.

We shall summarize these rules briefly using as an example
methylthiirane, from which it is easy to visualize cis and transdimethylthiirane simply by combining two half molecules as indicated below.

Enantiomeric choice

The chiral initiator generally used resulted from the reaction of an organometallic compound ($ZnEt_2$, $CdMe_2$...) with a chiral alcohol or a glycol. Such an initiator is composed of monosubstituted and disubstituted alkoxide species and corresponds to the general formula $(R-Met-OR^*)_x(R^*O-Met-OR^*)_y$.

In a first "coordination-stereoelection" step one of the enantiomers is preferentially chosen by the chiral initiator.

Let us consider the spatial representations of the monomer and the chiral component of the initiator, a monosubstituted 1,2 diol for instance :

Both asymmetric centers represented above may be considered as "homochiral" since they bear analogous substituents at homologous positions.

When the chemical composition of the initiator corresponds to a ratio of alkylalkoxide/dialkoxide species $x/y < 1$ the initiator chooses preferentially the enantiomer which has an homochiral configuration. Such type of choice is called "homosteric".

In this particular case, using the Cahn, Ingold, Prelog rule one can say that an initiator derived from a 1,2-diol with R configuration will preferentially choose the R enantiomer of methyl-thiirane. However, one must be careful in the application of the (R,S) rule since this is not always equivalent to our concept of homochirality.

It was found, on the other hand, that when using initiators with predominance of monoalkoxide species ($x/y > 1$) the opposite choice of the heterochiral configuration occurs. This process is called "antisteric". It was suggested that the reason of such a behaviour is a consequence of the external "overall" chirality of the aggregated initiator species.

From a practical point of view homosteric initiators ($x/y < 1$) are generally obtained when using $ZnEt_2$ as organometallic derivative, while heterosteric initiators ($x/y > 1$) are easily prepared from $CdMe_2$.

Ring-opening reaction

The usual alkoxide initiators produce almost exclusively a
α-scission of the ring. Therefore the configuration of monomeric
units in the polymer chain is the same as in the reacting monomer

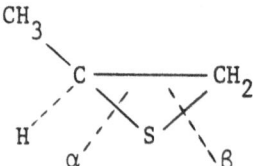

However when initiators derived from 1,2 diols with two vicinal asymmetric carbons of the same configuration were used, ring-
opening occured on both sides leading to polymers with a configuration opposite to that predicted from the former enantiomeric
choice rule.

For example, when racemic methylthiirane is polymerized with
$ZnEt_2$-(R,R) 1,2 - diphenyl 1,2-ethanediol initiator system, polymer and unreacted monomer are both of predominant S configuration,
which means that a high amount of α-scission occured with an inversion of the asymmetric homochiral carbon attacked[15].

Kinetics of the process

It was shown that for most monomers the rate of the enantiomer consumption obeys to a first order law :

$$d \, |R|/dt = k_R \, |R| \qquad\qquad d|S|/dt = k_S \, |S|$$

$$d|R|/d|S| \; = k_R/k_S \, |R|/|S| \qquad\qquad\qquad (1)$$

$r_R = k_R/k_S$ is the stereoelective coefficient expressing the pre-
ference of the initiator for a given enantiomer. It was found to
be constant during the polymerization process. The integration of
equation (1) gives the first-order relationship :

$$(1 - x)^{r-1} = \frac{1 \; + \; \alpha/\alpha_0}{(\, 1 \, - \; \alpha/\alpha_0)^r}$$

where x is the conversion and α/α_0 the optical purity of unreacted monomer. This relationship fits well with experimental data
for several monomers, e.g. methylthiirane polymerized with zinc and
cadmium chiral diolates[10]. For monomers with bulky substituents,
e.g. t-butyl thiirane, this relation must be replaced by the

so-called second-order law[10] derived from the equation :

$$d|R|/d\ |S| = \rho\ |R|^2/|S|^2$$

We shall now discuss these various stereochemical aspects for disubstituted cyclic monomers and particularly for trans and cis di-methylthiiranes.

POLYMERIZATION OF TRANS DIMETHYLTHIIRANE

Trans dimethylthiirane (trans-DMT) is a racemic monomer, mixture of (R,R) and (S,S) stereoisomers bearing two asymmetric carbons of identical configuration. It was shown by Vandenberg[2,5] that whatever the initiator used, polymerization proceeds with complete inversion of configuration of the asymmetric carbon involved in the ring-opening process. Therefore the polymers obtained are always optically inactive i.e. composed of -RS- or -SR- structural units. The polymerization with cationic and coordinated achiral initiators were studied by Vandenberg[5]. The polymers obtained showed no or very little crystallinity which was attributed to the low stereoregularity of these polymers (table 1). We have polymerized trans-DMT with a series of optically active initiators and we report here on the kinetics of the resolution reaction and on the structure of the polymers.

Kinetics of the resolution reaction

We used first as initiator the standard system obtained by reacting diethylzinc with R(-)-3,3-dimethyl-1,2-butane-diol (R(-) DMBD) which was known at that time to be the most efficient system for the stereoelective polymerization of thiiranes.

The plotting of experimental data (optical purity of unreacted monomer $(\alpha/|\alpha_o|)$ versus conversion (x)) shows interesting features. The sign of residual monomer changes at a definite conversion which means that the two enantiomers polymerize according to different kinetic laws. Such a particular behaviour is observed for the first time in the case of cyclic monomers. Several possibilities were tested in order to find a theoretical equation fitting with the experimental data.

The best correspondance for a large range of conversions is obtained for the following relationship of the relative consumption rates of the stereoisomers[13]:

$$\frac{d\ |RR|}{d\ |SS|} = \gamma|RR|^2 \qquad (1) \qquad\qquad \gamma = \frac{K_{RR}}{K_{SS}}$$

which can be integrated, giving :

$$\frac{1}{|RR|_o} - \frac{1}{|RR|} = \gamma (|SS| - |SS|_o)$$

After introduction of the experimental data α/α_o and x, one obtains, when the starting monomer is racemic ($|RR|_o = |SS|_o = \frac{|M|_o}{2}$) the following equation :

$$\frac{1}{(1 + \alpha/\alpha_o)(1 - x)} = \frac{\gamma}{4} |M|_o^2 (1 - \alpha/\alpha_o)(1 - x) + \frac{\gamma|M|_o^2}{4} + 1 \quad (2)$$

This equation fits satisfactorily with the theoretical curve up to 60 % conversion, then a large scattering is observed. Other equations (for example $\frac{d|RR|}{d|SS|} = \gamma|RR|$) don't give better results. It is possible that a change of mechanism occurs at high conversion. Nevertheless, for the first time, a change of the sign of residual monomers in the course of polymerization is observed (fig. 1).

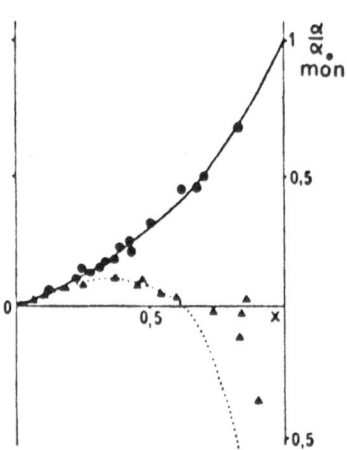

Experimental data : ● methylthiirane ▲ transdimethylthiirane

Fig. 1.   Stereoelective curves corresponding to the polymerization of methylthiirane and transdimethylthiirane using $ZnEt_2$-R(-)DMBD initiator system.

Temperature has some effect on the kinetics : at higher temperature, e.g. 50°C, the change of the sign of the unreacted monomer occurs for a much lower conversion (10 % conversion versus 60 % at 25°C).

The nature of the chiral component used in the initiator is also deeply affecting the stereoelectivity of the process. Results obtained with four different chiral 1,2 diols all bearing asymmetric centers of R configuration are compared in table 2.

Due to the small number of experiments it is particularly difficult to say at the present time if the same mechanism is operative for all the four cases with only a shift of the crossover point. The variations observed seem to be much more important than in the case of methylthiirane with the same initiators[10,15].

Polymer properties and structure

The properties of the poly trans DMT were examined. The main part (85-90 %) of polymers is soluble in toluene (fraction A) with a minor part insoluble in toluene but soluble in chloroform (fraction B). Intrinsic viscosities (dl/g), measured in chloroform, varied from 1.0 to 2.2 . Most of the polymers, except those prepared with DPED initiator, were crystalline as shown by DTA measurements. Several peaks are observed, located near 47, 65 and 90°C. The melting points corresponding to the disappearance of the last crystal were respectively found at 55,72 and 120°C.

Two features should be mentionned. The peak at 100-120°C which is the broadest one, is much larger in polymers obtained at low conversions. On the contrary the peak at 47°C which is much sharper seems to increase with conversion. On the other hand the former peak (at 120°C) seems to be larger in fractions B (fig. 2).

Cationic degradation studies were carried out in cooperation with Professor Goethals's group. In this method the reaction proceeds by elimination of butene molecules, and, as a result, a mixture of isomeric trithiepanes is obtained. The polymers we have examined gave a ratio of cis-cis-cis/cis-trans-cis trithiepane isomers as high as 7.6. This means that the fraction of the diisotactic trithiepane cis-cis-cis is equal to 0.88 i.e. that the original polymer is highly stereoregular. A similar result was found for cationic polymers prepared with the Et_3OBF_4 initiator system[16].

The analysis of ^{13}C NMR spectra of polytrans DMT reveals that both types of carbons are stereosensitive. The methyl carbon gives a pattern of more than four peaks located around 21.3 ppm, while the methine carbon at 48.8 ppm is much less stereosensitive showing dyad effects (only a shoulder on the main peak is observable).

Table 2. Polymerization of racemic trans dimethylthiirane
with various diethylzinc-R_1-CHOH-CHOH-R_2 chiral
systems.

R_1-CHOH-CHOH-R_2			Polymeriz. time (days)	Yield (%)	Unreacted monomer α_D^{25} (neat)
abrev.	R_1	R_2			
R(-)DMBD	tBu	H	4.7	24	- 11.9
			5	37	- 15.4
R(-)PD	Me	H	3.8	16	+ 0.03
			14	28	+ 0.55
R,R(-)BD	Me	Me	14	5	- 0.65
			33	11	- 1.80
R,R(-)DPED	Ph	Ph	6	35	+ 0.09

DMBD : 3,3 dimethyl 1,2 butanediol   PD : 1,2 propanediol
DPED : 1,2 diphenyl 1,2 ethanediol   BD : 2,3 butanediol

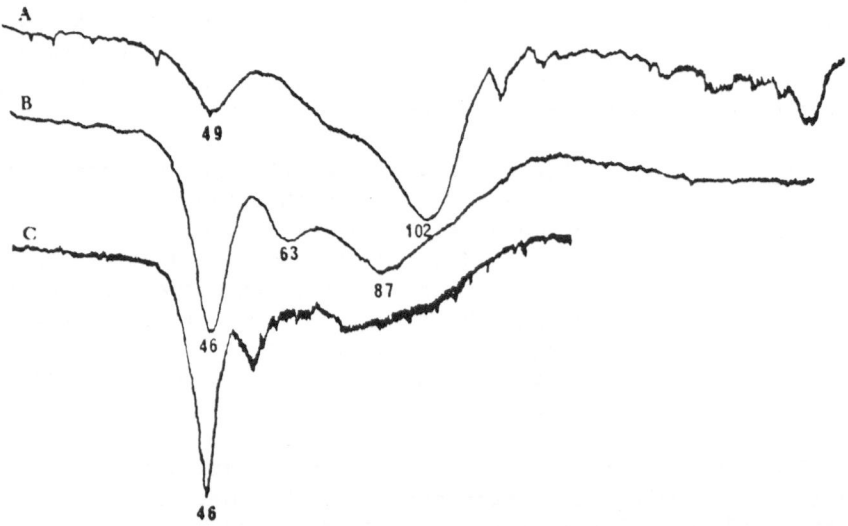

Fig. 2. DTA diagrams of poly-trans-DMT prepared at different
conversions using ZnEt$_2$-R(-)DMBD initiator system.
A : 5 %   B : 37 %   C : 44 %

An examination of the spectra indicates that in the methyl region the lowfield signal is the most intense for polymers isolated
at low conversions, while the highfield signal increases with the
conversion. At complete conversion both signals are practically
equivalent (fig. 3).

From the kinetic data, the isomer RR is preferentially chosen
at the beginning of the reaction. This should give an excess of
-SR- configuration units in the polymer. It is observed that the
lowfield signal of methyl carbon in the 25.15 MHz spectrum contri-
butes indeed 52 % of the overall pattern at a conversion of 37 %.
When conversion increases, this peak becomes smaller and the highfield one increases. At 90 % conversion both peaks are practically
equivalent (35 and 32 % respectively). The assignments of the peaks
are not yet completely established. Each corresponds probably to a
triad effect of configurational units. For example an SR-SR-SR enchainment may be called an isotactic triad of SR units and corresponds probably to the lowfield peak.

There is a correlation between the variations observed in ^{13}C
NMR spectra and the crystallization properties measured by DTA.

The presence of multiple peaks in ^{13}C spectra substantiates
the presence in most of the products of a structure much more complex than diisotactic.

POLYMERIZATION OF CIS-DIMETHYLTHIIRANE

Cis-dimethylthiirane is an achiral monomer of meso configuration
containing two vicinal asymmetric carbons of opposite configuration.
The polymerization of such a monomer is interesting from both stereochemical and mechanistic points of view. The configurational structure of the polymer depends on the ring-opening reaction (opening
in α or in β) i.e. on the choice of attack on one asymmetric
carbon of a particular configuration.

Vandenberg had demonstrated that the ring-opening occurs with
complete inversion of configuration of the asymmetric carbon attacked.
In the polymer obtained from cis-DMT one should therefore finds configurational units of RR and SS type. Using different achiral cationic and coordination initiators Vandenberg had prepared crystalline polymers for which he proposed a diisotactic structure i.e.
polymer chains containing only RR or SS units. The evidence for
such a structure was the finding of an optically active polymer,

$|\alpha_p|_D^{25}$ = + 5.8, (CHCl$_3$), when using diethylzinc-1-menthol initiator system [5].

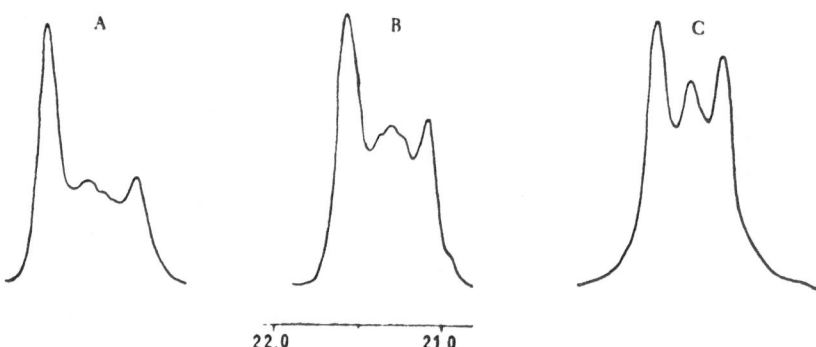

22.0   21.0

Fig. 3. ^{13}C NMR 25.15 MHz of polytrans DMT prepared with ZnEt$_2$-(R)-DMBD initiator system. Methyl pattern Variation with conversion A. 37 % B. 53 % C. 90 %.

Several years ago we undertook systematic studies of these reactions using various chiral and achiral initiators. First, we have performed some experiments with anionic, cationic and coordination achiral initiators. It was found that cis-DMT polymerizes very rapidly with cationic (EtO$_3$BF$_4$) and anionic (tBuOK/DMSO) initiators (in a few minutes quantitative yields are obtained), but more slowly with coordination type initiators such as ZnEt$_2$-H$_2$O or cadmium tartrate (48 hours at least are necessary for completion of the reaction). All the polymers obtained were crystalline with almost the same m.p. located around 80-84 °C.The polymer prepared with cadmium tartrate showed however several other m.ps. at higher temperatures (93, 110°C).

Cationic degradation studies indicate that in all these polymers there is some excess of diisotactic dyads but their fraction is less than 0.70[16].

Cis-DMT was then polymerized with different chiral initiators. Polymers with high optical activity were obtained in most cases. The rotatory power and the nature of the products depended on the initiator used.

First, we have tried to polymerize cis-DMT in bulk with our standard initiator ZnEt$_2$-R(-)DMBD. Some typical results are given

in table 3. The overall polymer obtained is dextrorotatory
($|\alpha_p|$ = + 42 in CHCl$_3$) and it can be fractionated giving a frac-
tion A, soluble in toluene, and a fraction B, insoluble in toluene
but soluble in chloroform. The latter is generally of higher optical activity than the former. When polymerization is carried out a
long time disulfide linkages appear in the polymer as shown by NMR
and CD studies. In some case fraction B is almost insoluble in all
solvents. The polymers prepared with ZnEt$_2$-R(-)DMBD system are crystalline. DTA diagrams of fractions B show melting peaks near 125°C
and sometimes a peak located near 196°C.

Polymerizations were also performed in solution using different solvents such as toluene, tetrahydrofuran or limonene. In the
two last solvents only fraction B was obtained. The rotatory power
was the highest when using limonene as a solvent.

When CdMe$_2$ was used with R(-)DMBD instead of ZnEt$_2$ the resulting polymer was levorotatory in CHCl$_3$, which shows, as seen below,
that -SS- units are predominant in the polymer. This is typical
for an "antisteric" process as it may be observed for example for
methylthiirane polymerization with the same initiator.

Polymerizations were also carried out using other chiral components of the initiator.

As may be seen in table 4, several unexpected results were obtained.

The main fraction of the polymer prepared with the diethyl
zinc-R,R(-)BD initiator system is levorotatory which means that
SS type units are predominant in the chain. This indicates that antisteric ring-opening occurs with this initiator. On the other hand,
a small fraction of a dextrorotatory polymer with a high rotatory
power (+ 129) and a high melting point (190°C) was isolated.

With DPED and BN based initiators only fractions A were obtained. Some typical DTA diagrams of various polymers are given in
fig. 4. The molecular weights of these polymers are lower than those
obtained with DMBD based initiators. They are all dextrorotatory
which again seems to support an antisteric attack. Polymers from
DPED initiators are crystalline while those from BN initiators are
amorphous.

According to NMR and CD studies polymers from BD and BN initiators contain a substantial amount of disulfide linkages while
the latter are absent in DPED type polymers.

A complete understanding of the physical and structural properties of these polymers necessitates detailed studies of chiroptical behaviour and stereoregularity which are discussed in the
next sections.

Table 3. Polymerization of cis-DMT by $M_t R_2$ (-)DMBD (1:1)

$M_t R_2$	Solvent	Polym. time (days)	Conversion %	$[\alpha_p]$ whole polymer (a)	Fraction A(b)		Fraction B(c)		$[\eta]$ (a)
					%	$[\alpha_p]$ (a)	%	$[\alpha_p]$ (a)	
ZnEt$_2$	bulk	0.63	54	+ 42	35	+ 30	65	+ 55	
"	bulk	0.63	97	+ 42	65	+ 59	35	ins.	
"	toluene	2	61	+ 42	100	+ 42	–	–	0.44
"	THF	1.7	43	+ 58	–	–	100	+ 58	3.0
"	limonene	1	59	+ 73	–	–	100	+ 73	1.3
"	limonene/toluene = 2.4	5	49	+ 55	25	+ 33	75	+ 65	3.9
CdMe$_2$	bulk	5	72	– 45	100	– 45	–	–	1.75

(a) in chloroform   (b) soluble in toluene   (c) ins. in toluene, sol. in chloroform

Table 4. Polymerization of cis DMT using $ZnEt_2$-dihydroxychiral compound (1:1) initiator systems

Polymerizations carried out at room temperature [C]/[M] = 4-5 moles %

Dihydroxy compound	Solvent	Polym. time (days)	Conversion (%)	$[\alpha]_p$ whole polymer (a)	Fraction A (a)		Fraction B (b)		m.p. (°C)	M_n osmo. (d)
					%	$[\alpha]_p$ (e)	%	$[\alpha]_p$ (c)		
R(−)DMBD	bulk	0.63	54	+ 42	35	+ 30	65	+ 55	125/195(f)	34 000
R(−)DMBD	bulk	0.63	97	+ 42	65	+ 59	35	ins.	162(f)	34 000
	toluene	2	61	+ 42	100	+ 42	−	−	−	43 000
R,R(−)BD	bulk	2	88	− 8(e)	88	− 27	12	+ 129	190(f)	32 000(g)
S,S(−)DPED	bulk	6	94	+ 154	100	+ 154	−	−	71	24 000
	toluene	21	90	+ 117	100	+ 117	−	−	65/119	25 500
S(−)BN	bulk	3	37	+ 41	100	+ 41	−	−	−	21 000
	toluene	6	52	+ 74	100	+ 74	−	−	−	14 000

BN : 2,2' dihydroxy 1,1' binaphthyl − (a) soluble in toluene − (b) ins. in toluene, sol. in $CHCl_3$ − (c) in chloroform − (d) in toluene − (e) recalc. value − (f) fraction B − (g) fraction A

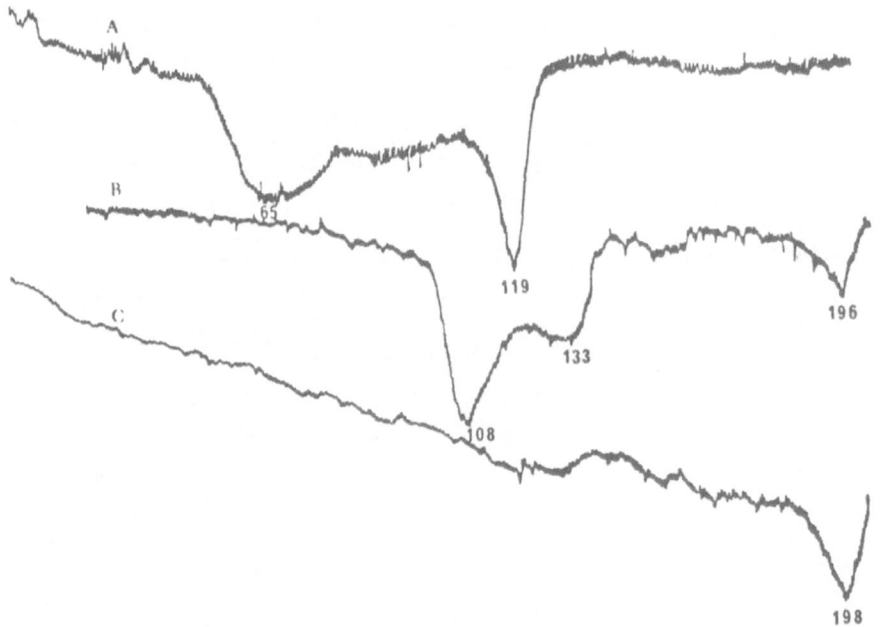

Fig. 4. DTA diagrams of poly-cis DMT prepared using various chiral initiaator systems. A: $ZnEt_2$-(S,S)DPED (fr. A) – B : $ZnEt_2$-(R) DMBD (fr. B) – C : $ZnEt_2$-(R,R)BD (fr. B)

Chiroptical properties

ORD and CD spectra of various samples of polycis DMT were examined. The polymers prepared with initiators of R-configuration are
generally dextrorotatory in usual solvents. As expected the polymer
prepared with cadmium initiator was levorotatory which is in agreement with an "antisteric" type process. The rotatory power is generally increasing with the tacticity of the polymers which corresponds to an enrichment in one of the enantiomeric units. We have
found in CD spectra of all polymers prepared from R type initiators
a negative Cotton effect located at 225-240 nm. Previous studies indicated that all monosubstituted polythiiranes bearing asymmetric
carbons of R configuration, showed a negative Cotton effect located
near 235 nm which was attributed to the thioether chromophore[4].
Therefore it seems reasonable to assign the negative Cotton effect
observed at 225-230 nm in poly cis-DMT to the thioether chromophore
corresponding to R configuration. As seen in table 5, the polymer
prepared with cadmium initiator shows a positive Cotton effect as
expected for the predominance of S configuration.

In polymers prepared with the standard initiator $ZnEt_2$-R(-)DMBD
in usual conditions one finds another positive Cotton effect located
below 210 nm. This transition was already observed in open-chain
sulfides[17]. The situation is much more complex in polymers prepared
with other initiators derived from (S,S)-DPED, (R,R)-BD and
(S)-binaphthol. According to elemental analysis some polymers contain an excess of sulfur and one might expect the presence of disulfide linkages. Indeed, negative Cotton effects located between 270
and 290 nm were found in all of them. The magnitude of these effects
seems to depend on the amount of S-S linkages which was the highest
for polymers prepared with the binaphtol initiator system.

It was reported that the S-S chromophore gives transitions located in this region as observed on some model compounds[18]. We have
also previously observed in polymethylthiirane containing S-S linkages a Cotton effect located at 275 nm[19]. The dichroic intensity
of the latter was much smaller than that of the effect at 235 nm
and was opposite in sign. In bis (2-methyl butyl) disulfide effects
of opposite sign were also found located at 270 and 238 nm, but in
this model compound the S-S chromophore is linked to methylenic carbon[18].

Interesting information is obtained from the trithiepane prepared by degradation of poly (R,R) cis-DMT. With this compound which
is a cyclic model of poly cis-DMT's bearing S-S linkage one finds a
negative Cotton effect located at 290 nm, a small positive one near
250 nm, a strong negative one at 230 nm, followed by a negative one
at 213 nm and then an unidentified positive one.

Table 5. Chiroptical properties of different poly cis DMT's and related products CD spectra run in cyclohexane (C) and hexafluoroisopropanol (H)

Sample	$[\alpha]_D^{25}$ (CHCl$_3$)	Solvent	Cotton effects ($\Delta\varepsilon$, λ nm)		
			290–260	250–225	220–200
ZnEt$_2$-(R)DMBD	+ 22	C		− 0.31 (229)	positive
tetramer[a]	− 36	C		− 1.51 (232)	− 1.82 (211) + 8.04 (203)
trithiepane[a]	+ 18	C	− 0.05 (290)	− 0.25 (230)	+ 0.14 (213) negative (<210)
CdMe$_2$-(R)DMBD	− 45	C	− 0.018 (270)	+ 0.072 (242)	negative
ZnEt$_2$-(S)BN	+ 74	H	− 0.143 (283)	+ 2.57 (239) − 1.97 (226)	+ 0.57 (213)
ZnEt$_2$-(S,S)DPED	+ 154	H	− 0.008 (281) + 0.016 (260)	− 0.156 (232)	positive

[a] degradation product from a polymer $[\alpha_p]$ + 66 (CHCl$_3$)

It is important to underline that the finding of identical Cotton effects near 230 nm representative of the thioether chromophore in poly cis-DMT and in its cationic degradation products, e.g. trithiepane and tetramer, support the hypothesis that configurations of asymmetric centers are not changed during the degradation reaction. Finally another model compound a (+)bis |(S) 1 methyl propyl| disulfide in which the S-S chromophore is directly linked to an asymmetric carbon shows a positive effect near 250 nm and a negative one at 225 nm[20]. It appears therefore that the effect located at 270-290 nm should be assigned to the S-S chromophore. The situation is not as simple in the 210-260 nm region where Cotton effects due to S-S and S-linkages overlap. Further studies are necessary to complete the identification of observed transitions.

One must add that polymers prepared with (S,S)DPED zinc initiators show a negative Cotton effect at 232 nm meaning the predominance of RR configurational units. The same behaviour was observed in methylthiirane polymerization and it was explained by substantial ring-opening in the α-position[15]. The rotatory powers in CHCl$_3$ observed for these polymers are abnormally high, and they vary with solvent. The presence of residues coming from the initiator is suspected and could explain the transition observed at 260 nm.

In conclusion, one should notice that if rotatory power can be used for identification of configurations in simple case, one must be very careful when irregularities such as S-S linkages appear in the polymer chain.

ENANTIOMERIC DISTRIBUTION IN THE POLYMER CHAIN

The enantiomeric distribution |RR|/|SS| in the polymer chain, which represents the selectivity of the polymerization process, could be approached by several techniques.

Cationic degradation

The degradation of poly cis-DMT using methyl triflate or Et$_2$OBF$_4$ leads to a mixture of two isomers of trithiepane (I) and (II) and[16,21] some amount of a cyclic tetramer.

(I)   (II)

Isomers (I) of trithiepane have all their asymmetric carbon atoms of the same configuration that is a RR-RR or SS-SS structure while isomer (II) contains two units having asymmetric atoms of opposite configuration (RR-SS structure). Thus, only isomers (I) are optically active, while isomer (II) is achiral.

The ratio of trithiepanes I/II was followed in the degradation products of optically active poly cis DMT which contains an excess of RR units prepared with a $ZnEt_2$-R(-)DMBD initiator.

As seen from table 6, the ratio I/II, taken at the beginning of the degradation reaction, is increasing with the optical activity of the polymers that is with an increase of RR content in the polymer chain in other words the tacticity. Ratio I/II can also be expressed in terms of isotactic dyads and we shall see from NMR studies that degradation and NMR technique give almost identical results concerning the enantiomeric composition, i.e. the stereoregularity of the chain. This confirms that the degradation reaction which involves the formation of a sulfur-sulfur bond does not change the configuration of the carbon centers.

Tacticity studies

Poly cis DMT's of different origins were studied by ^{13}C NMR. Two patterns are observed : one located around 45 ppm which corresponds to the methine carbon of the chain, the other one near 18 ppm is related to the methyl group. In the case of poly cis DMT's the methine carbon is more stereosensitive than the methyl group contrarily to what was observed for poly trans DMT's.

At 25.15 MHz the methine pattern is composed of four (or more) peaks, one of them, located at the lower field (45.8 ppm.) is well separated. This peak increases with the optical activity of polymers and it might therefore be assigned to enchainment of units containing carbons of the same configuration e.g. -RR-RR-RR isotactic type triads. Other peaks located at higher field are due to syndiotactic (RR-SS-RR)or heterotactic (RR-SS-SS and RR-RR-SS) triads of configurational units.

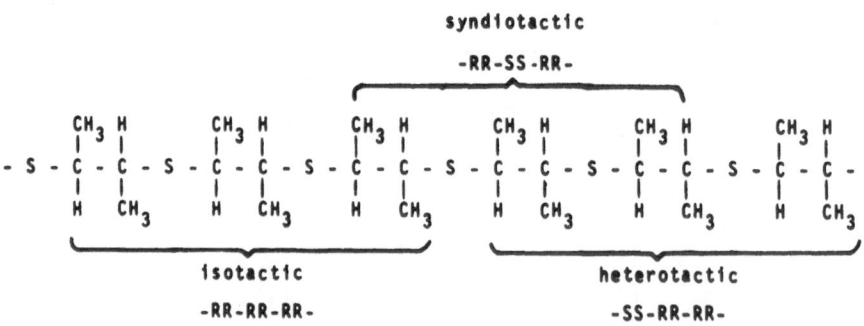

Table 6.   Tacticities of different optically active polycis DMT's
prepared with ZnEt$_2$-(R)-DMBD initiator system and ratio
of trithiepanes (I)/(II) obtained by cationic degradation of these polymers by methyl triflate in chloroform
at 60°C.

$\|\alpha_p\|$ (CHCl$_3$)	- 11	+ 32	+ 49	+ 59	+ 66	+ 84
i %	0.42	0.52	0.55	0.58	0.62	(0.76)
(I)/(II)	2.4	3.15	2.9	3.3	3.6	5.2

(I) RR-RR or SS-SS   (II) RR-SS

      The methyl carbon is less stereosensitive and gives at 25.15 MHz
only two peaks. The peak located at higher field increases with polymer rotatory power and is therefore due to isotactic enchaiments
(16.85 ppm).

      For polymers obtained with achiral initiators (tBuOK, BF$_4$Et$_3$0,
ZnEt$_2$-H$_2$O) the ratio of "isotactic" peak area to overall pattern
area is very close to 1:4, corresponding to a random type distribution. For optically active polymers this ratio increases with the
rotatory power as seen in table 6. It appears, however, that the ex-
trapolation of the experimental data to the limit, i.e. $\|\alpha_p\|$ = 0
for optically inactive polymer, leads to a value i = 0.41, much
higher than that found previously with achiral initiators. We concluded to the presence of several types of sites in the initiator
system : some of them are non stereospecific and produce polymers
with a random distribution, others are highly stereospecific and
lead to chains with long sequences of enantiomeric RR or/and SS
units. It is not clear presently if we are dealing with enantiomeric
blocks or a mixture of enantiomeric chains. The data may be satis-
factoraly explained by superimposition of a random bernouillian pattern and of a purely isotactic one, the latter being responsible for
the optical activity observed. The tacticity results are in good
agreement with cationic degradation results as seen from table 6.

      More detailed information on the structure could be obtained
from spectra run at high field ^{13}C NMR. At 90.52 MHz one finds
16 lines for the methine pattern, which can be interpreted in terms
of pentad effects for enantiomeric units, and approximatively
8 lines for the methyl carbon. At 100.62 or 125.77 MHz the stereosensitivity is again increased and one observes 32 lines for the

methine carbon. Studies are carried out presently in order to ex -
plain these spectra[22]. An example of a high field spectrum is given
in fig. 5.

Other interesting information is obtained from spectra of poly
cis DMT's prepared with (R,R)-butanediol, (S)-binaphthol type initiators, and (S,S)-diphenyl-1,2-ethanediol. These polymers, as we
have seen before, show unusual chiroptical properties and in the
first two ones -S-S- linkages were observed in CD spectra. The examination of ^{13}C NMR at 25.15 MHz confirms these findings. In figure 6 is given a spectrum of a polymer prepared with zinc-(S) binaphthol initiator. Many supplementary peaks appear in each pattern.

In the methine region one finds between 51.76 and 47.38 ppm
10-12 peaks divided into two patterns. Another set of peaks are
located at 43.87 to 43.10 ppm highfield from the normal methine
pattern (45-46 ppm).

In the methyl region at highfield from the normal double peak
(near 17.50 ppm) one finds a single peak at 15.94 ppm and several
strong peaks between 15.13 and 13.58 ppm.

Disulfide and polysulfide linkages have already been studied
by ^{13}C NMR in polymethyl thiiranes[19, 23]. A multiplet corresponding
to -S-S-CH linkages appeared at 45 ppm with a lowfield shift close
to 4-5 ppm from the normal -S-CH peak.

For poly cis DMT's the lowfield shift for methine carbons is
even higher in the presence of S-S linkages. One should notice that
supplementary peaks appear also at a lower field than the normal
methine position.

In the methyl region all the supplementary peaks are found at
a higher field than the normal peak. The pattern composed of 4 peaks
located between 15 and 14 ppm is very intense. The magnitude of these
peaks varies with the tacticity.

The overall amount of supplementary patterns observed in methine
and methyl region is increasing with the increase of % S in the polymer determined by elemental analysis. The % S, i.e. the amount
of disulfide linkages varies with the nature of initiator used and
is particularly high with binaphtol zinc initiator (table 7). In
the latter case polymer composition corresponds to a
$-CH(CH_3)-CH(CH_3)-S_{1.5}-$ repeating unit. It is on the other hand difficult to correlate the rotatory power with the excess of % S observed. More detailed studies on a series of polymers prepared with
different initiators are necessary.

It must be mentioned that in polymers prepared with DPED zinc

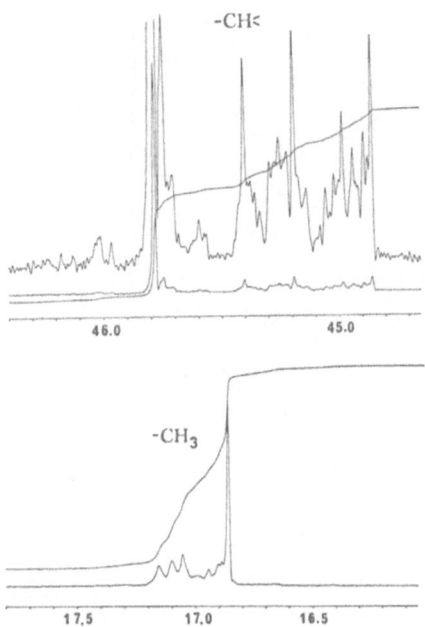

Fig. 5. ^{13}C 100.62 MHz NMR : poly cis-DMT $|\alpha_p|$ + 42 (CHCl$_3$) prepared with ZnEt$_2$-R(-)DMBD initiator system. Solvent : CDCl$_3$

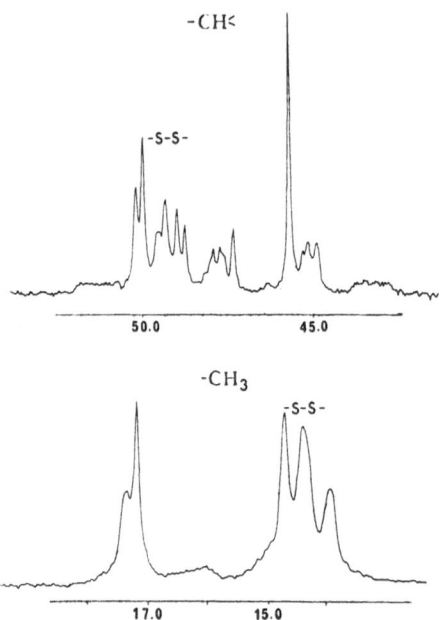

Fig. 6. ^{13}C 25.15 MHz NMR : poly cis-DMT $\left|\alpha_p\right|$ + 74 (CHCl$_3$) prepared with ZnEt$_2$-(S) binaphthol initiator system (in CDCl$_3$)

initiator, which do not contain an excess of sulfur, no supplementary peaks were observed in NMR spectra. These polymers have a high
rotatory power which is not in agreement with the tacticity observed. The presence of residues coming from the initiator could be
the origin of the enhancement of $|\alpha_p|$ observed.

Table 7.   Variation of excess of sulfur in poly
           cis-DMT's prepared with different chi-
           ral initiators.

Initiator	$\|\alpha_p\|$ (a)	% i (b)	% S (c)	$\dfrac{S_x}{mon.\ unit}$ (d)(e)
ZnEt$_2$-DMBD	+73	62	37	0,03
CdMe$_2$-DMBD	-45	48	41	0,22
ZnEt$_2$-BD	-27	29	42	0,27
	+126	40	43	0,32
ZnEt$_2$-BN	+41	27	44	0,37
	+74	56	47	0,55

(a) in chloroform - (b) isotactic triad - (c) from
elemental analysis (d)-Sx-S-CH(CH$_3$)-CH(CH$_3$)-

(e) for monomer unit C$_4$H$_8$S : S % = 36.4.

DISCUSSION

     From the results described above it appear that thiiranes with
two asymmetric carbons atoms in the cycle present interesting stereochemical behaviour in the presence of chiral initiators.

     This behaviour may be very often deduced from rules established
for methylthiirane, a model molecule of DMT's, but in other cases
it has original aspects.

     For the trans monomer, a resolvable compound, the intrinsic
choice with the usual chiral initiator prepared from ZnEt$_2$ corresponds to the homosteric choice. In the present case a R type initiator chooses preferentially the RR enantiomer. The kinetic equation describing this process is quite different from that of methyl
thiirane. Its originality consists in the change of the sign of un-

reacted monomer at a defined conversion. Thus during the same process one obtains a monomer that is first levorotatory and then dextrorotatory. The mechanism based on kinetic data seems to be complex and it is probable that two mechanisms which were not yet identified are superimposed in the process[13].

The polymers obtained are optically inactive, which confirms again the complete inversion of attacked asymmetric carbon atom during ring-opening as established by Vandenberg[2]. According to cationic degradation studies these polymers present some stereoregularity, which was confirmed by ^{13}C NMR studies. This stereoregularity changes with the conversion. Combination of triads of monomeric units -SR- and -RS- are present, but they could not be yet identified and require studies at higher fields. The polymerization of a pure enantiomer, RR for example, should provide very useful information. It is clear that in these polymers, the simple pure disyndiotactic RS-RS structure is not observed. The polymers are crystalline and present melting points which change with conversion. This behaviour is different from that of polymers prepared with aluminum initiators studied by Vandenberg[5] in which no melting points were observed even with the chelated initiator.

In the case of the <u>cis</u> monomer the results are particularly interesting from the stereochemical point of view. The meso monomer allows the possibility of ring-opening on two asymmetric carbon atoms of opposite configuration.

From our results it is clearly established that the principle of homochirality is again operative. A standard zinc chiral initiator bearing one asymmetric carbon atom such as $ZnEt_2$-(R) DMBD system attacks preferentially the molecule in β position, that is on the opposite side of the homochiral carbon atom. Due to inversion of configuration of the attacked carbon atom, predominant enantiomeric units of one configurational type are formed (-RR- in the case of the initiator cited), and the polymer is optically active. This enantiogenic process is a true asymmetric polymer synthesis. Although the optical purity of our products is not completely established, one may expect from some NMR measurements data that enantiomeric enrichment may reach 70 % in some cases[16].

When using $CdMe_2$-(R) DMBD initiator systems one obtains an enrichment in opposite enantiomeric units i.e. -SS-. This means that the heterochiral carbon atom was preferentially attacked. This result has to be compared with methylthiirane polymerization in which with the same initiator antisteric process was observed i.e. a preferential choice of the heterochiral carbon atom. Thus similarity in choice is observed. However, the result with cis-dimethylthiirane is more significant since it shows that the initiator is able to recognize the configuration of an asymmetric carbon atom in a molecule and not only to distinguish between enantiomers. It is a higher step in chiral recognition.

The situation with initiators derived from chiral components
with two asymmetric carbon atoms seems much more complex. It was
observed that when using (S,S) DPED and (R,R)BD type initiators
with methylthiirane the polymers obtained were enriched in enantiomer
of configuration opposite to that predicted by the homochiral choice
This result was explained by a predominant ring-opening in α position i.e. on the asymmetric atom.

In the case of cis-DMT one observes the same situation. Although with (S,S)DPED predominant SS units should be observed according to the homochiral rule, one finds a dextrorotatory polymer
which is enriched in RR units. This is substantiated by an examination of the CD spectrum.

The polymers prepared from (R,R)BD and especially from (S)-BN
initiators contain an excess of sulfur as shown by elemental analysis. In CD spectra one finds a Cotton effect at 290 nm characteristic of the disulfide linkages. These linkages are also observed,
but in a small amount, in $CdMe_2$-(R) DMBD type polymers and even in
some $ZnEt_2$-(R) DMBD type polymers.

The formation of disulfide linkages in thiiranes polymerization
was already described by Aliev et al[24] in the case of anionic polyme
rization with organolithium compounds, by Goethals et al[21] in the
case of cationic polymerization and by ourselves[19] when aluminum
initiator are used. It involves the formation of the corresponding
olefin which was also observed in the present case. In anionic polymerization with lithium compounds a relay mechanism was preposed
which gives polymers with polysulfide linkages. In cationic polymerization the reaction may go on until complete degradation into
cyclic products of defined structures.

In the case of cis DMT the reaction is more limited and gives
in the case of the binaphthol initiator a maximum amount of one
S-S linkage per two monomeric units. It is not possible to say now
if these linkages are formed during the propagation or through a
degradation reaction. Probably both of these processes are involved
and various mechanisms may be proposed[19,24,25].

Poly cis-DMT prepared with our systems are crystalline and present some stereoregular structure. The melting points of optically
inactive polymers obtained from achiral initiators are different
from those prepared with chiral initiators. Difference in crystalline structures as observed in the case of poly t-butylthiirane[26].
may be considered and studies are progressing in this field. The
study of stereoregularity by ^{13}C NMR revealed the complexity of the
chain structure. An extreme stereosensitivity of the various carbons

to tacticity was observed. At high field NMR more than heptad effects for configurational units could be observed. This complexity has not yet allowed us to propose a defined structure. The peak corresponding to isotactic n-ades only was well assigned and a correlation with optical activity established. In the case of optically active polymers the chain could be formed by isotactic blocks followed by random sequences, or by a mixture of isotactic chains and heterotactic chains. In the case of optically inactive polymers the heterotactic chain might have a particular arrangment which allows the crystallization of the products. The polymers containing a large amount of disulfide linkages did not show any melting point up to now. It is interesting to notice that in the case of methylthiirane the zinc-binaphthol initiator leads to very different results absence of S-S linkages and very high stereoelectivity[27].

The results presented in this paper contribute to the understanding of the stereochemistry of the polymerization of disubstituted thiiranes and emphasize new properties and new behaviour of these polymers. Many questions are however still unsolved and particularly the problem of chain structure. In addition very similar results to those observed with cis DMT were obtained with cyclohexene sulfide. These results will be reported elsewhere.

REFERENCES

1. C.C. Price and M. Osgan, J. Amer. Chem. Soc. 78, 4787 (1956)
2. E.J. Vandenberg, J. Polymer Sci., A1, 7, 525 (1969)
3. N. Spassky and P. Sigwalt, Bull. Soc. Chim. 4617 (1967)
4. N. Spassky, P. Dumas, M. Sépulchre and P. Sigwalt, J. Polymer Sci., Symposium n° 52, 327 (1975)
5. E.J. Vandenberg, J. Polymer Sci., A1, 10, 329 (1972)
6. S. Inoue, T. Tsuruta and J. Furukawa, Makrom. Chem. 53, 215 (1962)
7. T. Furukawa, N. Kawabata and A. Kato, J. Polymer Sci., B, 5, 1073 (1967)
8. N. Spassky and P. Sigwalt, Comptes Rendus Ac. Sc. 265, C, 624 (1967)
9. T. Tsuruta, J. Polymer Sci., D, 179 (1972)
10. M. Sépulchre, N. Spassky and P. Sigwalt, Israel J. of Chem., 15, 33(1976/77)
11. N. Spassky, A. Momtaz and M. Sépulchre in Preparation and Properties of stereoregular Polymers, Ed. by R.W. Lenz and F. Ciardelli - D. Reidel Publ. Co. p. 201 (1979)
12. A. Momtaz, N. Spassky and P. Sigwalt, Nouveau J. Chimie, 3, 669 (1979)
13. A. Momtaz, N. Spassky and P. Sigwalt, Polymer Bulletin, 1, 267 (1979)
14. N. Spassky, A. Leborgne, A. Momtaz and M. Sépulchre, J. Polymer Sci., Polymer Chem. Ed., 18, 3089 (1980)

15. N. Spassky, A. Momtaz, M. Reix and M. Sépulchre, Preprints
    Makro Mainz, v. 1, 265 (1979)
16. E.J. Goethals, R. Simonds, N. Spassky and A. Momtaz, Makrom.
    Chem., 181, 2481 (1980)
17. P. Salvadori, Chem. Comm. 1203 (1968)
18. J.P. Casey and R.B. Martin, J. Amer. Chem. Soc. 94, 6141 (1972)
19. P. Dumas, N. Spassky and P. Sigwalt, J. Polymer Sci., Polymer
    Chem. Ed. 14, 1015 (1976)
20. E. Chiellini, private communication
21. W. Van Crayenest and E.J. Goethals, Eur. Polym. J., 12, 859
    (1976)
22. W.E. Hull, A. Momtaz and N. Spassky, to be published
23. A.D. Aliev, S.L. Alieva and B.A. Krentzel, Vysokomol. Soed. XXII
    A, 1171 (1980)
24. A.D. Aliev, B.A. Krentsel, G.M. Mamediarov, I.P. Solomatina and
    E.P. Tiurina, Eur. Polym. J., 7, 1721 (1971)
25. A.D. Aliev, I.P. Solomatina and B.A. Krentzel, Macromolecules
    6, 797 (1973)
26. H. Matsubayashi, Y. Chatani, H. Tadokoro, P. Dumas, N. Spassky
    and P. Sigwalt, Macromolecules, 10, 996 (1977)
27. M. Sepulchre, K. Hintzer, V. Schurig and N. Spassky, C.R. Acad.
    Sci., 291, C, 267 (1980)

STEREOCHEMICAL ASPECTS OF THE CATIONIC POLYMERIZATION OF CHIRAL

ALKYL VINYL ETHERS

Emo Chiellini, Roberto Solaro, and Francesco Masi

CNR Center of Stereordered Optically Active Macromolecules, Institute of Industrial Organic Chemistry
University of Pisa, Italy

SYNOPSIS
        A survey of the stereochemical aspects relevant to the cationic
polymerization of chiral alkyl vinyl ethers has been presented. Accordingly the routes to the preparation of optically active poly(alkyl vinyl ether)s by starting both from optically active monomers
and racemic ones have been presented. In particular, to approach the
mechanistic aspects of the stereospecific heterogeneous catalytic
systems based on alkyl or alkoxy aluminum and sulphuric acid, the
activity of soluble organoaluminum sulphates has been investigated.
The kinetic trend of the polymerization of some optically active
vinyl ethers has been determined simply by optical rotation measurements. On the basis of the presented data a coordinate polymerization mechanism has been proposed for the multicomponent catalyst introduced by Vandenberg.

INTRODUCTION

        Previous to the intuitive and tentative rationalization of the
stereospecific polymerization of α-olefins by anion coordinate catalyst[1] it was known that under suitable conditions *iso*butyl vinyl
ether could be polymerized at low temperature in the presence of a
conventional Lewis acid to a tacky or moderately hard thermoplastic
material[2].
        Starting from the mid 50's and for more than one decade a great
deal of interest was devoted by several industrial and academic research groups[3] to the stereospecific polymerization of alkyl vinyl

ethers in a side by side development with the α-olefin polymerization that in the meantime had found fruitful industrial outlets.
Even though the efforts in that direction were not crowned by any
wide-scale practical application, nevertheless they contributed heav-
ily to the general understanding of the stereospecific polymerization of vinyl monomers.

In the present paper we wish to highlight the major results of
the contributions afforded by the use of chiral vinyl ethers in
helping to solve some aspects of the stereochemistry of the cationic
polymerization of alkyl vinyl ethers.

Alkyl vinyl ethers readily undergo cationic polymerization
thanks to the presence of the alkoxide group which is directly bound
to the vinyl group and is able to release electrons thus making the
π-bond active toward an electrophilic attack with consequent stabilization of the positive charge on the carbon in the α-position. A
large variety of hard and soft, either conventional or modified
Lewis-acid type cationic initiators have been tested in the polymerization of alkyl vinyl ethers both under homogeneous and heterogeneous conditions at ambient and low temperatures[3] (Table 1).

Table 1. Catalytic systems active in the polymerization
of alkyl vinyl ethers.

A. FRIEDEL-CRAFTS TYPE	D. METAL SULPHATES
B. ZIEGLER-NATTA TYPE	E. GRIGNARD REAGENTS
C. METAL ALKOXIDES, METAL OXIDES & HALOGEN DERIVATIVES	F. STABLE CARBOCATIONS

Among those particularly attractive appear the "multicomponent
systems" based on metal alkyls, metal alkoxides and metal sulphates
in combination with sulphuric acid and suitable activators as described in several patents assigned to the Hercules Powder Co.[4-8]
to which a substantial contribution was offered by E.J. Vandenberg.

In relation to this we are going to present some recently obtained data which should help in giving a better insight into the
reaction mechanism relevant to their chemical and stereochemical
activity.

The utilization of chiral alkyl vinyl ethers, since they provide[9-12] a powerful means for a staightforward semiquantitative

evaluation of the stereospecificity of the different catalytic systems and polymerization conditions has been particularly valuable
for this investigation.

1. POLYMERIZATION OF CHIRAL ALKYL VINYL ETHERS

Chiral alkyl vinyl ethers, like every chiral monomer, may give
rise to optically active polymers both by homopolymerization and
copolymerization procedures. By starting either from monomers enriched in one enantiomer or from a racemic mixture, optically active
homopolymers of alkyl vinyl ethers can be obtained by using chirality
sensitive catalysts. These should be able to discriminate between
the two enantiomers of the starting monomer (*Stereoselective* and
stereoelective polymerization*) without altering the stereochemical
requirements of the preexisting chiral centers (Scheme 1).

Scheme 1. Synthesis of Optically Active Polymers.

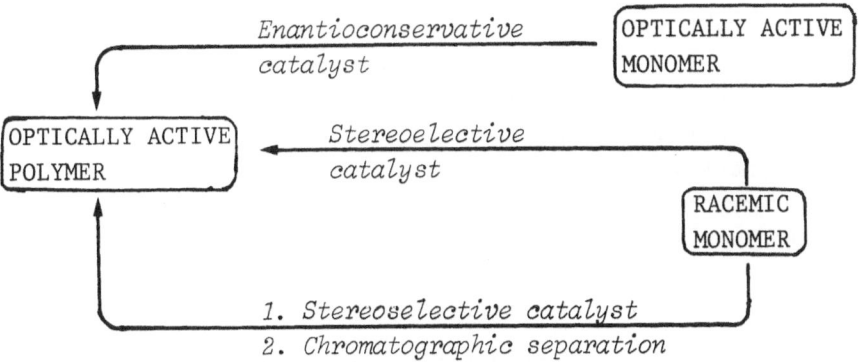

The first route is the only reliable one for the synthesis of
optically active poly(alkyl vinyl ether)s on preparative scale,
whereas the other two starting from racemic monomer have been found
useful only for speculative deductions. By copolymerization processes optically active poly(alkyl vinyl ether)s are obtainable from
an optically active precursor monomer which can be combined in copolymer macromolecules along with a prochiral or racemic alkyl vinyl
ether (Scheme 2).

---

* According to a recent IUPAC nomenclature proposal the terms
stereoselective and **stereoelective** should be substituted with
enantioselective and *enantioelective* respectively.

Scheme 2. Synthesis of Optically Active Polymers.

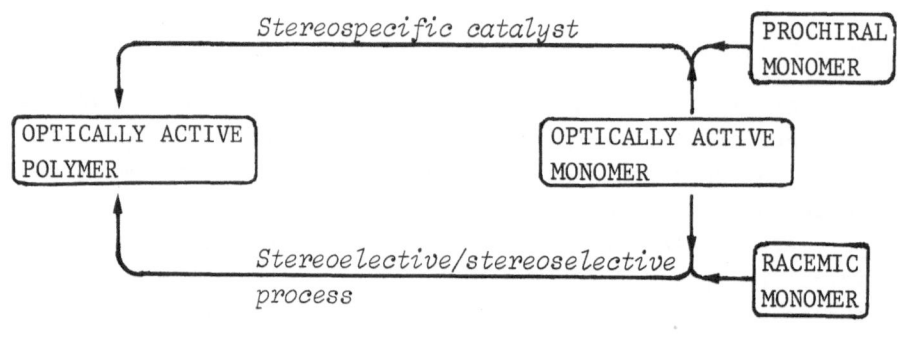

In the former case, when isotactic copolymer macromolecules are obtained, the counits of the prochiral monomers may assume an induced preferential conformation and give a substantial contribution to the overall polymer optical rotation. In the latter case a discrimination by the reactants of the two antipodes of the racemic monomer toward the copolymerization can lead at the same time to a dynamic enantiomeric enrichment of the non-polymerized monomer and to a polymerization product whose optical rotation is substantially affected by the presence of optically active monomeric units derived from the racemic monomer.

1a. Polymerization of Optically Active Monomers

Isotactic polymers from optically active monomers are readily produced by polymerization with different catalytic systems under suitable experimental conditions. In Table 2 are collected the most significant examples known up to now and in particular the correlation between the optical rotation of the starting monomer and the polymer derived therefrom is shown.

With only the exception of the polymer obtained from (S)-2--methylbutyl vinyl ether, where the asymmetric carbon atom is in the β position with respect to the ethereal oxygen and the optical rotation of the polymer is comparable to that of the starting monomer, in all other cases a substantial enhancement of the optical rotation is observed in going from the monomer (or low molecular weight model compound) to the polymer.

This enhancement of optical rotation, observed also in structurally analogous poly(α-olefin)s[13] obtained by anionic coordinate polymerization, is explained as being due to the existence of a peculiar conformational discrimination leading to helical structures with a prevailing screw sense.

Table 2. Synthetic Optically Active Semicrystalline Poly(Alkyl Vinyl Ether)s.

Monomer			Polymer	Ref.
Type	Absolute configuration	$[\phi]_D^{25}$	$[\phi]_{D_{max}}^{25}$	
$C_2H_5-CH-O-CH=CH_2$ $\quad\vert$ $\quad CH_3$	S	+ 14.4	+328	10
$tC_4H_9-CH-O-CH=CH_2$ $\qquad \vert$ $\qquad CH_3$	S	− 13.1	+319	30
$nC_6H_{13}-CH-O-CH=CH_2$ $\qquad\quad \vert$ $\qquad\quad CH_3$	S	− 1.4	+150	
$C_6H_5-CH-O-CH=CH_2$ $\quad\vert$ $\quad CH_3$	R	+ 78.8	+120	20
(cyclohexyl with isopropyl)-O-CH=CH$_2$	1R,3R,4S	−134.0	−405	19
	1R,3S,4S	+ 16.7	+148	18
	1R,3S,4R	+ 23.0	+ 32	
$C_2H_5-CH-CH_2-O-CH=CH_2$ $\quad\vert$ $\quad CH_3$	S	+ 7.7	+ 5.6	10

These structures are stable in the solid state and are cooperatively maintained in solution[14-16]. The degree of prevalence and hence the increase of the optical rotation of the polymer are positively affected by the isotacticity of the polymer. Therefore, considering that during the polymerization a complete retention of the configuration of the asymmetric carbon atom both in alkyl vinyl ether of secondary and primary alcohol is maintained[17,18], the optical rotation can be taken as a suitable measure for a qualitative evaluation of the configurational regularity of the main chain in optically active macromolecules of the reported alkyl vinyl ethers, and in turn may be used as a tool to evaluate the stereospecificity of different catalytic systems.

In Tables 3 and 4 data are reported relevant to the two typical sets of information attainable by the use of optically active poly-(alkyl vinyl ether)s, such as (-)-menthyl vinyl ether and (+)-(S)--secbutyl vinyl ether with optical purity larger than 90%.

Table 3. Optical Rotation of Poly(Menthyl Vinyl Ether) Samples Obtained under Different Conditions.

Polymerization conditions			Polymer		Ref.
Catalyst	Temp. (°C)	Solvent	$[\Phi]_D^{25}$ a	Isotactic diads (%)	
$Mn-MnO_3/H_2SO_4$	0	$n-C_6H_{14}$	-256	isotactic	18
$[(iC_3H_7O)_2Al]_2SO_4$	25	C_6H_6	-365	70	30
$C_7H_7SbCl_6$	30	CH_2Cl_2	-368	70	
	0		-375	70	19
	-30		-380	70	
$C_2H_5AlCl_2$	-78	$C_6H_5CH_3$	-405	80	

aUnfractionated sample in hydrocarbon solution.

In both cases, polymer with fairly high molecular weight and high optical rotation can be prepared by using either homogeneous or heterogeneous catalytic systems at room temperature and below 0°C. It is worth noting that, at least for polymerization runs carried out under comparable conditions, a modest rôle is played by the operating conditions in affecting the optical rotation and hence the isotacticity of the poly(menthyl vinyl ether) samples, whereas a more dramatic effect is detectable in the case of poly(secbutyl vinyl ether) samples[19].

In the former case the structural geometry of the monomer with three different asymmetric carbon atoms must be claimed as responsible for the rather high isotacticity degree observed in polymeric samples obtained under free cation propagation conditions[19].

More generally, the best catalytic systems with respect to the stereochemical control appear to be the modified Lewis acids operating at -78°C and the multicomponent insoluble catalyst based on

Table 4. Optical Rotation of Poly (S)- Butyl Vinyl Ether Samples
          Obtained under Different Conditions.

Polymerization conditions				Polymer		Ref.
Catalyst	Temp.	Solvent	$\|\phi\|_d^{25a}$	IR^b regularity		
	(°C)			index		
	30		+137	–		
C H SbCl	0	CH Cl	+157	–		19
	-30		+174	–		

$Al(O-iC_3H_7)_3$ and H_2SO_4[10,11]. This last system is practically incapable of polymerizing (-)-menthyl vinyl ether[20], which on the contrary does polymerize with the soluble bis-(diisopropoxyaluminum) sulphate.

Table 4. Optical Rotation of Poly[(S)-secButyl Vinyl Ether] Samples Obtained under Different Conditions.

Polymerization conditions			Polymer		Ref.
Catalyst	Temp. (°C)	Solvent	$[\Phi]_D^{25}$ [a]	IR[b] regularity index	
$C_7H_7SbCl_6$	30	CH_2Cl_2	+137	-	
	0		+157	-	19
	-30		+174	-	
$[(iC_3H_7O)_2Al]_2SO_4$	25	C_6H_6	+195	±	30
$Al(iC_4H_9)Cl_2$	-78	$C_6H_5CH_3$	+260	+	9-10
$Al(O-iC_3H_7)_3-H_2SO_4$	25	AcOEt	+277	+	

[a]Unfractionated sample in hydrocarbon solution.
[b]Evaluated as D_{827}/D_{771}.

The utilization of optically active monomers with low optical purity[21] does not appreciably influence the tacticity degree of the polymer and the overall molar rotation can be generally determined by linear extrapolation, the only exception being the case of poly-(secbutyl vinyl ether) samples, for which a small positive deviation from the linear trend has been reported[17,21]. This result allows one to mention the convenience of using alkyl vinyl ethers characterized by an even modest enantiomeric excess for making deductions valid in the entire range of enantiomeric purity of the system used.

1b. Polymerization of Racemic Monomers

Generally speaking the stereospecific polymerization of a racemic monomer may be realized in the presence of a catalyst prepared from either achiral or chiral precursors characterized by a preferential chirality. In the former case (Scheme 3a), the polymerization process can be approached, from the point of view of the stereochem-

ical features of the initial monomer, as a true copolymerization of
the two monomeric antipodes, and two limiting cases consisting of
the production of either a homopolymer mixture or a random copolymer
of the two antipodes are possible. That behaviour is normally caused
by the intrinsic nature of the catalyst either able to selectively
distinguish (*stereoselective catalyst*) or not (*non-stereoselective
catalyst*) between the two opposite steric form of the two monomeric
enantiomers. In the latter case (*stereoelective polymerization*)
(Scheme 3b) the presence of a chiral catalyst with a preferential
chirality able to discriminate between the two monomeric enantiomers
should make possible the polymerization of the two enantiomers at a
different rate. This leads, as a limiting case (stereoelectivity
100%), to the recovery of the pure homopolymer of one enantiomer on
the one hand and on the other, the unconverted enantiomer.

Scheme 3. Polymerization of a Racemic Monomer.

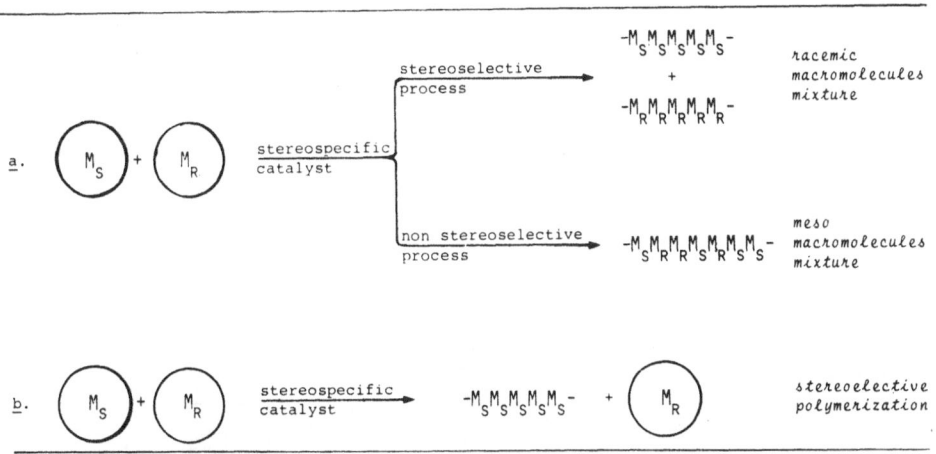

1b. 1. Stereoselective polymerization. Alkyl vinyl ethers along
with α-olefins[22] are the only vinyl-type monomers able to give
stereoselective polymerization in the presence of heterogeneous multicomponent catalysts, that have been separated by elution chromatography, after adsorption on an optically active insoluble support,
in fractions characterized by an opposite sign of optical rotation.

In Table 5 are represented the structures of partially isotactic semicrystalline poly(alkyl vinyl ether)s obtained by polymerization of the corresponding racemic monomers with $Al(O-iC_3H_7)_3/H_2SO_4$
stereospecific catalyst[21].

Table 5. Chromatographic Separation of Poly(Alkyl Vinyl Ether)s[20,21].

$$\begin{array}{cc}
+CH_2-CH+ & +CH_2-CH+ \\
\quad\quad O & \quad\quad O \\
\quad *CH-CH_3 & \quad\quad CH_2 \\
\quad\quad C_2H_5 & \quad *CH-CH_3 \\
& \quad\quad C_2H_5
\end{array}$$

Supports

$$\begin{array}{cc}
+CH_2-CH+ & *\\
\quad *CH-CH_3 & +C-CH-O+ \\
\quad\quad C_2H_5 & O\ \ CH_3
\end{array}$$

After supporting them on either poly[(S)-3-methyl-1-pentene] or
poly[(S)-lactide] and subsequent elution with suitable solvents,
stereoselectivity was shown to occur only in the case of poly(secbutyl vinyl ether). In Fig. 1 a typical profile is represented of
the fractionation by elution with solvents of this polymer supported
on poly[(S)-lactide].

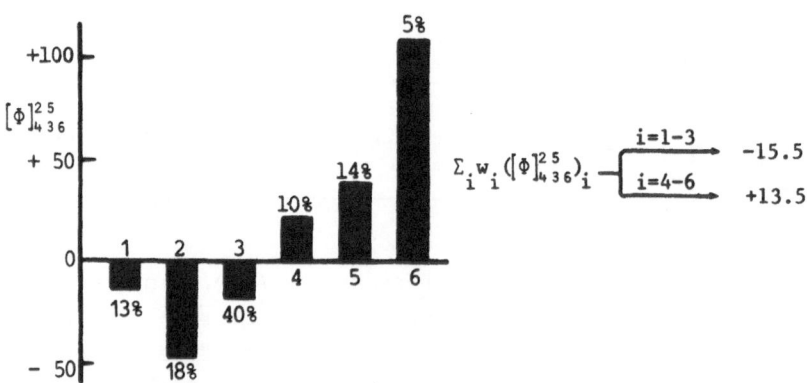

Figure 1. Chromatographic Separation of Poly[(R)(S)-secButyl Vinyl
          Ether][21].

In both cases no stereoselectivity was observed for samples prepared under homogeneous conditions with conventional cationic catalysts. In keeping with the results obtained in the case of poly-(α-olefin)s[22] we stress that stereoselectivity in a polymerization process depends on: 1) structure of the monomer, 2) nature of the catalyst, and 3) isotacticity degree of the polymer, which depends in turn on the nature of the catalyst and polymerization conditions

1b. 2. <u>Stereoelective polymerization</u>. Very little has been done on the polymerization of racemic alkyl vinyl ethers and the results reported up to now are indeed not very encouraging. Since the early seventies[23] we started to study the polymerization of racemic *sec*-butyl vinyl ether in the presence of Vandenberg-type catalysts, based on optically active alcoholates or alkylalcoholates (metal: Zn, Al or Ti and alkoxy: menthoxy and 1-phenylethoxy), and H_2SO_4, which are active in the polymerization at room temperature. More recently it has been claimed that stereoelective polymerization of racemic *sec*-butyl vinyl ether and *cis* and *trans*-*sec*butyl propenyl ethers can be achieved by using the optically active menthoxy aluminum dichloride modified Friedel-Crafts catalyst[24]. No appreciable stereoelectivity was observed by using (S)-*sec*butoxy aluminum dichloride and 2-methyl-butoxy aluminum dichloride. By using the first mentioned catalytic system, when the conversion was lower than 70%, we were able to obtain both optically active unreacted monomer and polymer fractions with optical rotation consistent with a prevalent polymerization of the enantiomer of opposite configuration to that of the unreacted monomer. From the reported optical data of both polymerized and non-polymerized monomer, enantiomer excesses lower than 1% are cal-culable (Table 6), thus indicating the very low potential of the stereoelective polymerization in preparative applications of optical-ly active poly(alkyl vinyl ether)s.

No optical rotation of the residual monomer was apparently ob-served in a more recent paper[31], thus making the claimed stereo-electivity very disputable, in consideration also of the extremely low value of the specific rotatory power of the polymer. We have moreover to mention that no significant enhancement of the stereo-regularity of the prepared polymers is noticed by using optically active catalysts in place of $Al(O-iC_3H_7)_3/H_2SO_4$ system. This result becomes particularly interesting in light of the fact that the sys-tem based on titanium tetramenthoxide is completely soluble in the reaction medium, thus for the first time permitting one to stress that the steric control at the level of the main chain is the same in homogeneous and heterogeneous polymerization processes of alkyl vinyl ethers.

Table 6. Polymerization of Racemic *sec*Butyl Vinyl Ether in the
Presence of Optically Active Catalysts[23].

Catalyst $R_mMe(OR)_n/H_2SO_4$		Polymer	Unreacted monomer	Preferentially polymerized enantiomer	
$R_mMe(OR)_n$	Absolute config.	$[\Phi]_{436}^{25}$	$[\Phi]_{365}^{25}$	Absolute config.	Optical purity (%)
$C_2H_5ZnO\text{-Ment}^*$	R	-0.8	$+0.01$	R	0.13
$Ti(O\text{-Ment}^*)_4$	R	-4.1	$+0.04$	R	0.32
$Ti(O\text{-}\overset{*}{C}H\text{-}CH_3)_4$ $\quad\;\; C_6H_5$	S	-9.1	-0.02	S	0.47
$Al(O\text{-}\overset{*}{C}H\text{-}CH_3)_3$ $\quad\;\; C_6H_5$	S	-1.4	$+0.02$	R	0.38

$-\text{Ment}^*$ = (1R, 3R, 4S)-1-Methyl-4-*iso*propylcyclohexyl-3- .

2. COPOLYMERIZATION OF OPTICALLY ACTIVE ALKYL VINYL ETHERS WITH NON-OPTICALLY ACTIVE ONES

In the late sixties it was accidentaly found that optically
active α-olefins could be copolymerized with racemic α-olefins[25]
and prochiral ones[26] to give optically active copolymers whose optical rotation denoted the presence in copolymer macromolecules of
optically active counits derived from the racemic monomer and from
the prochiral one. In the former case it was also possible to isolate the unconverted monomer, characterized to a certain extent by
an enantiomeric excess (*Stereoelective copolymerization*).

At the same time the fractionation of the polymeric product by
extraction with boiling solvents caused a substantial separation of
the optically active homopolymer fraction along with another fraction consisting essentially of the copolymer of one of the two antipodes with the optically active monomer (*Stereoselective copolymerization*) (Scheme 4).

Scheme 4. Copolymerization of a Racemic Monomer with an Optically
Active One.

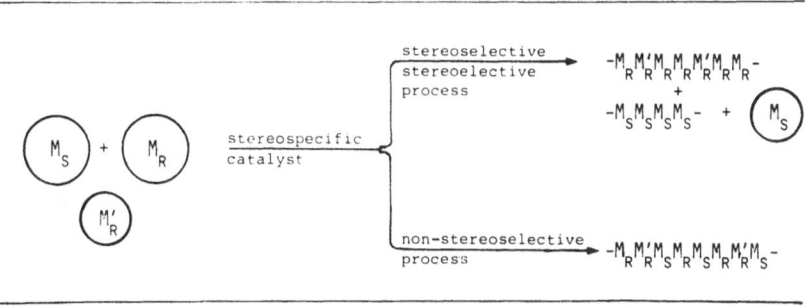

The extension of these concepts to the copolymerization of optically active alkyl vinyl ethers led to the conclusion that the
stereochemical control exerted by the Ziegler-Natta type catalyst
in the copolymerization processes was not an exclusive peculiarity
of those catalytic systems and may occur also in the absence of a
transition metal component in the catalytic system.

Another interesting aspect of the copolymerization of optically
active alkyl vinyl ethers with prochiral monomers was the potential
of tailormaking polymers with a strong asymmetric induction even by
small amounts of optically active monomeric precursors.

2a. Stereoselective and Stereoelective Copolymerization

Racemic secbutyl vinyl ether was copolymerized in the presence
of $Al(O-iC_3H_7)_3/H_2SO_4$ catalyst with (-)-menthyl vinyl ether[20].

In all the examined cases the copolymerization process was demonstrated to be stereoelective and stereoselective on the basis of
the optical rotation of the recovered non-polymerized monomer (optical purity 1-3%) (Tables 7 and 8).

That result made possible the determination of the absolute
configuration of the preferentially polymerized enantiomer.

In Table 7 are summarized the relationships between sign of the
optical rotation and absolute configuration of the starting optically active monomer. For the examined cases it was observed how the

the matching of the different enantiomers was exclusively realized
on the basis of the sign of the optical rotation rather than on the
absolute configuration.

Table 7. Stereoelectivity in the Copolymerization of *sec*Butyl Vinyl
Ether [20].

Optically active comonomer			Preferentially polymerized antipode	
Type	Sign	Configuration	Sign	Configuration
1-Phenylethyl	−	S	−	R
vinyl ether	+	R	+	′S
(−)-Menthyl vinyl ether	−	R	−	R

The optical rotation of the unconverted *sec*butyl vinyl ether
in the stereoelective copolymerization is summarized in Table 8.

Table 8. Stereoelective Copolymerization of (R)(S)-*sec*Butyl Vinyl
Ether (BVE) [20].

Optically active vinyl ether (VE)	Conversion BVE (%)	Non-polymerized BVE $[\alpha]_D^{25}$
(−)-Menthyl vinyl ether	50	+0.10
(+)(R)-1-Phenylethyl vinyl ether	38	−0.27
(−)(S)-1-Phenylethyl vinyl ether	20	+0.28

Molar ratio BVE/VE = 3–6.

The fractionation with boiling solvents of the crude copolymerization products brought, as expected, a separation of copolymer
fractions with different content of the enantiomer units more akin
to the starting optically active comonomer. In particular the last
fraction extracted had an optical rotation entirely consistent with
all the previous deductions and an unequivocal proof of the stereo-

selective-stereoelective copolymerization was stated (Table 9).

Table 9. Fractionation of Copolymers of (R)(S)-*sec*Butyl Vinyl Ether
with (-)-Menthyl Vinyl Ether (MVE) and (R)-1-Phenylethyl
Vinyl Ether (PEVE)[20].

Fractions extracted with	Copolymer with MVE		Copolymer with BEVE		
	wt-%	$[\alpha]_D^{25}$	wt-%	PEVE-%	$[\alpha]_D^{25}$
1. Methanol	4.0	−29.0	7.4	22.0	+ 19.8
2. Acetone	8.0	−46.6	77.4	20.5	+ 28.3
3. Diethyl ether	70.4	−53.4	12.8	10.0	+ 26.1
4. Benzene	5.4	+ 7.3	2.4	1.7	−123.0

Poly(MVE): $[\alpha]_D^{25}$ −220; poly(BEVE): $[\alpha]_D^{25}$ + 70.

Major effects were observed in copolymerization experiments
carried out with optically active 1-phenylethyl vinyl ether[20].
That behaviour can be associated with the presence on the surface
of the heterogeneous catalyst of chiral sites in enantiomeric bal-
ance which in turn can be saturated by selective interaction with
the preexisting enantiomer. It can also be related to the presence
of symmetric catalytic sites which become chiral after complexation
with monomer and retain a sort of memory of the shape of the former-
ly complexed monomer even after insertion into the growing chain.

2b. Copolymerization with Prochiral Monomers

Menthyl vinyl ether and a few other chiral alkyl vinyl ethers
have been copolymerized with prochiral alkyl vinyl ethers such as
benzyl vinyl ether[27,28], *p*-phenylbenzyl vinyl ether[27], methyl vinyl
ether[27] and *iso*propyl vinyl ether[29,30]. The optical rotation of the
prepared copolymers was equal to or higher than that expected on a
compositional basis and using the additive principle of the optical
rotation. A substantial contribution to the optical rotation was
observed in particular in the copolymers with the mentioned aryl
vinyl ether[27,28] and *iso*propyl vinyl ether[30].

This behaviour resembles in its trend that reported for α-olefin
copolymers[22] and acrylic and methacrylic ester copolymers[31-33],
whose potential in the field of synthetic optically active polymers

has been demonstrated to some extent. In this phenomenon it seems
that the larger the effect the closer are the nature and shape spa-
cings of the structural units derived from the prochiral monomers
as compared to those derived from the optically active ones[29,34,35].
This procedure has been very satisfactorily and conveniently extended to cross copolymerization between a vinyl monomer different from
alkyl vinyl ethers, readily polymerizable by a cationic initiator,
such as polynuclear aromatic and heteroaromatic derivatives[36-38],
for whom in particular there exists a definite chemical composition
at which the induced optical rotation is maximized. Unfortunately
in this last case the stereospecific catalysts based on aluminum
alcoholates and sulphuric acid are not able to give copolymers in
high yields.

In conclusion of this first part we gave a general idea of the
potential of optically active vinyl ethers in helping to understand
the many aspects of the polymerization of alkyl vinyl ethers and to
suggest the most convenient routes leading to structurally and stereochemically definite synthetic macromolecules characterized by the
presence of the ethereal function in the side chains.

In the remaining part we wish to present the most recent data
obtained in our experiments undertaken to ascertain the way in which
the interesting and complex catalytic system introduced by Vandenberg is working.

3. CONSIDERATIONS ON THE MECHANISTIC ASPECTS OF THE MULTICOMPONENT CATALYTIC SYSTEMS USED IN THE ALKYL VINYL ETHER POLYMERIZATION

Analysis of the great deal of information published on the alkyl
vinyl ether polymerization indicates that several polymerization
mechanisms have been proposed with respect to any particular catalytic system, reaction condition and monomer. It is practically impossible to formulate a rational mechanism suitable to cover all
experimental findings.

With this limitation in mind, we considered that, among the
large variety of the catalytic systems used, those based on aluminum
alkyls or alkoxides, in combination with the right proportion of
sulphuric acid give apparently amorphous heterogeneous slurries and
are extremely suitable for providing isotactic poly(alkyl vinyl
ether)s.

In this context it appeared worth undertaking an investigation
to make a soluble mixed sulphate able to reproduce a *"simple model"*
of the microchemical environment present in the heterogeneous systems mentioned above (Fig. 2).

Figure 2. "Model" Catalytic Systems.

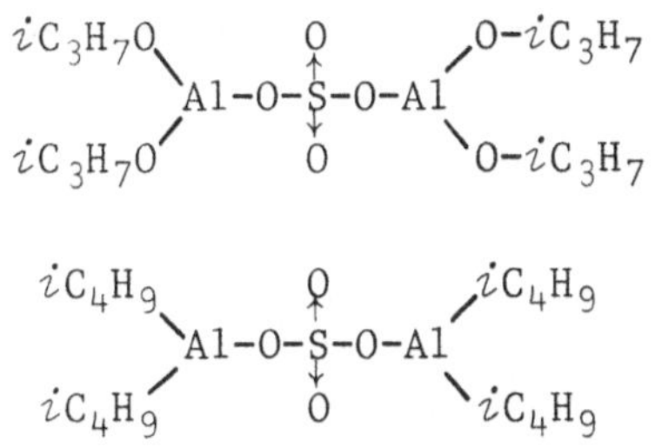

$$iC_3H_7O \quad\quad O \quad\quad O-iC_3H_7$$
$$\diagdown Al-O-\overset{\uparrow}{\underset{\downarrow}{S}}-O-Al \diagup$$
$$iC_3H_7O \diagup \quad\quad O \quad\quad \diagdown O-iC_3H_7$$

$$iC_4H_9 \quad\quad O \quad\quad iC_4H_9$$
$$\diagdown Al-O-\overset{\uparrow}{\underset{\downarrow}{S}}-O-Al \diagup$$
$$iC_4H_9 \diagup \quad\quad O \quad\quad \diagdown iC_4H_9$$

By reaction of concentrated sulphuric acid with a 5/1 molar excess of an aluminum alkyl or alkoxide, an equivalent amount of alcohol or hydrocarbon was developed with the presumed formation of a not well defined alkyl or alkoxy aluminum sulphate. In Schemes 5 and 6 are represented respectively the reactions leading to bis-(diisobutylaluminum)sulphate and bis(diisopropoxyaluminum)sulphate which can be isolated as pure solid products by crystallization from hydrocarbon solvents.

Scheme 5. Preparation of bis(diisobutylaluminum)sulphate.

$$2 \begin{array}{c} iC_4H_9 \\ \diagdown \\ iC_4H_9 \diagup \end{array} Al-Cl + CH_3O-\overset{\overset{O}{\uparrow}}{\underset{\underset{O}{\downarrow}}{S}}-OCH_3 \xrightarrow[40-90°C]{toluene} \begin{array}{c} iC_4H_9 \\ \diagdown \\ iC_4H_9 \diagup \end{array} Al-O-\overset{\overset{O}{\uparrow}}{\underset{\underset{O}{\downarrow}}{S}}-O-Al \begin{array}{c} \diagup iC_4H_9 \\ \\ \diagdown iC_4H_9 \end{array}$$

$$+$$

$$2 \ CH_3Cl$$

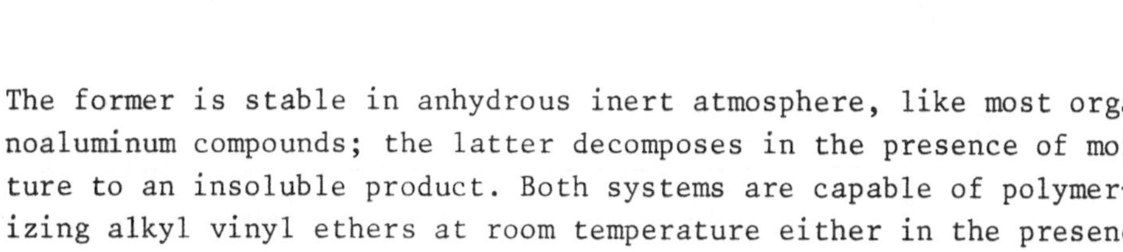

The former is stable in anhydrous inert atmosphere, like most organoaluminum compounds; the latter decomposes in the presence of moisture to an insoluble product. Both systems are capable of polymerizing alkyl vinyl ethers at room temperature either in the presence or absence of an equimolar amount or an excess of "activator" such as aluminum alkyls.

In a proportion of 1-2 mole % with respect to the monomer, the aluminum alcoholate based system has been tested in the polymerization of optically active (-)-menthyl vinyl ether in the presence and in the absence of triisobutylaluminum under homogeneous conditions.

Scheme 6. Preparation of bis(di$i$sopropoxyaluminum)sulphate.

---

1. $Al(O-iC_3H_7)_3 + CH_3\underset{\overset{\|}{O}}{C}-Br \xrightarrow[25°C]{benzene} Al(O-iC_3H_7)_2Br + CH_3\underset{\overset{\|}{O}}{C}-O-iC_3H_7$

2. $2\ Al(O-iC_3H_7)_2Br + (CH_3O)_2SO_2 \xrightarrow[80°C]{benzene} [(iC_3H_7O)_2AlO]_2SO_2 + 2\ CH_3Br$

In Figure 3 is reported a plot of $(\alpha - \alpha_0)/(\alpha_\infty - \alpha_0)$ versus time, where α_0, α_∞ and α are the optical rotations at time 0, ∞ and t respectively.

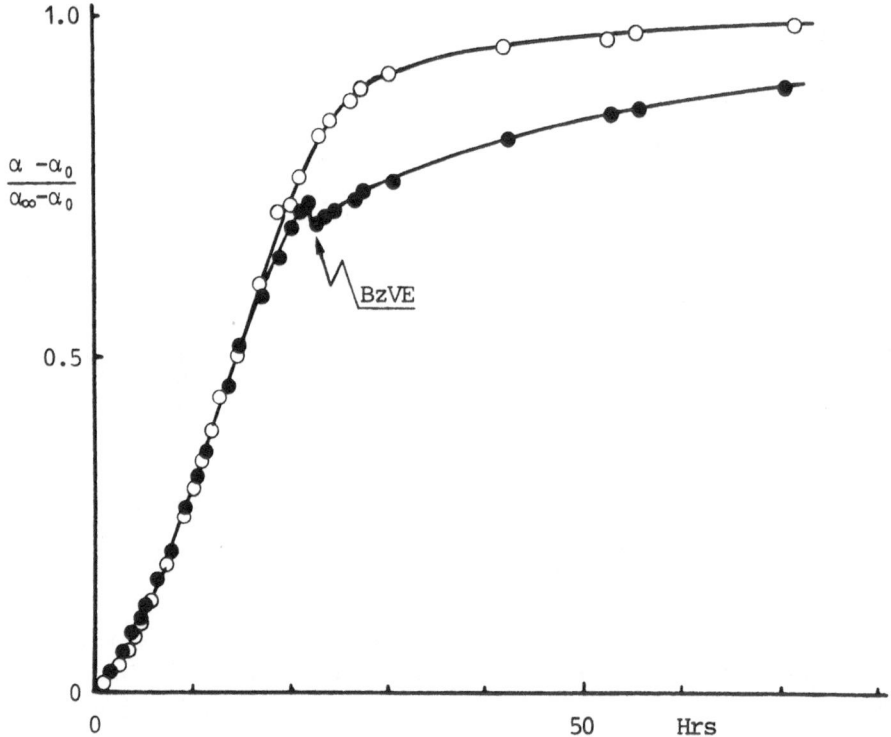

Figure 3. Variation of Optical Rotation During the Polymerization of
(-)-Menthyl Vinyl Ether in the Presence of Bis(di$i$sopropoxy
aluminum)sulphate as Catalyst.
–O– No activator.   –●– Activator $Al(iC_4H_9)_3$
(BzVE = Benzyl vinyl ether)

Thanks to the rather low activity of this system compared to that of conventional ones, working either in homogeneous or heterogeneous conditions, it is possible to make a profile of the kinetic trend of the polymerization by a polarimetric technique which is rather sensitive because of the marked enhancement of the optical rotation in going from the monomer to the polymer (Table 2).

The kinetic curves of Fig. 3 indicate:

1) No substantial rôle seems to be played by the $Al(iC_4H_9)_3$ added as an "activator". The specific activity of the catalyst in the two cases is practically unaffected.

2) A definite induction period is observable in both cases.

3) The addition, at about 70% menthyl vinyl ether conversion, of an equimolar amount of benzyl vinyl ether (BzVE) compared to the initial content of menthyl vinyl ether does not cause the cessation of the polymerization (only a depression due to dilution effect is observable). The added benzyl vinyl ether as shown by the analysis of the polymerization product is incorporated into macromolecules containing menthyl vinyl ether units.

4) The reaction conversion reaches about 95% in more than 50 hr.

5) The reported profile is reproducible within very narrow limits of error.

The mutarotation curves reported in Fig. 4 show a marked effect on the reactivity by structurally different alkyl vinyl ethers. Within the limits of the available data it seems plausible that the structure of the monomer affects the polymerization mechanism. Consequently the order of reactivity appears to hold, for the analysed cases in the reported conditions, as follows:

secbutyl vinyl ether > menthyl vinyl ether >> 2-octyl vinyl ether

In any case it is worth mentioning that 1) menthyl vinyl ether can be polymerized with the soluble system, whereas practically no polymer can be obtained by the corresponding insoluble one. 2) The molar optical rotation of the final products from the soluble system is only somewhat lower than that of the samples obtained in an insoluble slurry. 3) The ^{13}C-NMR spectra of poly(menthyl vinyl ether) obtained by the soluble system indicate 70% isotactic dyads and are rather similar to those of samples obtained at low temperature with a modified Friedel-Crafts catalyst and identical to that obtained with tropylium salts.

Research is in progress to verify if the systems under investigation are classifiable as a "living-type" cationic process[39]. Beyond the possible implications of this interesting point, we have reported here for the first time an example of a kinetic study of a

cationic polymerization reaction using a simple non destructive technique based on a "mutarotation" phenomenon.

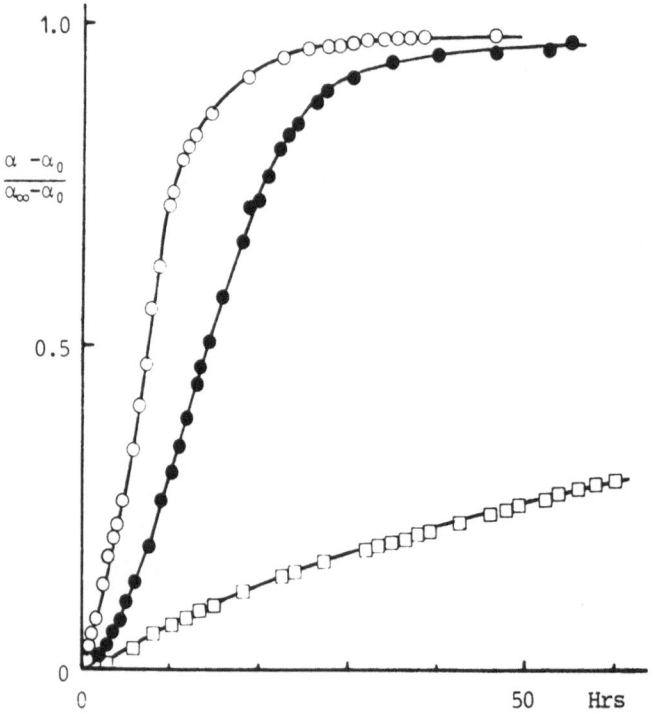

Figure 4. Variation of Optical Rotation During the Polymerization of
Some Optically Active Alkyl Vinyl Ethers.
  –○– (+)(S)-*sec*Butyl vinyl ether
  –●– (-)-Menthyl vinyl ether
  –□– (-)(S)-2-Octyl vinyl ether

In keeping with the results presented above we wish to propose,
as a working hypothesis, a polymerization mechanism (Fig. 5) that,
involves a six-center concerted approach, which should be classified
as a "coordinate cationic" one in agreement with the reaction mechanism proposed several years ago by Vandenberg[40] and Furukawa[41].
    Accordingly, this proposal excludes the intervention of any free
cationic species both as promoter and/or propagating moiety. This is

distinct from the "multicenter" and "coordinate" mechanisms previously proposed (Fig. 6 and 7) since no loose or tight ion pair is involved in the initiation or in the propagation step.

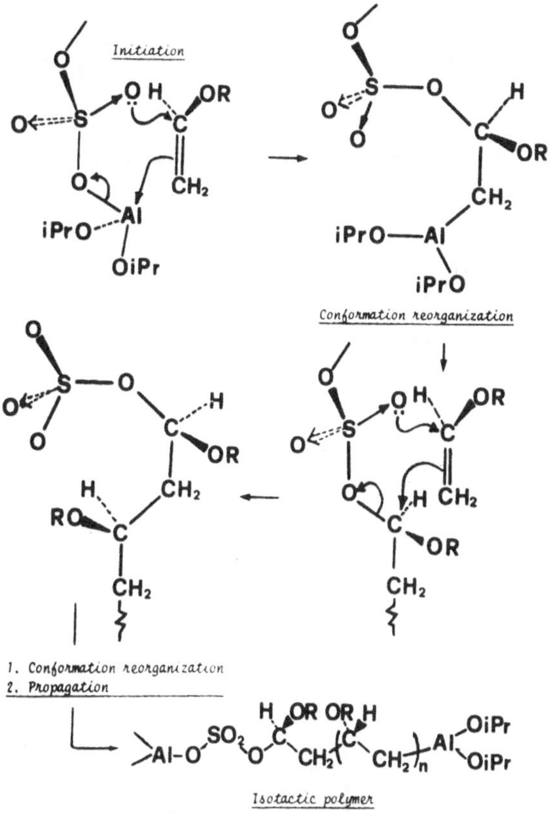

Initiation

Conformation reorganization

1. *Conformation reorganization*
2. *Propagation*

Isotactic polymer

Figure 5. Concerted "Cationic Coordinate" Mechanism.

In the initial step a presentation of the monomer should take place, with the enantioface tending to minimize steric effects, that is the olefinic methine carbon protrudes towards the oxygen atom of the sulphate group which is bound to the central sulphur atom by a covalent dative bond. Correspondingly the olefinic methylene atom is facing the aluminum atom which should in turn become bound to it during the monomer insertion reaction. From one side this gives rise to the opening of a σ Al-O bond and a π C=C bond and from the other side to the formation of an Al-C and a C-O σ bonds with a consequent overall thermodynamic destabilization of the system of about 30 kcal/mole. This energy loss should properly account for the observed induction period in the kinetic curves[*].

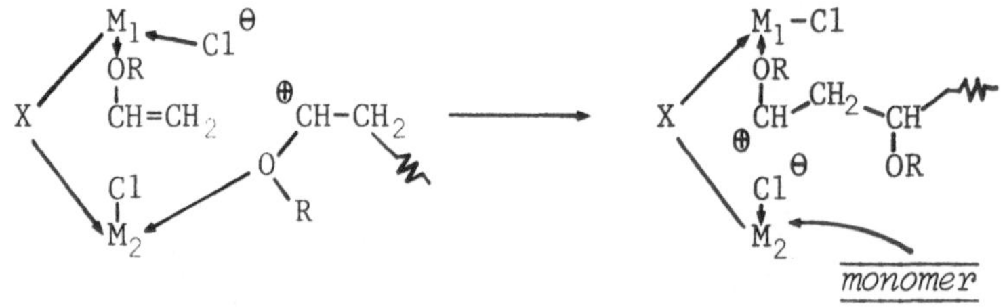

Figure 6. Cationic Coordinate Polymerization Mechanism According to
Vandenberg[40].

After a conformational reorganization, the previous structural
situation, modified only by the presence of a tertiary asymmetric
carbon atom replacing the aluminum position, should lead to the insertion of the second monomer molecule.

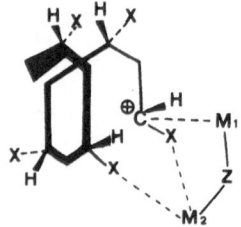

Figure 7. Multicenter Polymerization Mechanism According to Furukawa[41].

This insertion produces the configurational inversion of the first
introduced asymmetric carbon atom. Successive stereochemically controlled propagation steps should lead to an isotactic molecule. Considering however that in the absence of the conformational constraint

---

[*] The autors are grateful to Prof. C.C. Price for drawing this point
to their attention.

of being bound to any solid surface, it is rather hard to imagine a rigid stereochemically controlled reaction pathway under these conditions. Therefore it is conceivable that this system should provide, with respect to the insoluble system, polymer macromolecules characterized by a lower degree of isotacticity.

This scheme should be consistent with the observed induction period that might be attributed to a rate limitation in the first monomer insertion step. The propagation step, consisting of the insertion of an activated double bond onto a hemiacetal is supported by the experimental fact that dialkyl sulphates promptly polymerize alkyl vinyl ethers.

Analogously the very poor stereoelective polymerization observed in the presence of an insoluble catalyst obtained from optically active alcoholates should be entirely attributed to the chiral centers present on the metal components, that are now situated far away from the active sites after the first monomer insertion, thus making any asymmetric influence vanishingly small.

REFERENCES

[1] G. Natta, *J. Polymer Sci.*, *16*:143(1955); G. Natta, P. Pino, F. Danusso, E. Mantica, G. Mazzanti, and G. Moraglio, *J. Am. Chem. Soc.*, *77*: 1708 (1955).

[2] C.E. Schildknecht, A.O. Zoss, and C. McKinley, *Ind. Eng. Chem.*, *39*: 180 (1947).

[3] A.D. Ketley, *"The Stereochemistry of Macromolecules"*, A.D. Ketley Ed., M. Dekker Inc., New York, N.Y., 1967, *vol.* 2, p. 37.

[4] R.F. Heck, (Hercules Incorporated), *U.S. Pat. 3,014,013*, Dec. 1961; *Chem. Abstr.*, *57*: 1080e (1962).

[5] R. Chiang, (Hercules Incorporated) *U.S. Pat. 3,025,281*, March 1962; *Chem. Abstr.*, *57*: 4884h (1962).

[6] D.L. Christman and E.J. Vandenberg, (Hercules Incorporated), *U.S. Pat. 3,025,282*, March 1962; *Chem. Abstr.*, *57*: 2415d (1962).

[7] R.F. Heck, (Hercules Incorporated), *U.S. Pat. 3,098,061*, July 1963; *Chem. Abstr.*, *59*: 10314b (1963).

[8] R.F. Heck and E.J. Vandenberg, (Hercules Incorporated), *U.S. Pat. 3,025,283*, March 1962; *Chem. Abstr.*, *57*: 20415f (1962).

[9] P. Pino, G.P. Lorenzi, and E. Previtera, *Rend. Acc. Naz. Lincei*, *29*: 562 (1960).

[10] P. Pino, G.P. Lorenzi, and E. Chiellini, *Ric. Sci.*, *7*: 193 (1964).

[11] G.P. Lorenzi, E. Benedetti, and E. Chiellini, *Chim. Ind. (Milan)*, *46*: 1474 (1964).

[12] P. Pino, G.P. Lorenzi, and E. Chiellini, *J. Polymer Sci.*, *C*, *16*:

3279 (1968).

[13] P. Pino, *Adv. Polymer Sci.*, *4*: 393 (1965).

[14] P. Pino, P. Salvadori, E. Chiellini, and P.L. Luisi, *Pure Appl. Chem.*, *16*: 469 (1968).

[15] P. Pino, F. Ciardelli, and M. Zandomeneghi, *Ann. Rev. Phys. Chem.*, *21*: 561 (1970).

[16] P. Pino, P. Salvadori, G.P. Lorenzi, E. Chiellini, L. Lardicci, G. Consiglio, O. Bonsignori, and L. Lepri, *Chim. Ind. (Milan)*, *55*: 182 (1973).

[17] F. Ciardelli, E. Chiellini, and C. Carlini, in " *Charged and Reactive Polymers*", E. Selegny Ed., D. Reidel Publisher, Dordrecht Holland, 1979, *vol. 5*, p. 83.

[18] A.M. Liquori and B. Pispisa, *Chim. Ind. (Milan)*, *48*: 1045 (1966).

[19] A. Ledwith, E. Chiellini, and R. Solaro, *Macromolecules*, *12*: 240 (1979).

[20] E. Chiellini, *Macromolecules*, *3*: 527 (1970).

[21] E. Chiellini, G. Montagnoli, and P. Pino, *J. Polymer Sci.*, *B, 7*: 121 (1969).

[22] P. Pino, A. Oschwald, F. Ciardelli, C. Carlini, and E. Chiellini, in *"Coordination Polymerization"*, J.W. Chien Ed., Acad. Press, New YorK, N.Y., 1975, p. 25.

[23] E. Zucchi, *Thesis*, University of Pisa, 1971.

[24] T. Higashimura and Y. Hirokawa, *J. Polymer Sci.*, *Polymer Chem. Ed.*, *15*: 1137 (1977).

[25] F. Ciardelli, E. Benedetti, G. Montagnoli, L. Lucarini, and P. Pino, *Chem. Comm.*, 285 (1965).

[26] C. Carlini, F. Ciardelli, and P. Pino, *Makromol. Chem.*, *119*: 244 (1968).

[27] H. Yuki, K. Ohta, and N. Yajima, *Polymer J.*, *1*: 164 (1970).

[28] R. Solaro and E. Chiellini, *Gazz. Chim. Ital.*, *106*: 1037 (1976).

[29] G. Natta, G. Allegra, I.W. Bassi, C. Carlini, E. Chiellini, and G. Montagnoli, *Macromolecules*, *2*: 311 (1969).

[30] E. Chiellini and R. Solaro, unpublished results.

[31] H. Yuki, K. Ohta, Y.Okamoto, and K. Hatada, *J. Polymer Sci.*, *Polymer Lett. Ed.*, *15*: 589 (1977).

[32] K. Ohta, K. Hatada, Y. Okamoto, and H. Yuki, *J. Polymer Sci.*, *Polymer Lett. Ed.*, *16*: 545 (1978).

[33] Y. Okamoto, K. Suzuki, K. Ohta, K. Hatada, and H. Yuki, *J. Am. Chem. Soc.*, *101*: 4763 (1979).

[34] C. Carlini and E. Chiellini, *Makromol. Chem.*, *176*: 519 (1975).

[35] E. Chiellini and C. Carlini, *Makromol. Chem.*, *178*: 2545 (1977).

[36] E. Chiellini, R. Solaro, M. Palmieri, and A. Ledwith, *Polymer*, *17*: 641 (1976).

[37] E. Chiellini, R. Solaro, O. Colella, and A. Ledwith, *Europ. Polym. J.*, *14*: 489 (1978).

[38] E. Chiellini, R. Solaro, and F. Ciardelli, *Makromol. Chem.*, in press.

[39] T. Higashimura, H. Teranishi, and M. Sawamoto, *Polymer J.*, *12*: 393 (1980).

[40] E.J. Vandenberg, *J. Polymer Sci.*, *C*, *1*: 207 (1963).

[41] J. Furukawa, *Polymer*, *3*: 487 (1962).

TETRANEOPHYLZIRCONIUM AND ITS USE

IN THE POLYMERIZATION OF OLEFINS

R. A. Setterquist, F. N. Tebbe and W. G. Peet*

E. I. du Pont de Nemours & Company
Central Research and Development Department
Experimental Station
Wilmington, Delaware 19898

SYNOPSIS

Olefin polymerization catalysts based on tetraneophylzirconium and its reaction products with submicroscopic fumed alumina are discussed. (Neophyl is the 2-methyl-2-phenylpropyl group). Neophyl-zirconium aluminate or hydridozirconium aluminate supported on alumina are extremely active catalysts for high temperature (300°C) poly-merization of ethylene. The high catalyst efficiency and the innocu-ous nate of the catalyst residues eliminate the need for deactivation or removal of the catalyst from the resultant polymer. The effect of catalyst structure on activity and oplymer properties is discussed.

INTRODUCTION

Since the discovery in the fifties of transition metal halide catalysts for the preparation of high density polyethylene (HDPE), innocuous catalyst active at high temperature has been a goal of research. Residues of certain transition metal halide catalysts (titanium and vanadium chlorides, for example) if not removed from the HDPE or deactivated, may yield HDPE or poor stability, corrosive-ness and unacceptable color. Initial processes included catalyst deactivitation (1), or removal by extraction (1), or filtration (2) and solvent purification (1) that increased the complexity and cost. Recently, there have been reported a number of "Second Generation" processes for HDPE manufacture not requiring catalyst removal.

A catalyst system utilizing tetraneophylzirconium (TNZ) and submicroscopic fumed alumina has been developed which is highly

* Contribution No. 2954

active in ethylene polymerization, is innocuous and need not be removed from resultant polymer. In addition purification of the polymerization medium for recycle and reuse in particle or solution polymerization processes is unnecessary.

RESULTS AND DISCUSSION

<u>Tetraneophylzirconium</u>

Tetraneophylzirconium, otherwise identified as tetrakis (2-
methyl-2-phenylpropyl) zirconium or TNZ, is prepared in 97% yield by
reaction of a Grignard reagent, neophyl magnesium chloride, in ether/
hexane solvent with zirconium tetrachloride at $33°C^{4,5}$.

The ^{1}H NMR spectrum of TNZ and by-products found in the reaction
mixture from a TNZ synthesis is shown in figure 1. By-products include dineophyl, tert-butylbenzene, and di- and tri-neophyl zirconium
chloride. Values for the ^{1}H NMR chemical shifts of protons in TNZ
and in dineophyl zirconium dichloride isolated from a TNZ preparation
are shown in Table 1.

Tetraneophylzirconium, a white crystalline solid melting at
69°C, is stable for days at 25°C and years at -35°C in the absence
of air, moisture, or light. Thermal decomposition of TNZ at temperatures above 100°C yields primarily t-butyl-benzene and a dark
residue readily oxidized to a white zirconium oxide by air. Crystalline TNZ reacts with air but is not highly pyrophoric. Dry oxygen
reacts to yield a neophyl zirconate which on hydrolysis produces
neophyl alcohol. With hydrogen at 80°C, hydrocarbon solutions of
TNZ produce a hydrocarbon insoluble, reduced zirconium hydride.

The thermal stability of Group IV transition metal neophyl
compounds was found to decrease in the order Hf>Zr>Ti. Comparison
of ^{1}H NMR spectra taken during aging of TNZ and tetraneophylhafnium
(TNHf) in C_6D_6 shows TNNf to be the most stable, although both decom-
pose to tertbutylbenzene and broad spectrum impurities at a rate of
ca. 10% per week at 25°C. At the end of two weeks, the TNZ sample
was characterized by E.S.R., and a small amount of paramagnetic
material was observed.

The ^{1}H NMR spectra of TNZ and other alkyl Group IV transition
metal compounds are shown in Table 1. In accord with the decreased
electronegativity in the series Ti, Zr, Hf, a decrease in chemical
shift is noted for methylene portons of the respective neophyl
compounds. Substitution of chloride on the metal atom likewise
reduces electron density of the methylene carbon resulting in a
greater chemical shift of the methylene protons.

Since the NMR spectrum of TNZ shows no broadening of the peak
due to phenyl protons at δ7.01 ppm, it appears no interaction between

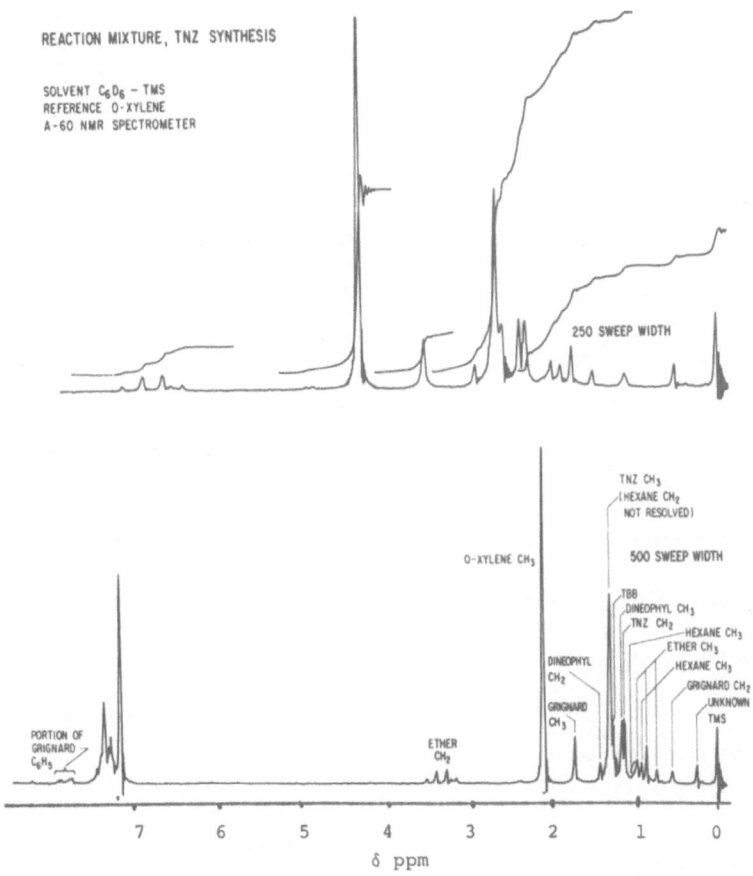

Figure 1. H' NMR spectrum of a TNZ reaction mixture taken
at the end of the reaction showing by-products.

TABLE I

^{1}H NMR Spectra of Alkyl Group IV B Compounds

Compound	$-CH_2-$[a]	$-CH_3$	Aromatic H[b]
$Zr[CH_2(CH_3)_2C_6H_5]_4$[c]	1.13	1.27	7.10
$Zr[CH_2C_6H_5]_4$[d]	1.5		6.4 (ortho) 7.1 (m,p)
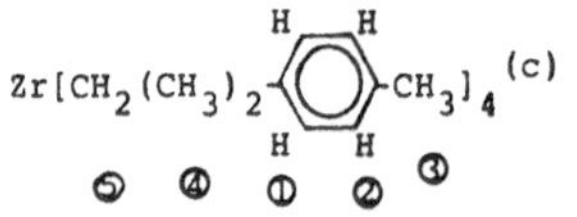 $Zr[CH_2(CH_3)_2$ ⬡ $-CH_3]_4$[c] ⑤ ④ ① ② ③	1.12 ⑤	1.30 ④ 2.20 ③	7.05 ② 7.20 ①
$Zr[CH_2(CH_3)_2C_6H_5]_2Cl_2$[c]	2.19	1.34	7.1
$Ti[CH_2(CH_3)_2C_6H_5]_4$[e]	1.8	1.2	7.1
$Hf[CH_2(CH_3)_2C_6H_5]_4$[e]	0.5	1.2	7.1

(a) Sharp Singlets

(b) Broad Multiplets

(c) 220 MH$_z$

(d) 100 MH$_z$ (Ref. 4)

(e) 60 MH$_z$

the phenyl ring and zirconium occurs in this compound, distinguishing
it from tetrakis (π-ally) zirconium and tetrabenzylzirconium[6,7]. The
absence of such interaction in TNZ provides more carbanion character
to the methylene carbon (less chemical shift), less cationic character
to the zirconium, and consequently, lower activity with nucleophiles
than tetrabenzylzirconium. Also, the stability of TNZ is thought to
be associated with the bulkiness of the neophyl group and the absence
of a hydrogen atom on the carbon atom beta to the metal; this reduces
the tendency for decomposition via β-hydrogen transfer and olefin
elimination[8]. The presence of bulky neophyl groups about zirconium
also may inhibit β-elimination of other alkyl groups (e.g., an
attached polyethylene polymer chain).[9]

Zirconium oxide, like other zirconium compounds, is generally
recognized to possess a low order of toxicity by most routes.[10]
Results of feeding a 33% by weight solution of TNZ in corn oil to
rats indicates an acute lethal dose of over 15,000 mg/kg. Accordingly
the residues from deactivation of TNZ with water and air (zirconium
oxide, tert-butylbenzene, and neophyl alcohol) would not be considered very toxic on a short-term basis. As an added benefit the TNZ
catalyst system should provide a process of reduced ecological
problems compared to many catalysts which must be deactivated and
disposed of with great caution to prevent environmental problems.
The added cost of disposal of many "spent" catalysts exceeds the
original cost of the catalyst.

TNZ alone is a poor catalyst for ethylene polymer polymerization.
Based on thermolysis studies of organometallics reported in the literature[11] and observations of the dark color developing with TNZ
catalyst in the polymerization vessel at temperatures above 150°C,
the conversion of the zirconium alkyl to Zr-H and reduced species
may be a key part of the polymerization mechanism. Ethylene appears
to enhance the decomposition of TNZ probably by insertion between Zr
and stabilizing neophyl groups, providing facile conversion to a
reduced metal hydride catalyst.

TNZ reacts with water or hydroxyls present on alumina with the
partial or total elimination of the neophyl groups as t-butylbenzene.
If the elimination of neophyl groups is not complete, the products
of the reactions are susceptible to reduction and are active catalysts
for olefin polymerization. Ballard reported that compounds formed
by the reaction of aklyl or benzyl zirconium compounds with alumina
or silica were active catalysts for polymerization of 1-olefins.[12]

The hydrides of zirconium (ZrH_x) formed by heating the metal
with hydrogen have been used with a metal halide to catalyze ethylene
polymerization, but the rate of polymerization was low.[13] We found
that a zirconium hydride species prepared by pyrolysis of TNZ in
hydrogen was a catalyst for the polymerization of ethylene at 150°C.
However, when the zirconium hydride was fixed on a high surface area

metal oxide by decomposition, olefin elimination or hydrogenation of zirconium alkyls fixed on the oxide surface, the activity in ethylene polymerization increased over 100 fold. Similarly, Zakharov found that heat treatment or hydrogenation of $Zr(C_3H_5)_4$ deposited on silica gel was attended by the formation of Zr hydrides which reacted with ethylene during polymerization forming propagation sites of the type $(SiO)_2Zr(H)(CH_2CH_3)$[14,15].

Typically the ZrH_x/Al_2O_3 catalyst was prepared in our work by hydrogenation of the reaction product of alumina and a Group IV transition metal organometallic such as tetraneophylzirconium, tetra-benzylzirconium, tetraneophyltitanium, tetraneophylhafnium, or zirconium borohydride for sufficient time to strip the alkyl groups from the transition metal.

Description of Hydridozirconium Aluminate Catalyst

An especially active catalyst for 1-olefin polymerization consists of the hydrogenated reaction product of tetraneophylzirconium (TNZ) and fumed submicroscopic alumina such as pyrogenic Alumina C obtained from the Degussa Corporation. The alumina is activated by heating at $700°$ to $1100°C$, rehydrated, and dried at $400°C$ in nitrogen until the hydroxyl content on the alumina is ca. 0.5 - 1.0 mM OH/g-Al_2O_3. The catalyst is prepared by contacting a suspension of the hydroxylated alumina in a high viscosity inert hydrocarbon medium with an inert hydrocarbon solution of TNZ. In general the proportion of TNZ employed is 0.05 - 0.35 millimoles per gram of alumina. Upon mixing the alumina and TNZ a reaction occurs involving the HO-groups on the alumina surface whereby Zr-O-Al bonds are formed with the elimination of some neophyl groups as described in Equation (1).

$$(1) \quad n\text{-}(C_{10}H_{13})_4Zr + Al_2O_3(OH)_a \longrightarrow$$
$$(HO)_{(a-ny)} Al_2O_3 (O)_{ny} Zr_n (C_{10}H_{13}) (4-y)$$

NZA

$$(2) \quad NZA \xrightarrow{H_2} (HO)_{a-ny} Al_2O_3 \{O\}_{ny} Zr_n H_x$$

HZA

Kinetics for the reaction of TNZ with alumina of ca. 1mM OH/g at $25°C$ and $50°C$ in hexane were obtained by isolating the catalyst at different times, washing, and analyzing the ashed product for deposited zirconium using x-ray fluorescence. The specific rate constant decreased as hydroxyls became less available during reaction possibly caused by slower diffusion of TNZ into hydroxyls blocked by deposited TNZ. Slow migration of both OH and TNZ as the reaction proceeds does occur since neophylzirconium aluminate (NZA) after washing with hexane to remove unreacted TNZ slowly decreases in

neophyl content with time, while aging of NZA in the presence TNZ
slowly increases the neophyl content of the solids.

The reactions involved in the formation of HZA catalyst by
treatment of NZA with hydrogen are shown in Equation 2.   In addition
to substitution of the neophyl groups by hydrido groups and the
formation of tertbutylbenzene, reductive elimination to yield reduced zirconium species and hydrogenation of aromatic compounds (such
as tertbutylbenzene) in the system also occurs. The ease of elimination of neophyl groups from neophyl Group IV metal aluminates was
determined in hydrocarbons to decrease in the order Ti>Zr>Hf where
Ti and Zr are hydrogenated at about the same rate while Hf takes at
least three times as long under the same conditions to react to the
same extent.

The calculated time for 1/2 the Zr-neophyl groups to be removed
was 4.5 min at $100^\circ C$ and 0.2 min at $200^\circ C$ under these conditions.
The activation energy was 11 Kcal.   The rate of reduction of the
transition metal could be qualitatively estimated by the rate of
darkening of the neophyl compound as colored, reduced metal species
formed.   The Ti and Zr compounds darkened almost immediately at $100^\circ C$
in H_2, whereas the Hf compound showed little color change.   However,
in N_2 the reduction was much slower even at the same extent of
neophyl elimination.

In view of the many partially complete reactions occurring
during the formation of HZA, it is probable that a complex assortment
of active species exist in the catalyst. A depiction of the catalyst
in Figure 2 shows several types of active species resulting from
different ligands and oxidation states of zirconium, as well as
unreacted hydroxyl groups, adsorbed TNZ, and aluminum alkyl bonds
formed by transalkylation.

Olefin Polymerizations in the Presence of HZA

Although NZA is a catalyst for 1-olefin polymerizations,[16]
the actual catalyst may be the hydrido derivative (HZA), formed in
situ. Formation of a metal hydride by B-hydrogen transfer from a
growing polymer chain Equation (3) or hydrogenolysis of the Żr-C
bond Equation (4) may be involved as considered by Ballard[15] or
Zakharov[17].

(3) $(Al_2O_3)Zr(CH_2CH_2)_nR \longrightarrow (Al_2O_3)ZrH +$
$\qquad\qquad\qquad\qquad\qquad\qquad\qquad CH_2=CH(CH_2CH_2)_{n-1}R$

(4) $(Al_2O_3)Zr(CH_2CH_2)_nR \xrightarrow[H_2]{} (Al_2O_3)ZrH +$
$\qquad\qquad\qquad\qquad\qquad\qquad\qquad H(CH_2CH_2)_nR$

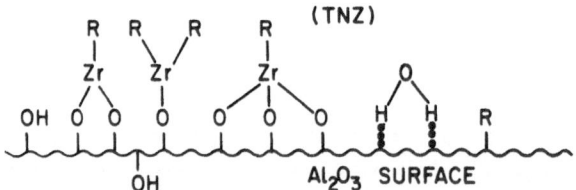

R = POLYMER CHAIN, ALKYL GROUP, HYDROGEN

Figure 2.   Depiction of Species on HZA catalyst.

The former method yields vinyl-ended polymer, and the latter reaction yields methyl-ended polymer.

We theorized that a more active catalyst could be obtained by formation of the hydrido metal catalyst prior to introduction into the polymerizer, especially in polymerizations utilizing short hold-up time in the polymerization vessel. This would provide the hydride initially in high concentration eliminating the induction period associated with incidental reactions otherwise required for conversion of NZA to the active hydride species. The observed high activity of the HZA catalyst is in agreement with our theory[18]. In batch ethylene polymerizations, as described in the Experimental Section, the hydrogenated catalyst was generally observed to be more active than unhydrogenated catalyst. For example, as shown in Table II, the activity of partially hydrogentated NZA, NTiA, and NHfA catalysts increased with removal of neophyl groups.

In continuous polymerization of ethylene at 1.5×10^7 Pa and short hold-up time (1.5 min), the HZA catalyst was observed to be 20% more active at $250°C$ and 55% more active at $175°C$ than NZA catalyst. In continuous polymerization of propylene at $85°C$, an HZA efficiency of 167,000 g polymer/mole Zr vs. 20,000 g polymer/mole Zr for NZA catalyst was obtained using a 4.8 min hold-up time.

In addition to ethylene and propylene, HZA catalyst provided high molecular weight polymers from other α-olefins such as butene-1 and octene-1. The relative comonomer activity by butene-1 in ethylene copolymerizations was approximately 1.7 times greater than for a Ti/V/Al alkyl catalyst at the same conditions. Thus, lower comonomer concentrations is required, and the penalty in catalyst consumption because of comonomer will be lower for 1-olefin/ethylene processes catalyzed by HZA.

A drawing of the continuous high density polyethylene (HDPE) process utilizing HZA is shown in Figure 3 in which a solution of the tetrahydrocarbyl transition metal compound from feed tank (6) and of the slurry of alumina in inert hydrocarbon from feed tank (7) are passed through mixing valve (15) to premix vessel (1) agitated by mixer (12). After a preselected hold-up time in the premix vessel, the slurry is passed to hydrogenation vessel (2) through mixing valve (18) where a solution of H_2 in the inert liquid hydrocarbon from reservoir (8) is added to the stream. The slurry is hydrogenated in vessel (2) which is stirred by mixer (13). After a preselected hold-up time in the hydrogenation vessel, the slurry of transition metal hydride aluminate supported on alumina is passed out through control valve (16) to a polymerization vessel (3) which is thoroughly mixed by mixer (14) to provide a uniform, constant environment. Ethylene, or other olefin feed, from reservoir (9) is passed through mixing valve (17) where it is admixed with recycling polymerization medium from polymer separator (5) and then fed

TABLE II

Effect of Hydrogenolysis of Neophyl Group IV Metal
Aluminates on Activity in Ethylene Polymerization

Catalyst Type	mM Metal gAl$_2$O$_3$	% Hydrido Ligand	Transition Metal mM	Polymer Recovery g/min
NZA	0.2	0	0.01	0.68
HZA	0.2	100	0.01	1.02
NHfA	0.3	0	0.03	0.7
HHfA	0.3	100	0.03	1.0
NTiA	0.3	0	0.03	.2
HTiA	0.3	60	0.03	.3

Run Conditions: 40 PSI C$_2$H$_4$
1 min HUT at 150°C
Decalin® Solvent

Hydrido catalysts made by hydrogenolysis at 100°C for 20 min
under 20 PSI H$_2$.

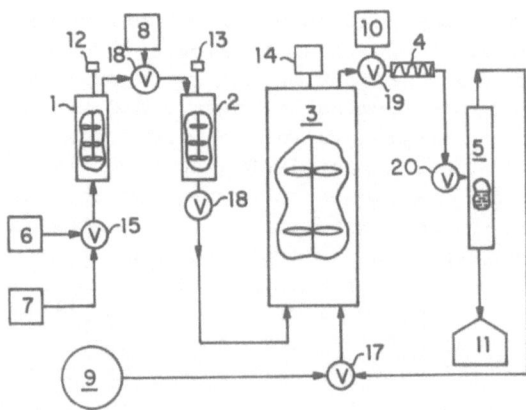

Figure 3.   Process for the preparation of HDPE
using HZA catalyst.

into polymerizer (3). After preselected hold-up time in the polymerization vessel, the reaction mixture is passed through mixing
valve (19) where catalyst deactivator from reservior 10 is added,
and the mixture passed through a tubular, turbulent mixer (4) and
through pressure let-down valve (20) to the polymer separator (5).
From the separator the polymer stream is passed to a polymer recovery
system (11) while the liquid polymerization medium is passed back
as recycle to the polymerizer (3).

The catalyst deactivator from reservoir (10) can be steam or
other polar compounds reactive with the transition metal catalyst.
The traces of deactivated catalyst are removed from the polymer
separator incorporated with the polymer from which they need not be
removed because of the low level and inert nature of this innocuous
catalyst residue.

Alumina Support

The supporting structure for the zirconium hydride catalyst may
be any of a number of pure or natural high surface area metal oxides.
Of the many supporting materials scouted, the most active catalysts
were derived from so-called fumed aluminas prepared by the Degussa
and Cabot Corporations by flame hydrolysis of $AlCl_3$.

Fumed alumina is submicroscopic in particle size (100-300 Å
platelets) and has a moisture content of 2-6% as received. Drying
at elevated temperatures in nitrogen removes bound water with little
effect on porosity or surface area. Not all aluminas lose water in
this fashion in dried nitrogen. "Hydral" 710, alumina trihydrate
($Al_2O_3 \cdot 3H_2O$) slowly loses water below $150°C$ with formation of many
micro-pores, and a structure displaying a high surface area (300
m^2/g). As with fumed alumina, water is liberated continuously above
$250°C$; however, the sizes of the micro-pores increase, and surface
area decreases in the porous alumina.

The type of catalyst formed by the reaction of TNZ with fumed
alumina differs from that formed with porous alumina. Fumed alumina
is composed of 100-300 Å nonporous platelets connected in loose
macro-porous agglomerates (1000Å) which have an average pore diameter in the range of 200Å as measured by B.E.T. methods. These
particles are unlike the less active alumina from alpha-alumina trihydrate which is composed of cubic and hexogonal platelets, 3000-
10,000Å across, filled with cylindrical pores (ca. 50Å). Reaction
of the organometallic compound with hydroxyl groups within the pores
of dried alumina trihydrate and removal of polymer formed at catalytic sites within the pores is not as favorable as comparable reactions on the surface of nonporous alumina plates. The greater activity in ethylene and propylene polymerizations of zirconium catalysts
made from fumed alumina compared to porous alumina is shown in

Table III. Fumed alumina dried in nitrogen-containing hydrogen
provided the most active catalysts.

The beneficial effect of a more open, porous catalyst support
was confirmed in continuous ethylene polymerizations in which an
increase in pore diameter of a beta-alumina from 49Å to 70Å doubled
the activity of HZA catalyst.

During the reaction with alumina, TNZ because of its bulky
groups may separate widely on the alumina surface. Separation sites
should improve stability and activity since destruction of active
sites by reaction between neighboring zirconium species is less
probable.

A TNZ/Al_2O_3 catalyst was coated lightly with polethylene at
25°C in a slurry polymerization and examined by electron microscopy
at 350,000-500,000 magnification. The formation during polymerization of a regular network of humps or fibrils at a distance of 20Å
apart on the surface of the catalyst was observed. Tentatively,
the humps are thought to be polymer formed at transition metal
centers. The volume of polymer should reflect the activity of that
site. The humps were largest at the edges of the catalyst particle
from which it might be concluded that the exposed sites are most
active.

Finally, the open structure and small particle size of fumed
alumina are desirable support features which provide ease in monitoring the catalyst and rapid dispersion and distribution of the small
quantities of catalyst through the viscous medium in the polymerizer.
The loose agglomerates of particles are easily disintegrated to a
size which is optically and physically innocuous in the final polymer.

Polymerization of Ethylene with HZA

Ethylene was polymerized using HZA catalyst at 150-300°C in a
process similar to that shown in Figure 3 in a continuously well-
stirred reactor. The kinetic equations used to compare catalyst
activity assumed a constant environment and second order kinetics
which could be translated to yield an overall polymerization rate
constant expressed as:

$$K = \frac{Q}{1-Q} \times \frac{1}{\tau} \times \frac{1}{C_f - C_p}$$

Where K = Overall polymerization rate constant (min^{-1} concentra-
         $tion^{-1}$)
      Q = Fractional ethylene conversion
      τ = Hold-up time in polymerizer (min^{-1})

TABLE III

Comparative Activity of Catalysts on Porous and Nonporous Al$_2$O$_3$

BATCH RUNS

Catalyst (1)	Alumina (2) Support	Catalyst-Millimoles Zirconium	Olefin (3)	Pressure Atm.	Polymerization Temperature °C	Time Min.	Polymer Yield Grams	Catalyst Activity g/mole/hr/atm
NZA	Nonporous	0.27	P	7.0	65	60	42.7	22.6
NZA	Porous	0.54	P	8.2	50	120	3.3	0.4
NZA	Nonporous	0.015	E	2.4	150	1	1.04	1615.
NZA	Porous	0.015	E	2.4	150	1	0.13	218.
NZA	Nonporous	0.0112	E	2.4	150	1	0.8	1780.
NZA	Nonporous	0.0075	E	2.4	150	1	0.7	2330.

CONTINUOUS RUNS

Catalyst	Alumina Support	Zirconium Concen.-Moles/l	Olefin (4)	Polymerization Temperature °C	Holdup Time Min.	Polymer Production g/hr	Polymerization Specific Rate Constant(5) Min.$^{-1}$ mole^{-1}
NZA	Nonporous	2 x 10^{-5}	E	250	2.56	177	1.48 x 10^5
BZA	Porous	18.6 x 10^{-5}	E	250	2.69	154	0.067 x 10^5

(1) NZA = neophylzirconium aluminate, BZA: benzylzirconium aluminate

(2) Nonporous = Degussa fumed alumina, dried at 400°C to 0.6% H$_2$O; surface area = 100m^2/g. Porous = Hydral-710 dried at 600°C to 0.5% H$_2$O: surface area = 187m^2/g.

(3) E = ethylene, P = propylene.

(4) 7% E in hexane at 2250 psi.

(5) $a = \frac{1}{\tau} \frac{Q}{1-Q} \frac{1}{C}$

where a = specific rate constant
τ = holdup time in polymerization
Q = ethylene conversion
C = catalyst concentration

C_f = Feed catalyst concentration
C_p = Concentration of poisons

As shown in Figure 4 the activity of HZA catalyst decreases with increased hold-up time or temperature. Presumably this decrease in activity is due to decay of the active catalyst species.

Activity was affected by the ratio of Zr to alumina. The optimum ratio depended upon impurities in the system and the hydroxyl content of the alumina. Alumina having a hydroxyl content of 0.6-1.0 mMOH/g alumina provided optimum activity at 0.20-0.27 mMZr/g alumina. In the range of 0.20-0.27mMZr/g alumina the activity was little affected by ratio. However, at 0.35 mm/g the catalyst required for 90% ethylene conversion might be 2-3 times that needed at lower ratios. Destruction of the catalyst by mutual reaction of sites in close proximity on the support surface may occur at high Zr/Al_2O_3 ratios during hydrogenolysis or polymerization.

Molecular Properties of HZA Polymer

The molecular weight of the polyethylene product was affected by parameters well-known in polymerization processes. Molecular weight of the product decreased with increasing temperature, increased ethylene conversion (reduced ethylene concentration), and increased concentration of transfer agent (H_2). These general affects are illustrated in Figure 5 which shows the affect of temperature and ratio of hydrogen to ethylene on polymer melt index (M.I.) which is related to polymer weight average molecular weight (M_w) as follows:

$$M.I. = \frac{\beta}{M_w}$$

By controlling the hydrogen/ethylene ratio and the temperature of the polymerization, polymer having any desired melt index within the range of useful commercial HDPE could be obtained. Useful high molecular weight polymer could be made at temperatures as high as $300°C$.

The molecular weight distribution (MWD) of HDPE is practically assessed by Gel Permeation Chromatography, (G.P.C.) and Size Exclusion Chromatography, (S.E.C.). The melt rheology of a polymer is assessed by measuring the flow of polymer melt in a Melt Indexer using two points of a shear/stress relationship. A power law equation having the form $\overset{\circ}{\gamma} = K\tau^n$ (where $\overset{\circ}{\gamma}$ = shear rate, τ = shear stress, K = constant and n = Stress Exponent(SE)) is used to determine how non-Newtonian a melt is. A value of 1 for SE refers to a Newtonian fluid, while values greater than 1 refer to increasing non-Newtonian behavior. The SE value is related directly to the molecular weight distribution and the amount of long chain branching in the polymer.

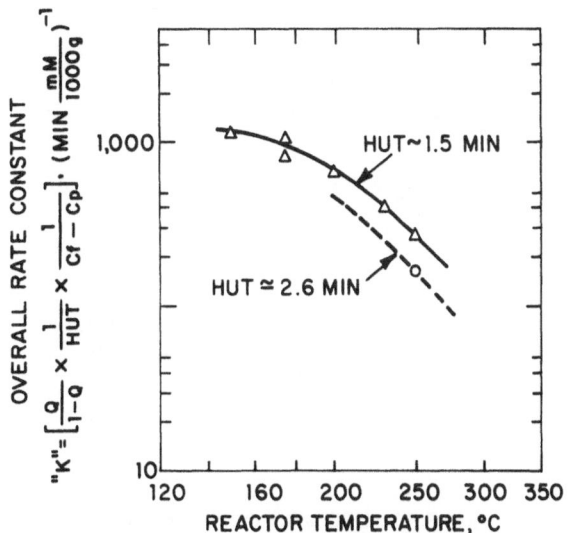

Figure 4. Effect of polymerization temperature and
hold-up time on the activity of HZA catalyst
in ethylene polymerization.

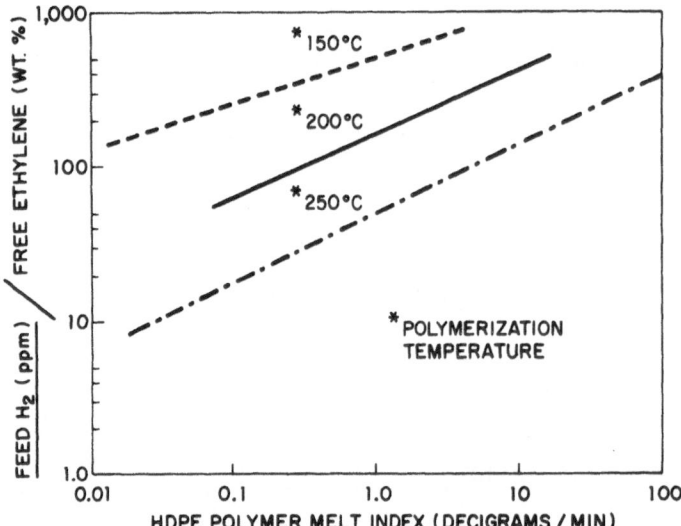

Figure 5.   Effect of polymerization temperature and Hydrogen (ppm)
to final ethylene concentration (wt%) ratio on the melt index
(ASTM-1238-GST Condition E) of HDPE made with HZA catalyst.

Three observations were made relating the rheological properties and molecular characteristics of HZA polymer.

o   Considerable skewing of the molecular weight distribution towards a higher molecular weight fraction occurs with HZA resins. (Figure 6).

o   For the same weight average molecular weight (as determined by Melt Index-M_w Correlations) the HZA resins have a higher Z average molecular weight (a higher molecular weight moment) than many commercial resins.

o   Many HDPE polymers show a good correlation between SE and M_w/M_n. The HZA resins show a better correlation between SE and M_z/M_w than M_w/M_n (correlation coefficients of 0.82 vs. 0.49, respectively).

These results of melt elasticity measurements and correlations between light-scattering molecular weight and molecular weight by S.E.C. indicated the amount of long chain branching of HZA resins is not different from most commercial HDPE.

Therefore, differences in the behavior of HZA resins can be explained in part by a disproportionate amount of high molecular weight fraction present.

Several process parameters which were found to affect the stress exponent of the HZA polymer included:
o   Temperature: direct effect.
o   Hydrogen concentration in the polymerizer: direct effect.
o   Hold-up time in the polymerizer: direct effect.
o   Conversion of ethylene: direct effect.
o   Reaction uniformity: as expected temperature gradients and concentration gradients within the reactor increased the S.E. of HZA polymer.
o   Deactivation of the catalyst: different catalyst deactivation agents (n-propanol, water) had similar effects.
o   Catalyst modifications: minimizing the multiplicity of different catalyst species arising during catalyst synthesis and in polymerization is a fundamental item of concern for heterogeneous catalysts in general. Agents which modify the catalyst, such as impurities in the system, unreacted alumina hydroxyls, and unreacted TNZ must be minimized. Effective disintegration and dispersion of the support during cata- lyst preparation and use are essential.

Other Physical and Chemical Properties of HZA Polymers

HDPE polymers and 1-olefin copolymers of invariant white color

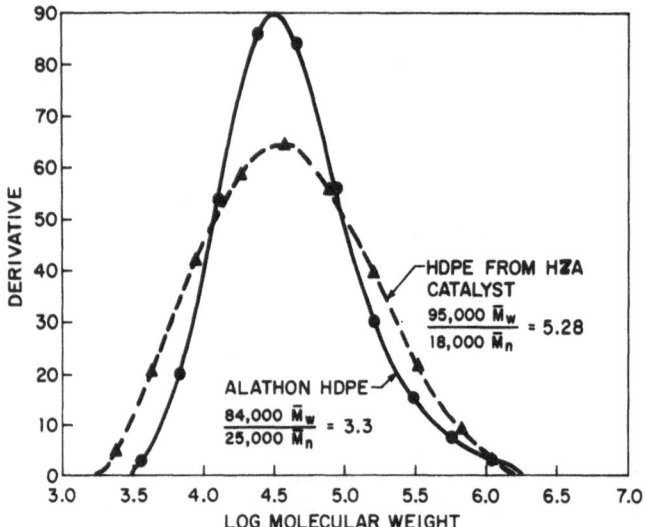

Figure 6.   A gel permeation chromatography normalized frequency
distribution plot of HDPE made using HZA catalyst
compared with a commercial Alathon® HDPE.

and excellent optical properties were obtained with HZA catalyst.
Physical properties of the resins were as expected for polymer of the
same molecular weight dispersity.

HZA polymers contain a low ratio of vinyl to methyl end groups
as shown in Table IV. Accordingly, most termination of the growing
polymer chain occurs by hydrogenolysis at temperatures in the range
of $230°C$. However, in the case of an octene/ethylene copolymer made
at $290°C$, the predominant mode of termination was by olefin elimination.

CONCLUSIONS

A highly active, innocuous catalyst based on the hydrogenated
reaction product of tetraneophylzirconium and activated fumed alumina
was developed for the polymerization of ethylene. The catalyst is
typically represented by an overall stoichiometry of

$$(HO)_{0.05}(Al_2O_3)_{9.8}(O_2ZrH)_{0.2}$$

and may be depicted as the complex alumina surface structure shown
in Figure 1.

The catalyst is useful for preparation of high molecular weight,
high density polyethylene and copolymers in a process not requiring
catalyst removal or extensive recycle solvent purification. Safety
in handling and storage and ease in transfer and monitoring the
catalyst under commercial conditions were recognized.

EXPERIMENTAL

Since the organometallic compounds and catalysts used in this
study are subject to deactivation by O_2, H_2O, CO_2, and other reactive
substances, precautions were taken to maintain equipment free of
atmospheric contact. Solvents, hydrogen, and monomers were freed of
moisture or oxygen by use of conventional desiccating agents and
sodium.

The weight-average molecular weights of the polyolefin products
were determined from an established correlation between polymer melt
flow (ASTM 1238-GST condition E) and weight-average molecular weight
as determined by light scattering. Number-average and weight-average
molecular weight also were determined by gel permeation chromatography using established standard analysis in 1,2,4-trichlorobenzene
solvent at $135°C$ on Stryragel columns.

The ^1H NMR characterization was carried out according to general
procedure using 0.06 g of organometallic in 0.2 cc of 99.8% deutero-
benzene in a 5mm glass NMR tube at $42°C$.

TABLE IV

Properties of Polyolefins Made With HZA and HTiA Catalysts

Polym; Conditions	HZA HDPE[c]	HZA HDPE	HZA HDPE	HZA Ethylene/C$_8$ Copoly	HTiA HDPE[d]
Reactor H.U.T. (min)	4	1.5	1.5	1.5	1.5
Ethylene Conv. (%)	86	55	76	58	89
Reactor Temp. (°C)	230	230	160	290	230
Polymer Properties					
Melt Index (decigrams/min)[a]	1.1	1.34	0.31	0.83	0.04
S.E. [b]	1.42	1.35	1.31	1.45	3
$\bar{M}_n$ from G.P.C.	20,000	27,000	15,000		32,000
Vinylidene/poly. molecule	-	0.05	0.01	-	-
CH$_3$-/poly. molecule	1.79	1.65	1.66	1.2	1.93
Vinyl/poly. molecule	0.2	0.33	0.2	0.7	.07
Vinyl/CH$_3$	0.11	0.2	0.12	0.6	.04

(a) Melt Index (M.I.); ASTM Method 1238-GST, Condition E.

(b) Stress Exponent (SE) = 2.095 log ($\dfrac{\text{MI at 6840 g load}}{\text{MI at 2160 g load}}$)

(c) HZA = Hydrido Zirconium Aluminate Catalyst

(d) HTiA = Hydrido Titanium Aluminate Catalyst

Tetraneophylzirconium Synthesis

An essential factor in development of a viable alkyl zirconium catalyst was facile synthesis of the alkyl zirconium compound in attractive yields under commercially attainable conditions. The general laboratory procedure for tetraneophylzirconium (TNZ) proved to be fully adequate for scale-up to large scale (200 lb) manufacture. The general procedure is quite adaptable, and with minor modifications has been used to synthesize tetraneophyltitanium, tetraneophylhafnium, tetrapara-methyl-neophylzirconium, and tetrapara-tert-butyl-neophylzirconium.

Neophylmagnesium Chloride Etherate

A conventional Grignard reaction between magnesium and neophyl chloride in ca. 1/1 ether 'hexane at 40-45°C was used to prepare 2M neophylmagnesium chloride for use in the TNZ preparation. The yield of the Grignard is in the range of 95%. The major by-products of the Grignard reaction are bineophyl (5 mole %) and phenylisobutene (0.9%). Neophyl chloride of high purity obtained from Chemical Samples Co. was sparged with nitrogen to remove air and water prior to use.

The Grignard solution was assayed by titration with 1-butanol using 1,10-phenanthroline as an indicator, by titration with HCl and by GC anaylsis of it alcoholysis products. The ^{1}H NMR spectrum at 220 MHZ of neophylmagnesium chloride etherate in C_6D_6 shows aryl at $\delta7.63$ and $\delta7.19$, neophylmethyl at $\delta1.58$, and neophylmethylene at $\delta0.49$ relative to tetramethylsilane.

The Grignard is very stable. After months at 25°C in an inert atmosphere no evidence of decomposition was observed. Accordingly, large quantities of the Grignard could be synthesized and stored for TNZ manufacture in campaign operation.

Tetraneophylzirconium

A dry nitrogen-filled flask equipped with a mechanical stirrer and with a dry-ice condenser with nitrogen inlet was charged with neophylmagnesium chloride (149.35g, 0.774 mol), ether (115.13g), and hexane (189.14g). The slightly cloudy reaction mixture was cooled to 15.6°C, and zirconium tetrachloride (49.15g, 0.2109 mol. 9% excess) was added in portions over 5 min. to the briskly stirred reaction mixture. The mixture was cooled to room temperature, and the $MgCl_2$ was separated from the tetraneophylzirconium solution by filtration. The solids were washed with an additional 500cc of n-hexane. The combined filtrates were evaporated to dryness at 30° in vacuo to yield 117.9g ().189 mol) of tetraneophylzirconium (yield 97%).

Tetraneophylzirconium prepared by the described procedure was
analyzed, and typically found to contain 0.4-1.2% chloride, 2-4%
trineophylzirconium neophylate, and 2-4% dineophyl. The ^{1}H spectrum
of tetraneophylzirconium at 220 MHZ shows aryl at δ7.1, methyl at
δ1.27, and methylene at δ1.13.

Tetra p-methylneophylzirconium Synthesis

The p-methylneophyl chloride was synthesized by the reaction
of toluene and methallyl chloride in the presence of H_2SO_4 at 25°C.
The p-methylneophyl chloride (b.p. 82 C at 266 Pa) contained 92%
para and 8% ortho isomers.

The p-methylneophyl magnesium chloride was prepared by addition
of 1/1 p-methylneophyl chloride in hexane to magnesium shavings in
ether at 30°C. After the addition was complete,the temperature of
the reaction mixture was raised to 45°C for 1 hour. The yield of
Grignard reagent was 86%. The Grignard solution was separated from
unreacted magnesium, and zirconium tetrachloride (0.07 molar excess)
was added to the solution at -10°C. The reaction mixture was warmed
to 45°C and stirred for 2 hours. The resulting brown mixture was
cooled to 25°C, and the $MgCl_2$ separated and washed with hexane. The
combined filtrates were evaporated at 25°C under vacuum to yield a
brown oil. The oil was diluted in hexane and cooled to -35°C.
Crystals of crude tetrapara-methylneophylzirconium (70% yield) were
separated by filtration. After recrystallization from hexane, pure
tetrapara-methylneophylzirconium, melting point 58.5°C, was obtained.
An ^{1}H NMR spectrum at 22 MHz in C_6D_6 shows aryl at δ7.05 and δ7.2,
methyl at δ1.3 and δ2.2, and methylene at δ1.12. Elemental analysis
of the product gave 75.72% C, 8.61% H, 12.98% Zr. (Theory: 77.75%
C, 8.83% H, 13.42% Zr.)

Tetraneophyltitanium Synthesis

A dry nitogen-filled flask was charged with neophylmagnesium
chloride (0.4 mole) in ether (100g) and toluene (100g). The mixture
was cooled to -60°C, and titanium tetrachloride (19g, 0.1 mole) in
toluene (50g) added. The slurry was stirred for 1 hour at -60°C
and allowed to warm to 25°C. The slurry was filtered to remove
$MgCl_2$, and the filtrate evaporated to a brown oil. The oil was
diluted with 100g of hexane and cooled to -25°C. Yellow-green
crystals of tetraneophyltitanium separated. After recrystallization
from hexane, solid tetraneophyltitanium (36g, 0.06 mole, 60% yield)
was attained. Tetraneophyltitanium was found to melt at 82.5° by
D.S.C. melting point determination in N_2. Elemental analysis of
the product gave: 81.03% C, 9.0% H_2. (Theory for tetraneophyltitanium 82.7% C, 9.0% H). ^{1}H NMR spectrum at 60 MHz of tetraneophyltitanium in C_6D_6 at 42°C shows aryl at δ7.1, methyl at δ1.2, and
methylene at δ1.8.

Tetraneophylhafnium Synthesis

A dry nitrogen-filled flask was charged with neophylmagnesium chloride (0.3 moles) in ether (100g) and toluene (100g). The mixture was cooled to $-10°C$, and hafnium tetrachloride (25.6g, 0.08 moles) added. The slurry was filtered to remove $MgCl_2$, and the filtrate evaporated at $25°C$ to a brown oil. The oil was diluted with 50g of hexane and cooled to $-35°C$. The cream-colored crystals were separated, recrystallized from hexane, and dried to yield 36g (0.05 moles) of tetraneophylhafnium (63% yield). Tetraneophylhafnium was found to melt at $69°C$ by D.S.C. melting point determined in N_2. Elemental analysis of the product gave 63.3% C, 7.1% H. (Theory: 67.56% C, 7.32% H.) ^1H NMR spectrum of tetraneophylhafnium in C_6D_6 at $42°C$ shows aryl at $\delta7.1$, methyl at $\delta1.2$, and methylene at $\delta0.5$.

Procedure for Alumina Drying

Submicroscopic alumina having a surface area of 112 m^2/g (by BET analysis) was obtained from the Degussa Company, activated by heating in a sweep of nitrogen at $850°C$, followed by rehydration to ca. 2% water, and redrying to 0.48% water by weight by continuously conveying the alumina with a rotating screw through a stainless steel cylindrical dryer heated to $600°C$ under a counter-current dry nitrogen sweep. The moisture content of the alumina was found by determining the amount of methane evolved in reaction of the surface hydroxyl groups on the alumina with methyl magnesium bromide.

General Procedure for Preparation of Alkylzirconium Aluminate Catalyst

A toluene solution of tetraneophylzirconium (or other group IV metal catalyst component) was added to a 3.5% by-weight suspension of dried alumina in mineral oil containing 4.5% petroleum jelly as a thickening agent and mixed for 3 hours to allow complete reaction between the hydroxyl groups on the alumina and the tetraneophylzirconium. The suspension was analyzed by spearating the solids and the neophyl content obtained by determining the amount of tertbutylbenzene evolved upon alcoholysis using gas chromatography techniques.

Preparation of Hydridozirconium Aluminate Catalyst

A 0.000625 M neophylzirconium aluminate slurry in n-hexane (containing 0.3 millimole zirconium/g alumina) was fed at a rate of 480 cc/min into a 300cc stirred autoclave where it was hydrogenated by mixing with 54 millimoles H_2/hr. The hydrogenation autoclave was maintained at $125°C$, and the suspension of catalyst on alumina remained in the hydrogenation autoclave for an average hold-up time

of 5.3 minutes. Analysis of the hydride catalyst showed 95% of the
neophyl groups originally present were displaced by this treatment.

Procedure for Polymerization of Ethylene

     Batch Synthesis. To a vigrously stirred 0.4% solution of eth-
ylene in 360cc of decahydronaphthalene in 400cc pressure bottle under
autogenous pressure at 150°C was added a slurry of supported transition metal catalyst. After 1 minute the reaction was quenched with
n-butanol to destroy the activity of the catalyst. The reaction
mixture was cooled, and the solid polyethylene removed by filtration,
washed with cyclohexane, and dried in a vacuum oven at 80°C for 16
hours.

     Continuous Synthesis. The mineral oil suspension of the sup-
ported transition metal catalyst, fed at a rate of 0.8g catalyst/hr
of cyclohexane and continuously fed to a 253cc stainless steel agitated polymerization vessel maintained 2250 psi and an appropriate
reaction temperature (200-300°C) where it was contacted with an 8.3%
(by weight) ethylene solution in cyclohexane at a rate of 350g eth-
ylene/hour. The concentration of transition metal in the reactor
was in the range of 10^{-6} to 10^{-5} molar. In order to control the
molecular weight of the polyethylene, hydrogen was fed to the polymerization vessel at a rate of 30-1000 millimoles/hr as a 0.004
molar solution in cyclohexane. The hold-up time in the polymerization vessel was maintained at approximately 1.5 minutes by continuously withdrawing the polymerization mixture to a deactivation
chamber where the catalyst was deactivated to terminate the polymerization by reaction with water added at a rate of 2M/hr. The solution
of polymer was discharged through an automatic, controlled pressure
reducing valve into a product receiver at 50°C where the solid polyethylene was separated from the polymerization medium. After washing
with cyclohexane containing distearylthiodipropionate antioxidant
(1% on a polymer basis), the polymer was vacuum-dried at 80°C for
16 hours.

     Kinetic anaylsis of the reactions were based on the change in
the rate of polymerization (a) as shown by:

$$a = \frac{1}{(C)} \quad \frac{Q}{1-Q} \quad \frac{1}{\tau}$$

where (C) = catalyst concentration
        τ = hold-up time in the polymerization
        Q = ethylene conversion

Procedure for Polypropylene Synthesis

A 1 liter autoclave containing dry nitrogen was charged with 168g of propylene and warmed to 25°C. While stirring the catalyst slurry was introduced into the polymerization vessel, and the temperature of the reaction mixture controlled at 50-56°C for 1 hour at autogenous pressure (275-300 psi), after which the autoclave was vented and cooled. The product was stirred with cyclohexane to convert it to a gel, diluted with acetone, and the solid separated by filtration after addition of an antioxidant and dried.

ACKNOWLEDGMENTS

Acknowledgment is made of the contiributions of our colleagues R. C. Czerner, E. T. Pieski, D. Vassallo, P. Donohue, L. Firment, W. Huang, J. A. Crowther, and C. S. Naples who assisted in this work.

ABBREVIATIONS USED IN TEXT

C_f	Feed catalyst concentration
C_p	Concentration of poisons
G.C.	Gas Chromatography
HDPE	High density polyethylene
HHfA	Hydrido Hafnium Aluminate catalyst
HTiA	Hydrido Titanium Aluminate catalyst
HZA	Hydrido Zirconium Aluminate catalyst
K	Polymerization Rate constant
M.I.	Melt index of polymers at 190°C ($\frac{decigram}{min}$ ASTM method 1238-GST Cond. E)
M_n	Number average molecular weight
M_w	Weight average molecular weight
M_z	Z average molecular weight
M.W.D.	Molecular weight distribution
N.M.R.	Nuclear Magnetic resonance
NZA	Neophylzirconium Aluminate
Q	Ethylene Conversion
SE	Stress Exponent (measure of HDPE melt flow behavior)
SEC	Size exclusion chromatography
TNZ	Tetra (neophyl) zirconium
HUT	Hold-up time during polymerization

REFERENCES

1. D. B. Ludlum; To Du Pont U.S. Patent 3,308,112; (1967).
2. H. H. Hagemeyer, D. C. Hull and S. J. Park; To Eastman Kodak U.S. Patent 3,600,463 (1971).

3. Marshall Settig, "Polyolefin Production Processes; Latest
    Developments", 1976, Noyes Data Corporation, Park Ridge, New
    Jersey, 07656, Page 222.
4. R. Setterquist, To Du Pont U.S. Patent 4,017,525 (1977); U.S.
    Patent 3,932,307 (1976).
5. W. G. Peet, F. N. Tebbe and G. W. Parshall; Res. Discl., #16815,
    April 1978.
6. J. B. Becconsall, B. E. Job and S. O'Brien, J. Chem. Soc. A423
    (1967).
7. U. Zucchini, U. Giannini, E. Albizzoti and R. D'Angelo, J. Chem.
    Soc., Chem. Commun., 1174 (1969).
8. R. J. Davidson, M. F. Lappert, R. Pearce; Chem. Rev., $\underline{76}$ 219
    (1976). G. Wilkinson, Science, $\underline{185}$, 109 (July 12, 1974). R. R.
    Schrock, G. W. Parshall, Chem, Rev. $\underline{76}$, 243 (1976).
9. J. K. Kocki, Organometallic Mechanism and Catalysis; Academic
    Press; New York, N.Y.; 1978, Page 257.
10. F. A. Patty, Industrial Hygiene and Toxicology, Vol.II, 2nd Ed.
    (1963), Page 1191-4.
11. Reference 9, Page 237.
12. D.G. H. Ballard, XXIII I.U.P.A.C. Meeting, Boston, U.S.A. July,
    1971.
13. Gilbert Bo and Philippee Perras, U.S. Patent 2,985,639 (1961).
14. V. A. Zakharov, V. D. Dudchenko, A. M. Kolchin, Yu. I. Yermokov;
    Kinet. Katal. 16 (3) 808 (1975).
15. D. G. H. Ballard, J. Polym, Sci., Polym. Chem. Ed. $\underline{13}$ 2191
    (1975).
16. R. Setterquist, To Du Pont U.S. Patent 3,932,307, Jan. 1976;
    U.S. Patent 4,011,383 March, 1977.
17. V. A. Zakharov, Yu. I. Yermakov; Catal. Rev. Sci. Eng. 19 (1)
    67 - (1979)
18. R. Setterquist, To Du Pont U.S. Patent 3,950,269 (1976); U.S.
    Patent 3,971,767 (1976).

POLYMERIZATION OF ETHYLENE

BY USE OF AN ORGANOSCANDIUM COMPOUND[12]

Glenn A. Moser

Hercules Incorporated
Research Center
Wilmington, Delaware   19899

SYNOPSIS

Bis(cyclopentadienyl)scandium π-allyl (I) catalyzes the polymerization of ethylene in hexane solution in the dark at room temperature and ordinary pressure with no other components necessary. The polymer prepared from (I) and ethylene is essentially linear high density polyethylene. During the

$$\text{ScCl}_3 + \text{CpTl} \longrightarrow (\text{Cp}_2\text{ScCl})_2 \xrightarrow{\text{C}_3\text{H}_5\text{MgX}} \underset{(\text{I})}{\text{Cp}_2\text{Sc}\pi\text{C}_3\text{H}_5} \xrightarrow[\text{hexane}]{\text{C}_2\text{H}_4} \text{polyethylene}$$

ethylene polymerization, there is a fairly rapid catalyst-destroying termination reaction which limits mileage to about 1g mmole of Scandium. Compound (I) does not catalyze propylene polymerization unless subjected to ultraviolet radiation. Even under these conditions, propylene yields only fluid, colorless oligomers and (I) is completely converted to unidentified products. The results constitute a long-sought example that olefin catalysis can occur by repeated additions of olefin to a simple, monomeric transition metal alkyl and provide a starting point for the design of practical catalysts.

INTRODUCTION

The commercial importance and the academic interest in the chemistry of Ziegler-Natta polymerization of α-olefins has been well established (1). Commercial processes use titanium or other transition metal compounds supported on silica, alumina, or other inorganic materials. Even the heterogeneous polymerization using titanium trichloride can be viewed as a supported catalyst, since

193

Table 1. Supported Catalysts – Polymerization of Ethylene

Company	Catalyst			Polymerization Conditions			$\dfrac{g\ Polymer}{g\ TM}$
	Metal Alkyl	Metal (TM)	Transition Support	h	°C	atm	
Union Carbide Br 1,253,872 11/17/71	AlEt$_3$ + AlEt$_2$OEt	CrO$_2$Cl$_2$	SiO$_2$	1.5	92	20	32,000
Solvay & Cie Fr 1,448,320	Al–i–Bu$_3$	TiCl$_4$	Mg(OH)Cl	2	90	10	31,000
Mitsui Ind. U.S. 3,642,746 2/15/72	Al–i–Bu$_3$	TiCl$_4$	MgCl$_2$ + MeOH	2	90	3.5	35,000
Phillips U.S. 2,963,470 12/6/60	None	CrO$_3$	SiO$_2$·Al$_2$O$_3$	1	104	30	36,000
Conventional Heterogeneous	AlEt$_3$	TiCl$_3$	None	5	80	10	1000–3000

only a small number of the titanium atoms are active. Table 1
summarizes several of the important systems used for the
polymerization of ethylene.

    All of the supported catalysts require activation prior to
use in order to achieve optimum polymerization capability. Most,
although not all, of the catalysts are activated by group I, II or
III metal alkyls; exceptions are $CrO_3/Al_2O_3/SiO_2$ and
Ni/Carbon. The most frequent examples of compounds used for
activators are the alkyl aluminum compounds.

    Unfortunately, in addition to alkylating the transition
metal center, the group I, II or III alkyls enter into other
reactions which greatly complicate the overall chemistry of the
system. For many reasons, including the complexities introduced
by the activator and the experimental difficulties encountered
when dealing with the chemistry of surfaces, there are no supported catalytic centers whose active site and chemistry are
well defined. Therefore, our interest was focused on designing
potential catalysts whose solid state and solution structure
could be determined by spectroscopic techniques.

    In designing a model system, several variables were
considered. These variables included the choice of the transition metal,its oxidation state, the ligand environment around
the chosen transition metal, and the resulting physical properties of the catalyst.The transition metals that were considered
in the planning stages for this work are shown in Table 2. They
include most of the early transition metals and some lanthanides.
Table 2 shows a compilation of the highest oxidation state and
the corresponding ionic radii for these metals. It was assumed
that a good combination of physical properties for the transition
metal was a large atomic radius and an oxidation state which left
the transition metal without d-electrons. The absence of
non-bonding d-electrons was felt to be desirable based on
theoretical studies by Hoffmann (2).

Table 2. Candidate Transition Elements Considered

Atomic Radius, Å
(Oxidation State)

Sc	Ti	V	Cr	Mn
0.81 (+3)	0.68 (+4)	0.59 (+5)	0.52 (+6)	0.46 (+7)
Y	Zr	Nb	Mo	Tc
0.93 (+3)	0.80 (+4)	0.70 (+5)	0.62 (+6)	- -
La	Hf	Ta	W	Re
1.15 (+3)	0.81 (+4)	0.73 (+5)	0.68 (+6)	- -

Yb	Lu
0.94 (+3)	0.93 (+3)

It was known that good catalytic activity occurs with
$(\pi-Cp)_2M(IV)$, $(M=Ti,Zr)$, under conditions such that the
tetravalent transition metal is likely to be alkylated and
associated with an unknown fourth ligand or counter ion (3).
The titanium systems suffer complications due to their tendency
to undergo reductive decomposition; the zirconium systems, on
the other hand, are more difficult to reduce. Because Sc(III)
is similar in size to Zr(IV), would require no fourth ligand or
counter ion for electric neutrality, and would be stable against
reduction, it was the metal chosen.

Bis(cyclopentadienyl)scandium-alkyl is electrically neutral
and was expected to be soluble in inert, (i.e. non-donor),
organic solvents. However, the scandium is coordinately
unsaturated, hence likely to form dimers through alkyl bridging.
To avoid this complication, the alkyl was replaced with the
π-allyl ligand in the design. The pi to sigma allyl conversion is facile in many transition metal allyls.   Reaction of
the π-allyl with monomer would prevent reversion to π-allyl and
generate the postulated catalyst in situ.

The organometallic chemistry of scandium has been considered
since 1954 when bis(cyclopentadienyl)scandium was reported (4),
but there was little additional published attention given to
scandium until Hart et al. (5) reported on the triphenyl and
triphenylethynyl compounds in 1970. Both compounds were
extremely pyrophoric and satisfactory carbon and hydrogen
analyses were not obtained in that study. An early report dealing with the preparation of $ScEt_3$ (Et_2O) (6) could not be
reproduced by later workers (5). It was expected that alkyl and
even aryl derivatives of scandium would not be particularly
stable; however, if known stabilizing ligands such as
cyclopentadienyl groups could be incorporated, new organoscandium derivatives seemed possible.

Two similar methods for the preparation of
bis(cyclopentadienyl)scandium chlorides have been reported and
are shown below.

$$2Cp_2Mg + 2ScCl_3 \xrightarrow[50°]{THF} [Cp_2ScCl]_2 + 2MgCl_2 \qquad \text{(Equation 1)}$$

$$2CpTl + ScCl_3 \xrightarrow[\Delta]{THF} Cp_2ScCl \cdot THF + 2TlCl \qquad \text{(Equation 2)}$$

Wailes and coworkers (7) (Equation 1) reported Cp_2ScCl in
dimeric form to be a yellow-green moisture sensitive solid.
The main drawback of this route is the difficulty in preparing
Cp_2Mg. More recently, workers at Du Pont (8) have used
cyclopentadienyl thallium (Equation 2) to prepare organoscandium
compounds. They found from ^{1}H NMR data that the coordinated
THF molecule was retained on sublimation. A high resolution mass
spectrum, however, did show a parent ion for $(Cp_2ScCl)_2$. This
does provide a convenient route to the Cp_2ScCl moiety for
subsequent reaction.

Wailes (7) reported that allylmagnesium chloride reacts with
bis(cyclopentadienyl)scandium chloride to give an orange,
monomeric π-allyl derivative (Equation 3).

$$Cp_2ScCl + C_3H_5MgCl \longrightarrow Cp_2Sc\pi-C_3H_5 + MgCl_2$$

(Equation 3)

Because, bis(cyclopentadienyl)scandium π-allyl has the
components thought necessary for Ziegler type polymerization of
α-olefins, this work was undertaken to explore its utility for
this application.

RESULTS AND DISCUSSION

Both $Cp_2ScCl \cdot THF$ and $(Cp_2ScCl)_2$ have been prepared
(Equations 4,5) from cyclopentadienyl thallium and scandium
trichloride. Clearly, two products

$$2\ CpTl\ +\ ScCl_3 \xrightarrow{\text{THF}} Cp_2ScCl \cdot THF$$

(Equation 4)

$$2\ CpTl\ +\ 2ScCl_3 \xrightarrow{C_6H_6} [Cp_2ScCl]_2$$

(Equation 5)

were formed and their composition was a function of the solvent.
Both products were identified by NMR spectroscopy and the data are
shown in Table 3.

The NMR spectrum of $Cp_2ScCl \cdot THF$ was in agreement with that
reported by Manzer (8). When the reactions were carried out in
THF, there was no IR or ^{1}H NMR indication that bridging chloride
dimer was formed. Formation of dimer was reported by Coutts and
Wailes (7) when the preparation was performed in THF and when
Cp_2Mg was the source of the cyclopentadienyl ligand. Since
their ^{1}H NMR data were reported in THF, it is likely that they
were, in fact, measuring the THF adduct formed by breaking the
bridging chloride structure in situ to form $Cp_2ScCl \cdot THF$.

Table 3. Proton NMR Data of Organoscandium Compounds(a)

Compound	Solvent	Cp	H_α $-OCH_2-$	H_β $-O-C-CH_2-$	H_1	H_2	H_3
$Cp_2ScCl \cdot THF$(b)	C_6D_6	6.25	3.52	1.31	–	–	–
$Cp_2ScCl \cdot THF$	C_6D_6	6.20(s)	3.55(m)	1.32(m)	–	–	–
Cp_2ScCl(c)	THF	6.16	–	–	–	–	–
$[Cp_2ScCl]_2$	Toluene-d_8	6.12	–	–	–	–	–
$[Cp_2ScCl]_2$	THF	6.18	–	–	–	–	–
$Cp_2Sc^\pi C_3H_5$ (c)	C_6D_6	5.90	–	–	7.29(q)	–	3.05(d)
$Cp_2Sc^\pi C_3H_5$	C_6D_6	6.20(s)	–	–	7.45(q)	–	2.20(d)
THF	$CDCl_3$	–	3.5-4.0	1.6-2.1	–	–	–

(a) Reported in δ relative to TMS.   s = singlet, d = doublet, q = quartet, m = multiplet.
H_1 represents the hydrogen atom on the center carbon of the π-allyl ligand.
(b) Ref. 8.
(c) Ref. 7.

Carbon-13 NMR Data of $Cp_2 Sc^\pi C_3H_5$

Compound	Solvent	π-C_5H_5	π-C_3H_5
$Cp_2Sc^\pi C_3H_5$	Toluene-d_8	112.9 ppm	67.3 ppm

The solvent peaks of the THF (solvent) would easily obscure the presence of THF (coordinated) and normal solvent effects can easily account for the difference in the observed resonance signal of the cyclopentadienyl groups. Since a single crystal x-ray structure determination has been reported (9); clearly, in the solid state, the bridging chloride dimer was prepared by Wailes.

When known $(Cp_2ScCl)_2$, prepared in benzene (Equation 5), was dissolved in THF, reisolated, and the compound's ^{1}H NMR spectrum recorded in C_6D_6, the spectrum was identical to that recorded for $Cp_2ScCl \cdot THF$ in C_6D_6 prepared via Equation 4. In fact, when $(Cp_2ScCl)_2$ prepared in benzene, was dissolved in THF, the cyclopentadienyl resonance was observed at 6.18 δ. The strong solvent peak of THF precluded observing coordinated THF. One would have to believe that THF was a sufficient nucleophile to break the bridging chloride dimer and that Wailes, et al., were able to remove solvated THF in their purification procedure, whereas this work and Manzer's indicate that even under sublimation conditions, the THF was retained and remains coordinated to the scandium in $Cp_2ScCl \cdot THF$.

Bis(cyclopentadienyl)scandium chloride dimer reacted with allyl Grignard (Equation 3) to form bis(cyclopentadienyl)scandium π-allyl as an orange crystalline solid. It was soluble in THF, hexane, and benzene and was extremely air sensitive. When exposed to air, the compound turned white. It was not observed to be pyrophoric. The compound appeared to be identical to that described by Wailes (7).

Further characterization was achieved by ^{1}H and ^{13}C NMR. The upfield doublet and downfield multiplet are well known for the room temperature spectra of π-allyl compounds (10). The ^{1}H NMR spectrum of the compound would be expected to be temperature dependent and to be resolved into two upfield doublets and a downfield quintet as the temperature was lowered. Although the multiplet signal (7.45 δ) for H_1 (the hydrogen atom on the center carbon of the π-allyl ligand) was weak in the ^{1}H spectrum, assignments were unequivocally made using spin decoupling techniques. Upon irradiation of the sample at 7.45 δ, the upfield doublet centered at 3.20 δ collapsed into a singlet; likewise, irradiation at 3.20 δ caused the coalescence of the multiplet centered at 7.45 δ. No other signals were affected by these decoupling experiments.

The polymerization activity of $(Cp_2ScCl)_2$ and $Cp_2Sc\pi C_3H_5$ toward ethylene and propylene was investigated. Polymerizations were run for 3h at 25° in hexane under 4.22 atm. of olefin with the exclusion of light. The dimer, $(Cp_2ScCl)_2$, by itself was not active toward propylene or ethylene polymerization. When activated with either

Table 4. Ethylene Polymerization using Bis(Cyclopentadienyl)scandium π-Allyl(a)

$(\pi\text{-Cp})_2\text{Sc}-\pi\text{-C}_3\text{H}_5$ Amount (mmoles)	Concentration (mM)	Yield(b) Insoluble (g)	Mileage $(g \cdot mm^{-1})$	Activity(b) Insoluble $(g \cdot mm^{-1} \cdot atm^{-1} \cdot h^{-1})$
10	167	11.551	1.16	0.091
5	84	6.432	1.29	0.102
3	50	2.875	0.96	0.076
1	16.7	1.170	1.17	0.093
0.1	1.67	0.131	1.31	0.104
0.01	0.17	0.098	9.8	0.774
0.1	1.67	0.131	1.31	0.104
0.1(c)	1.67	0.142	1.42	0.129
3.0	50	2.875	0.96	0.076
3.0	7.5(d)	3.068	1.02	0.081

(a) All polymerizations run for 3 h in the dark at 25° and 4.22 atm of
     monomer in 60 ml hexane (unless noted).
(b) Insoluble is relative to hexane. Units are grams of
     polymer/(mmoles metal)(atmosphere)(hour).
(c) Polymerization run at 60° and 3.67 atm monomer.
(d) Polymerization run in 400 ml hexane.

triethylaluminum (TEAL) or diethylaluminum chloride (DEAC),
there was some activity. Propylene gave only small amounts
of mostly atactic polypropylene and ethylene yielded small
amounts of polyethylene.

Bis(cyclopentadienyl)scandium π-allyl was dissolved in
hexane and subjected to ethylene. The yellow-orange solution
became cloudy within one minute; polyethylene was formed.
As the reaction continued, the solution lost its original color.
On one occasion, after one hour, the stirring was stopped and the
supernatant hexane solution (nearly colorless) was transferred via
a stainless steel cannula to a second vessel. Only a very small
amount of additional polyethylene formed (0.01 g). The activity
of $Cp_2Sc\pi C_3H_5$ based upon isolated polyethylene appears to
remain constant through a wide range of concentrations
(167 mM to 1.67 mM, Table 4). The reason for the higher activity
$(0.774 \text{ g} \cdot \text{mm}^{-1} \cdot \text{atm}^{-1} \cdot \text{h}^{-1})$ for 0.17 mM
$Cp_2Sc\pi C_3H_5$ is not understood, however, the result was
reproducible.

The polyethylene prepared from bis(cyclopentadienyl)-
scandium-π-allyl was high density (0.973-0.981 g/cm) with
number average molecular weight ranging on different preparations from 160,000 to 190,000 and the weight average molecular
weight from 740,000 to 2,430,000. The intrinsic viscosity of the
polymer in decalin ranged from 9.0 to 12.5 dl/g. By DSC, the
second melt showed the polymer to be 54.1% crystalline with 132°C
melting point. This was a decrease from 74.8% crystallinity
observed during the first heating determination of the melting
point. The ^{13}C NMR showed all the polyethylene prepared to be
highly linear.

The allyl scandium system did not effect the formation of high
molecular weight polypropylene. When the propylene polymerization
mixture was irradiated with a sun lamp, some oligomers were formed
and identified by GC-mass spectroscopy (Table 5).

Table 5 - GC-Mass Spectroscopy Analysis
of Oligomers from $Cp_2Sc\pi C_3H_5$ and Propylene

Molecular Weight	Number of Propylene Units	Relative Intensity
126	3	1
168	4	1/2
210	5	1/4
252	6	1/3
294	7	1/6
336	8	small
378	9	very small

The need to irradiate the π-allyl scandium system to show
activity toward propylene may reflect either steric influence
of the propylene methyl group or its inductive effect, or both.

It was not clear why propylene could not be polymerized to
high molecular weight polymer with the scandium compound;
however, propylene does not destroy the potential activity of
bis(cyclopentadienyl)scandium π-allyl for ethylene
polymerization since propylene treated scandium π-allyl
compound, when subjected to ethylene, was active for the
preparation of polyethylene. There was no indication of any
co-polymerization with residual propylene.

The mechanism of the ethylene polymerization was not studied.
The simplest hypothesis to explain what was observed is that
initiation involves relatively slow insertion of monomer into the
Sc-C bond of the σ-allyl, and that propagation is repeated rapid
additions of the monomer to the resulting Scandium alkyl. It was
not determined whether or not all the scandium was active nor why
the polymerization stopped. The presence of impurities during
polymerization was a concern; however, good reproducibility was
obtained. Because yield of polyethylene per amount of scandium
did not vary with scandium concentration, it is probable that
impurities were not causing or affecting the activity.

SUMMARY

Bis(cyclopentadienyl)scandium-π-allyl was prepared and
characterized. Even though activity was low and there was limited
mileage, no activator was required to prepare linear high density
polyethylene. It is likely that polymerization was achieved by
the repeated addition of ethylene monomer to monomeric transition
metal compound. Propylene did not homopolymerize but did not
affect ethylene polymerization.

EXPERIMENTAL SECTION

General

All investigations were carried out under inert atmosphere
conditions using either Schlenk tube or crown capped pressure
bottle techniques. Transfers were made in glove bags using a
high flow rate of nitrogen that had been passed through activated
Ridox and molecular sieve (Linde 3A). There was no fuming by
titanium tetrachloride in this atmosphere. The compounds and the
polymerizations were extremely air and moisture sensitive and
required a high degree of care. Ethylene and propylene were
purified by passing them through activated Ridox and molecular
sieve (Linde 3A).

Tetrahydrofuran and diethyl ether were distilled from sodium naphthalide and sodium benzophenone, respectively. Hexane and benzene were refluxed under argon and were distilled immediately before use from calcium hydride. Anhydrous scandium chloride, $ScCl_3$, was obtained from ROC/RIC and was sublimed before use. Sublimation was performed at 0.1 mm Hg in a quartz tube heated to 825°C in a vertically mounted tube furnace. Allyl Grignard reagents were prepared by standard methods from allyl chloride (Aldrich) and bromide (Aldrich) in both ether and tetrahydrofuran using magnesium turnings obtained from Alfa-Ventron. Cyclopentadienyl thallium was prepared by a modification of a literature method (11).

Preparation of Cyclopentadienyl Thallium

Cyclopentadienyl thallium was prepared in a Waring blender using 30 g potassium hydroxide in 300 ml water, 6 ml of freshly cracked cyclopentadiene and 10.8 g of thallium chloride. The blender was charged with the above components and allowed to stir for 30 seconds. The solid formed was filtered through a Buchner funnel and allowed to air dry. The thallium compound was sublimed (130°/0.05 mm Hg) before use. The compound, when pure, was cream-colored or white. When brown cyclopentadienyl thallium was used in reaction sequences, reduced yields resulted.

Preparation of Dicyclopentadienyl Scandium Chloride Tetrahydrofuran

Under inert atmosphere conditions in a 500 ml round bottom flask equipped with a stopcock side arm and a stir bar, sublimed scandium chloride (9.8 g; 0.065 moles), sublimed cyclopentadienyl thallium (34.68 g; 0.129 moles) and freshly distilled tetrahydrofuran (500 ml) were allowed to reflux for three hours. The reaction mixture was filtered while hot (about 50°C) through a medium frit (25-50µ) in Schlenk equipment. The filtrate was cooled to room temperature and the solvent was removed at reduced pressure. The resulting straw colored solid was sublimed (180°, 0.1 mm Hg) to give $Cp_2ScCl \cdot THF$ (4.07 g; 0.015 moles; 23%). There was a small amount (0.11 gram) of a light pink solid that was more volatile than $Cp_2ScCl \cdot THF$. The pink substance was not identified. The identity of $Cp_2ScCl \cdot THF$ was established by ^{1}H NMR.

Preparation of Dicyclopentadienyl Scandium Chloride Dimer

In a degassed large Schlenk tube equipped with a water condenser and a stir bar, sublimed scandium chloride (15.5 g; 0.101 moles), sublimed cyclopentadienyl thallium (56.5 g; 0.210 moles) and distilled benzene (150 ml) were allowed to reflux for 48 hours. The light grey solid was allowed to settle (about 15 minutes) and while the reaction mixture was still warm, the supernatant benzene was filter transferred (under Argon) to a second degassed Schlenk tube. The benzene was removed at reduced pressure to give a pale yellow powder. Sublimation (150°, 0.1 mm Hg) gave a small amount of cyclopentadienyl thallium. This was removed from the cold finger and the temperature was raised 200°C. White solid $(Cp_2ScCl)_2$ (6.2 g, 0.0147 moles, 29%) was isolated. This compound was characterized by infrared and ^1H NMR spectroscopy. It is soluble in benzene, toluene and THF. Tetrahydrofuran breaks the bridging dimer and one records the ^1H NMR spectrum identical to that of known $Cp_2ScCl \cdot THF$.

Preparation of Dicyclopentadienyl Scandium π-Allyl

A sublimed sample of $(Cp_2ScCl)_2$ (0.57 gram; M.W. = 421.21; 1.35 mmoles) was dissolved in 200 ml of dry benzene. The benzene solution was heated to about 50° at which point allyl Grignard (2.0 ml, 1.63M in THF; 3.26 mmoles) was added dropwise to the benzene solution and an immediate orange color developed. Stirring continued four hours at 50°C. The orange reaction mixture was filtered, leaving behind magnesium salts which gave a positive test with silver ion. The benzene was removed under reduced pressure to give an orange solid which was transferred to a sublimation apparatus and sublimed (95°; 0.1 mm Hg) to give an orange crystalline solid of $Cp_2Sc-\pi$-allyl (0.22 g; 1.02 mmoles; 38%). On several occasions this reaction went through a light green color soon after the allyl Grignard was added. No attempt was made to isolate this possible intermediate and the isolated yield of dicyclopentadienyl scandium π-allyl seemed unaffected by the presence or the lack of this green transition color.

Polymerization Reactions

All polymerization studies were performed in crown capped pressure bottles (200 ml or 850 ml). The equipment was thoroughly cleaned, heated to 120° and allowed to cool under an inert atmosphere. After the solvent (hexane) was added to the scandium compounds, normal daylight was excluded by wrapping the polymerization vessels with aluminum foil. Solution transfers of the freshly distilled hexane solutions of triethylaluminum and diethylaluminum chloride were accomplished by syringe techniques. Purified ethylene and propylene were supplied to the catalyst via needle/Swagelok assemblies by puncturing the butyl rubber liners and pressurizing the polymerization vessel. The irradiation source was a sun lamp.

ACKNOWLEDGEMENT

The author gratefully acknowledges Dr. H. G. Tennent whose guidance, encouragement and patience helped make this work both possible and enjoyable.

BIBLIOGRAPHY

1. H. Sinn and W. Kaminsky, "Ziegler–Natta Catalysis" in Advances
    in Organometallic Chemistry, Vol. 18, p. 99, 1980, Academic
    Press;
    P. Pino and R. Mulhaupt, Angew. Chem. Int. Ed. Engl. 19, 857
    (1980);
    J. Boor, Jr., Ziegler–Natta Catalysts and Polymerizations,
    Academic Press, Inc., 1979.

2. J. W. Lauker and R. Hoffmann, J. Am. Chem. Soc., 98, 1729
    (1976).

3. D. S. Breslow and N. R. Newburg, J. Am. Chem. Soc., 79, 5072
    (1957); W. P. Long, J. Am. Chem. Soc., 81, 5312 (1959); W. P.
    Long and D. S. Breslow, J. Am. Chem. Soc., 82, 1953 (1960); W.
    P. Long and D. S. Breslow, Liebigs Ann. Chem., 1975 463; W.
    Kaminsky and H. Sinn, Liebigs Ann. Chem. 1975, 424; W.
    Kaminsky and H. J. Vollmer, 1975, 438.

4. J. M. Birmingham and G. Wilkinson, J. Am. Chem. Soc., 78, 42
    (1956).

5. F. A. Hart, A. G. Massey and M. S. Saran, J. Organometal.
    Chem., 21, 147 (1970).

6. N. M. Pletz, Compt Rend. URSS, 20, 27 (1938).

7. R. S. P. Coutts and P. C. Wailes, J. Organometal. Chem., 25,
    117 (1970).

8. L. E. Manzer, J. Organometal. Chem., 110, 291 (1976).

9. J. L. Atwood and K. D. Smith, J.C.S. Dalton, 2487 (1973).

10 J. K. Becconsall, B. E. Job and S. O°Brien, J. Chem. Soc. (A),
    423 (1967).

11. C. C. Hunt and J. R. Doyle, Inorg. Nucl. Chem. Letters, 2,
    283 (1966).

12. Presented to the Polymer Division at the American Chemical
    Society National Meeting, Atlanta, Georgia, March–April 1981.
    Hercules Contribution No. 1735.

STEREOSELECTIVE AND STEREOELECTIVE POLYMERIZATION OF RACEMIC

4-METHYL-1-HEXENE WITH MgCl$_2$ SUPPORTED TITANIUM CATALYSTS

Piero Pino*, Giovanni Fochi, Andreas Oschwald, Oreste
Piccolo, Rolf Mülhaupt and Umberto Giannini

Swiss Federal Institute of Technology
Department of Industrial and Engineering Chemistry
Universitätstrasse 6, 8092 Zurich, Switzerland

INTRODUCTION

Although highly active polymerization catalysts prepared from
TiCl$_4$ supported on MgCl$_2$ and aluminum alkyls have been used for
the stereospecific polymerization of propylene since the early
seventies,[1] no report has appeared up to now concerning the polymerization of racemic α-olefins with this type of catalytic
system.

Since the polymerization of racemic α-olefins can yield interesting new information on the nature of the catalytic centers[2]
we have investigated the polymerization of racemic 4-methyl-1-he-
xene (4MH) in the presence of non-chiral and optically active
Lewis bases.

The MgCl$_2$/TiCl$_4$/AlR$_3$/Lewis base catalytic systems have proved
to be stereoselective, indicating that the above catalysts, as do
the traditional TiCl$_3$ based catalysts,[3] have chiral active centers capable of distinguishing between the two antipodes of the
monomer. When optically active Lewis bases were used, an appreciable stereoelectivity[4] was found, particularly in the synthesis of highly stereoregular polymers.

STEREOSELECTIVITY IN THE POLYMERIZATION OF RACEMIC 4-METHYL-1-HEXENE

It is known [3] that the polymerization of 4MH with the traditional $TiCl_4$ or $TiCl_3$ based catalytic systems is stereoselective, the
prevalence of monomeric units arising from (S) or (R) monomer, in
the fractions having positive or negative rotation, respectively,
being larger than that expected for a statistical copolymer of
(S) and (R) monomers. (Scheme 1).

Scheme 1

$n[R]+n[S] \longrightarrow ..[R][R][S][R][R][R][S]..+..[S][S][S][R][R][S][S]..$
[R];[S] monomers containing an asymmetric center with (R) or (S) ab-
        solute configuration.

The stereoselectivity is, in general, small; for the most stereoregular fraction of poly(rac.)4-methyl-1-hexene (PMH), the excess
of monomeric units of a single type can reach 15%.

PMH prepared using a $TiCl_4$-(EA)/$MgCl_2$/Al$(C_2H_5)_3$/Ethyl Anisate
(EA) catalytic system has been fractionated by the usual boiling
solvents extraction technique (see experimental section, Table 5).
The acetone insoluble-diethylether soluble fraction and the diisopropylether insoluble, benzene soluble fraction have been further
fractionated using isotactic poly-(S)-3-methyl-1-pentene (P(S)MP)
as support to separate macromolecules containing a prevalence of
(R) or of (S) monomeric units [3]. A slightly modified technique was
used for the separation (see experimental section).

The efficiency [5] of the new method (E) was tested on a mixture
(1:1) of diethylether insoluble fractions of (+)poly-(S)-4-methyl-
1-hexene and (-)poly-(R)-4-methyl-1-hexene prepared from monomers
containing more than 93% of the (S) or (R) antipodes, respectively (Sample Z, Table 1). With the reasonable assumption of a complete separability of the (S) and (R) fractions (D=100%)[5], an
efficiency of about 50% was estimated (Table 1).

The results of the fractionation of the polymers obtained with
the catalysts supported on $MgCl_2$ show beyond any doubt that the
catalytic system $TiCl_4$-EA/$MgCl_2$/Al$(C_2H_5)_3$/EA is stereoselective.
In fact (Table 1) the separability (D) both for the stereoregular
fraction (sample C) and for the stereoirregular fraction (sample A)
is much larger than the separability (D_{st}) calculated for a statistical copolymer having the same molecular weight, a logarith-
mic normal distribution of molecular weights, and $\overline{M}_w/\overline{M}_n$ ratio of 3[6].

Table 1. Comparison Between Separation Degree (F)[a] and Separability (D)[n] of some PMH Fractions and the Separability (D$_{st}$)[p] of a Statistical Copolymer of (R) and (S) Antipodes.

Catalyst	Sample	$[\eta]$ dl/g	$\bar{M}_w$[s]	$\dfrac{\Sigma[\alpha]_i w_i(-)}{W_t}$[v,u]	$\dfrac{\Sigma[\alpha]_i w_i(+)}{W_t}$[v,u]	F[a] %	D[n] %	E[m,n] %	D$_{st}$[p] %
NS[b]	Z[d]	n.d.	$16\cdot10^5$	-68.5	$+67.3$	50	100[ℓ]	0.5[m]	-
S[c]	A[e]	0.9[q]	$2.3\cdot10^5$	-6	$+11$	4.3	13	0.33	2.5
NS[b]	B[f]	0.22[r]	$0.16\cdot10^5$ [i,t]	-5.43 [z]	$+5.0$ [z]	4.5	13.6	0.33	6.0
S[c]	C[g]	3.87[q]	$14\cdot10^5$	-21.1	$+25.4$	15.4	30.8	0.5	1.0
NS[D]	D[h]	1.52[r]	$2\cdot10^5$ [t]	-8.5	$+8.0$	6.1	18.5	0.33	2.75

[a] F has been calculated from $\Sigma[\alpha]_i w_i(-)/W_t$ to avoid errors arising from the presence of traces of support having positive optical rotation in the last eluted fractions. [b] Non-supported catalytic system of the type TiCl$_3$"ARA"/Al(iC$_4$H$_9$)$_3$. [c] Supported catalyst of the type TiCl$_4$-EA/MgCl$_2$/Al(C$_2$H$_5$)$_3$/EA. [d] Equimolecular mixture of Poly(S)4-methyl-1-hexene and Poly-(R)-4-methyl-1-hexene containing more than 93% of (S) or (R) monomeric units respectively. [f] Fraction insoluble in diethylether. [e] Fraction insoluble in acetone, soluble in diethylether. [f] Fraction insoluble in acetone, soluble in ethylacetate. See ref. 3, run 13. [g] Fraction insoluble in diethylether, soluble in benzene. [h] Fraction insoluble in diisopropylether, soluble in isoctane. See ref. 3, run 16. [i] $\bar{M}_n = 9400$. (See ref. 3, run 13). [ℓ] Separability has been assumed to be 100% in view of the molar composition of the fraction. See footnote d. [m] E = D/F. See ref. 5. [n] See ref. 5.

[p] Calculated under the assumption that the fraction has $M_w/M_n = 3$ and a logarithmic normal distribution of molecular weights. See ref. 6. [q] In methylcyclohexane at 60°. [r] In tetraline at 120°. [s] See footnote q in Table 2. [t] See footnote n in Table 2. [u] W$_t$=total weight of the sample. [v] Optical rotation measured at 589 nm and 20°C if not otherwise stated. [z] Optical rotation measured at 400 nm and 25°C.

As shown by the D values, the stereoselectivity is larger for the stereoregular than for the stereoirregular fraction. If we compare the fractions with similar stereoregularity, the stereoselectivity is larger with the MgCl$_2$ supported catalyst (S in Table 1) than when the TiCl$_3$"ARA" traditional catalyst (NS in Table 1) is used.

The highest prevalence of (S) or (R) monomeric units in the polymer chain can be estimated from the highest optical activity of the fractions obtained in the separation on P(S)MP. It is safer to consider only the fractions having negative optical rotations, as traces of the support, which has a positive optical rotation[3], could enhance the positive optical rotation measured for the last fractions separated by methylcyclohexane extraction. (Table 2)

Taking into account the highest negative values of the optical rotation experimentally obtained, the highest % of (R) monomeric units in the polymer fraction can be evaluated on the basis of the optical purity of the monomer yielding a polymer with the same stereoregularity and optical rotation[7]. In fact, as the specific rotation of stereoregular PMH is not a linear function of the optical purity of the polymerized monomer[8], the ratio between the optical rotation of the polymer considered and the optical rotation of a polymer with the same stereoregularity, obtained from an optically pure monomer, does not yield a reliable indication of the prevalence of (R) or (S) monomeric units. The procedure used can yield too low values for the excess of monomeric units of one type, as it implies an efficiency of 100% in the fractionation of the sample considered. However the excess found is larger than that calculated[6] for a statistical copolymer of (R) and (S) antipodes for the fraction having a high stereoregularity (Table 2, fraction C) and is the same in the case of the stereoirregular fraction (Table 2, fraction A).

Considering the average rotation for the fractions having negative or positive optical rotation, the excess of monomeric units of one type of chirality is, in the polymer fractions with positive optical rotation, definitely larger than that calculated for a statistical copolymer of the (R) and (S) antipodes of the monomer (Table 3 samples A and C).

The maximum excess of monomeric units arising from one of the antipodes of the monomer is not very high; ignoring the efficiency of the separation process, the maximum excess of (R) monomeric units experimentally found is 12% and 4.4% for the most stereoregular fraction of the polymers prepared with the MgCl$_2$ supported and with the non-supported TiCl$_3$"ARA" catalytic systems, respecti-

Table 2.   Highest prevalence [R]max of (R) monomeric units found in the fractionation of different PMH samples on P(S)MP as support.

Catalytic System	Sample ℓ	Weight Fraction	$[\eta]$ dl/g	$\bar{M}_w$ q	$[\alpha]^{20}_D$	$[R]_{max}$ %	$[R]^m_{St}$
S[f]	A	0.13	(0.9)[a,b]	$(2.3 \cdot 10^5)$[d]	-19.1	52.5	52.5 [e]
NS[g]	B[h]	0.10	(0.22^g)[c]	$(0.16 \cdot 10^5$[d,n]$)$	- 8.4	51.1	57.2 [p,e]
S[f]	C	0.65	3.25 [b]	$15.5 \cdot 10^5$	-72	56.0	51.5 [e]
NS[g]	D[i]	0.12	(1.52)[a,c]	$(2 \cdot 10^5)$[d]	-24.2	52.2	52.7 [e]

a Intrinsic viscosity of the sample before fractionation.   b In methylcyclohexane at 60°.
c In tetraline at 120°.   d Average molecular weight of the sample before fractionation.
e Assuming for the fraction the same average molecular weight as for the original sample.
f TiCl4–EA/MgCl2/Al(C2H5)3/EA.   g TiCl3(ARA)/Al(iC4H9)3.   h Run 13 in reference 3.
i Run 16 in ref. 3.   ℓ See footnotes e, f, g, h in Table 1.   m See Footnote p in Table 1.
n Calculated according to the equation $[\eta]=9.18 \cdot 10^{-5} \cdot \bar{M}_n^{0,8}$ given for polypropylene in tetraline
at 133°C (see ref. 18).   P Calculated for a statistical copolymer of (R) and (S) monomers
monodisperse with respect to molecular weight and having $\bar{M}_w=1.6 \cdot 10^{-4}$ (see ref. 19).
q Calculated from $\bar{M}_\eta$ assuming $\bar{M}_w/\bar{M}_n=3$ (see ref. 20).

Table 3. Average Prevalence of (R) and (S) Monomeric Units ($[R]_{av}$ and $[S]_{av}$ in the Fraction Having Negative or Positive Optical Rotation Obtained in the Fractionation of PMH on P(S)MP for Different PMH Samples.

Catalytic System [a]	Sample fractionated [a]			Fraction with prevailing (R) monomeric units				Fraction with prevailing (S) monomeric units			
		$[\eta]$ dl/g	$\overline{M}_w$ [e]	$\Sigma w_i(-)/W_t$ [b]	$[\bar{\alpha}]_D^{20}$	$[R]_{av}$ %	$[R]_{St}$ [c] %	$\Sigma w_i(+)/W_t$ [d]	$[\bar{\alpha}]_D^{20}$	$[S]_{av}$ %	$[S]_{St}$ [c] %
S	A	0.9	$2.3 \cdot 10^5$	0.66	-8.21	51.15	50.8	0.39	27.65	52.6	51.6
NS	B	0.22	$0.16 \cdot 10^5$	0.84	-6.41	50.5	51.2	0.15	31.7	54.0	56 [f]
S	C	3.87	$15.5 \cdot 10^5$	0.66	-31.65	52.6	50.5	0.33	76.35	56.4	53
NS	D	1.52	$2.0 \cdot 10^5$	0.86	-9.76	50.8	50.4	0.13	60.0	54.9	52.6

[a] For the meaning of S, NS, A, B, C, D see footnotes b,c and e-h in Table 1.   [b] $\Sigma[\alpha]_i w_i(-)/\Sigma w_i(-)$
[c] See footnote $\underline{p}$ in Table 1 and footnote $\underline{e}$ in Table 2.   [d] $\Sigma[\alpha]_i w_i(+)/\Sigma w_i(+)$.   [e] See footnote $\underline{q}$ in Table 2.   [f] See footnote $\underline{p}$ in Table 2.

vely. The excess is much lower for the stereoirregular fractions A and C (5% and 2.2% respectively) Table 2. The average excess (Table 3) in the fractions having positive or negative rotation is larger for the positive ones than for the negative ones, showing that the efficiency of the separation process is greater for the fractions in which the chirality centers of the lateral chains prevailingly have the same absolute configuration in the supported polymer as in the support.

The average excess for the fractions having positive optical rotation is 12.8% and 7.2% for the stereoregular and stereoirregular fractions obtained with the $MgCl_2$ supported catalyst respectively, well above the excess calculated for the same weight fraction of a statistical copolymer of (R) and (S) monomers.

STEREOELECTIVE POLYMERIZATION OF RACEMIC 4-METHYL-1-HEXENE

When 4MH is polymerized in the presence of a catalyst containing an optically active component, one of the antipodes is polymerized more rapidly than the other. Therefore if the polymerization is stopped at partial monomer conversion, the recovered monomer and the polymer are optically active (stereoelective polymerization (Scheme 2) [9].

Scheme 2

$$n[R] + n[S] \xrightarrow{Cat^*} ...[R][R][R][R] ...+n[S]$$

Optically active organometallic compounds[9], optically active ligands [10] initially bound to the transition metal and optically active Lewis bases [11] have been used as the optically active components of the catalyst.

With 4MH only a very low stereoelectivity has been found up to now, the highest ratio between the polymerization rates of the two antipodes evaluated from the optical activity of the recovered monomer being 1.02 [12].

The polymerization of 4MH with the catalytic system $TiCl_4/MgCl_2/AlR_3/$Lewis base has been carried out in the presence of 20 different optically active Lewis bases [13]. However in this paper we shall discuss only the results obtained using (-)menthylanisate ((-)MA) as Lewis base and the catalytic system of the type $TiCl_4$(-)MA/$MgCl_2$/Al$(iC_4H_9)_3$/(-)MA. The experiments were carried out at room temperature at different [(-)MA]/[Al]+[Ti] ratios, (r), and different conversions. The non-reacted monomer was recovered and the polymers were fractionated by boiling solvents extraction, as previously reported [14].

Table 4. Optical Rotation and Optical Purity of the Recovered Monomer and of the Most Stereoregular (E) and Most Stereoirregular (B) Polymer Fraction.

Exp.	$(\underline{r})$[a]	Producti-vity[b]	Conver-sion %	Recovered monomer $[\alpha]^{25}_{365}$	Prev. Chir.	o.p. %	Fraction B[c] %	$[\alpha]^{25}_{D}$ Prev. Chir.	o.p.[e] %	Fraction E[d] %	$[\alpha]^{25}_{D}$ Prev.[e] Chir.	o.p.[e] %
1	0.131	159	88	n.d.	n.d.	n.d.	14.7	+2.67 (S)	0.19	44.7	-1.25 (R)	0.68
2	0.293	n.d.	8	n.d.	n.d.	n.d.	14.6	+0.94 (S)	0.23	37.3	-16.6 (R)	1.72
3	0.293	n.d.	33	-0.041	(S)	0.4	14.1	+1.98 (S)	0.53	34.7	-8.8 (R)	1.43
4	0.293	n.d.	71	-0.138	(S)	1.3	12.0	+4.23 (S)	1.09	31.3	n.d. n.d.	n.d.
5	0.293	n.d.	90	-0.241	(S)	2.2	12.5	+5.51 (S)	1.50	29.6	-7.8 (R)	1.25
6	0.408	67	15	-0.033	(S)	0.3	6.6	+1.0 (S)	0.25	39.8	-20.7 (R)	3.33
7	0.501	24	24	-0.104	(S)	1.1	9.9	+2.2 (S)	0.55	48.1	-8.2 (R)	1.48
8	0.538	53	66	-0.253	(S)	2.3	4.5	+5.9 (S)	1.39	52.5	-10.1 (R)	1.62
9	0.796	4	43	-0.297	(S)	2.7	6.9	+2.5 (S)	0.63	57.6	-10.4 (R)	1.67

a $(\underline{r})=[(-)MA]/[Al]+[Ti]$. [b] gPolymer/gTi.h.mol$_{(mon)}$ ℓ^{-1} [c] Acetone insoluble-ethylacetate soluble fraction. [d] diisopropylether insoluble-cyclohexane soluble fraction. [e] For the relationship between the sign of optical rotation and prevailing chirality, see ref. 21.

In all cases the recovered monomer is optically active (Table 4) and the prevailing chirality of the recovered monomer is (S); its optical purity increases by increasing (r) and by increasing conversion.

The determination of the optical rotation of the fractions of the polymers synthesized at (r)>0.1 has shown that the most stereo-irregular fraction (B) (acetone insoluble, ethylacetate soluble fraction) contains an excess of monomeric units arising from the (S) monomer. The most stereoregular fractions (E) (diisopropylether insoluble, cyclohexane soluble fraction), have negative optical rotation; hence they contain an excess of monomeric units arising from the (R) antipode of the monomer (Table 4).

The optical purity of the polymerized monomer giving rise to each single fraction can be roughly evaluated from the optical activity of the polymer fractions by comparing it to the optical activities of the corresponding fractions obtained when 4MH with different optical purities is polymerized [7,8]. The average values thus obtained are consistent with those calculated on the basis of the optical activity of the recovered monomer and of the conversion.

From the optical rotation of the polymer fractions, it appears that the optical purity of the polymerized monomer yielding the different fractions is influenced both by (r) and conversion. An increase of conversion (exp. 2-5 Table 4) causes an increase of the optical purity of the recovered monomer; correspondingly an increase of the optical purity of the polymerized monomer in the fraction having a low stereoregularity (B) has been observed. For the fraction with high stereoregularity (E), the contrary has been noticed, the optical purity decreasing by increasing conversion. For the less stereoregular fractions (B), polymerized and recovered monomer surprisingly have the same prevailing chirality, and the optical purity of the recovered monomer is in general higher than that of the polymerized monomer. The more stereoregular fractions (E) have, as expected, prevailing chirality opposite to that of the recovered monomers. The conversion being similar, the optical purity of the polymerized monomer increases with increasing (r).

DISCUSSION

The results obtained in the separation of the highly isotactic PMH, prepared with the $MgCl_2$ supported catalyst, into two optically active fractions having opposite signs of rotation and optical purities up to 12%, show beyond any doubt that the new catalytic system supported on $MgCl_2$, as the conventional $TiCl_3$ based

catalytic systems [3],is able to distiguish between the two antipodes of the monomer and therefore contains highly stereospecific
chiral catalytic centers (C_ℓ and C_d) on which the two antipodes
are polymerized with different rates.

The ratio between the rates of polymerization of the two antipodes
of 4MH on a single catalytic center can reach values up to 1.3,
which is remarkable, taking into account the fact that the asymmetric carbon atom is in β position with respect to the double bond.
In the polymerization of 4MH in the presence of (-)MA as Lewis base
at (r) ratios >0.1, there are obtained amorphous, stereoregular,
ethylacetate soluble polymers, having a prevalence of (S) monomeric units together with crystalline, highly stereoregular, diisopropylether insoluble polymers with a prevalence of (R) monomeric
nits; this shows unequivocally that at least two classes of catalytic centers exist. The first class (C_ℓ and C_d) consists of centers which are stereospecific and stereoelective and become stereoelective in the presence of an optically active Lewis base; the
second class (C) produces stereoirregular polymers and have a low
stereoselectivity and a low stereoelectivity if any. In fact the
polymers produced at (r)>0.1 contain monomeric units which prevailingly arise from the antipode of the monomer which prevails in
solution. By increasing conversion the optical purity increases,
both in the non-polymerized monomer and in the polymerized monomer,
and at high conversions, it is lower in the polymerized monomer
than in the residuous monomer.

These results indicate that only a minimum value for the ratio
between the polymerization rate of the two antipodes can be calculated starting with the optical activity of the recovered monomer[9,14]
since polymers with opposite prevailing chirality of the asymmetric
centers in the lateral chains can be produced.

The polymerization of 4MH at different (r) indicated further
that the Lewis acidity of the centers, C, is larger than that of
the centers C_ℓ and C_d, as the relative amount of the stereoirregular fraction B decreases and that of the stereoregular fraction
(E) increases by increasing (r).

The productivity, referred both to the highly stereoregular
fraction (E) and to the stereoirregular fraction (B) decreases in
absolute value by increasing (r) (Table 4).

The results of the polymerization of 4MH with the catalytic
system $TiCl_4$-(-)MA/$MgCl_2$/Al$(iC_4H_9)_3$/ (-)MA at least for (r)>0.1

are well accounted for if we postulate the equilibria (1)-(3).

$$C + B*AlR_3 \underset{K_2}{\overset{K_1}{\rightleftharpoons}} CB* + AlR_3 \qquad (1)$$

$$C_\ell + B*AlR_3 \underset{K_3}{\overset{}{\rightleftharpoons}} C_\ell B* + AlR_3 \qquad (2)$$

$$C_d + B*AlR_3 \rightleftharpoons C_d B* + AlR_3 \qquad (3)$$

In equations (1)-(3), C, C_ℓ and C_d are the catalytically active centers; CB*, $C_\ell B*$ and $C_d B*$ are the catalytic centers reversibly poisoned by the optically active Lewis base (in our case (-)MA). K_1 is larger than K_2 and K_3; K_2 is different from K_3, as the two equilibria lead to the two diastereomeric catalytically inactive species $C_\ell B*$ and $C_d B*$ having presumably different stability.

A second possible explanation requiring that CB*, $C_\ell B*$ and $C_d B*$ also are catalytically active is less satisfactory in view of the large decrease in productivity by increasing (r), but cannot be completely rejected. Indeed the following preliminary results indicate that the model we propose that would explain our results is largely oversimplified:

i) In few cases the prevailing chirality of the polymerized mono-
    mer forming the stereoirregular fraction (B) is very near to
    or even higher than that of the recovered monomer; furthermo-
    re at (r) <0.1 and at low conversions, stereoirregular frac-
    tions (B) containing a low prevalence of (R) monomeric units
    have been obtained [15,16]
    These facts indicate that also some centers which are not
    highly stereospecific can show a small stereoelectivity which
    might be connected with the presence of the optically active
    base in the active catalytic center.

ii) These indications have been confirmed by the results of the
    stereoelective polymerization of 3,7-dimethyl-1-octene[13],
    clearly showing the existence of catalytic systems with stereo-
    electivity of opposite type in the same catalytic system.

In conclusion, it appears that the investigation of the polymerization of racemic α-olefins with the MgCl2 supported catalysts
yields a unique opportunity to acquire a detailed knowledge of
the structure of the catalytic centers on which the polymerization of the α-olefins takes place.

The results show that a large variety of catalytic centers
with greatly different steric structures, Lewis acidity and cata-

lytic activity are present in the catalytic system investigated.

EXPERIMENTAL PART

Materials

Monomers, solvents and the P(S)MP used as support for the separation of P4MH into fractions having opposite optical rotations have
been described in a previous paper[3].

Characterization of the polymers

Melting points were determined by DTA measurements using a Mettler
Mod. TA 2000.

Intrinsic viscosity for P4MH was determined in methylcyclohexane
at $60^{\circ}C$.

Optical activity was measured with a Perkin Elmer Spectropolarimeter
Mod. 141 in Methylcyclohexane (C = 1-5 mg/ml). The wavelengths at
which the measurements were made are reported in the Tables.

$MgCl_2/TiCl_4$ component of the catalytic system

The preparation of the $MgCl_2/TiCl_4$ catalyst was carried out by
milling the $TiCl_4$-ethylanisate or the $TiCl_4$-menthylanisate complex[13]
in a ball-mill[13] which will be described elsewhere[17] for 12-20 h.
The catalyst was stored under N_2, in the absence of H_2O.

Polymerization of racemic 4-methyl-1-hexene (4MH) with the
$MgCl_2TiCl_4EA/Al(C_2H_5)_3$/EA catalytic system

75.7 mg of a catalyst prepared by milling $MgCl_2$ with the $TiCl_4EA$-
complex (Ti 0.67%) were introduced into a glass reaction vessel in a
N_2 atmosphere. A solution of 2.2 ml of ethylanisate and 0.4 ml
distilled $Al(C_2H_5)_3$ in 10 ml of racemic 4-methyl-1-hexene was added at room temperature. The polymerization began immediately and was
stopped after 16 hours.

The polymer was suspended in methanol and purified with HCl in the
usual way [3].

After drying,the polymer was fractionated by boiling solvents
extraction, as previously described[3].

Separation of the polymer into fractions having optical rotations of opposite sign

The fractions used for the fractionation were characterized as shown in Table 5.

Table 5. Characterization of the Fractions Used for the Separation According to the Prevailing Chirality of the Macromolecule's Side Chains.

Sample	Extraction with boiling solvents	X-Ray examination	IR index [a]	Mp $^{\circ}$C [b]	Intrinsic Viscosity dl/g [c]
A	Acetone ins. diethylether soluble	very low crystallinity	0.8	125	0.9
C	Diethylether ins.,benz.sol.	crystalline	n.d.	216	3.87
Z [d]	Diethylether insoluble Benzene sol.	crystalline	n.d.	n.d.	- [d]

[a] D_{995}/D_{965} (D=optical density at the indicated wave numbers).
[b] From DTA measurements.   [c] In methylcyclohexane at 60°C.
[d] This fraction was obtained by dissolving in benzene equalweights of diethylether insoluble, benzene soluble fractions obtained from (S)4MH (optical purity 95.5%) and (R)4MH (optical purity 87.2%) polymerized with a TiCl$_3$/Al(iC$_4$H$_9$)$_3$ catalyst as previously described [3]. The first fraction had [η]=4.6 dl/g (in methyl cyclohexane at 60°) and $[\alpha]_D^{20}$ = 274. The second fraction had [η]=5.1 dl/g (in methyl cyclohexane at 60°) and $[\alpha]_D^{20}$=-260. The polymer mixture was precipitated with CH$_3$OH from the benzene solution and dried in vacuum.

Fractionation of sample A

127.4 mg of the sample were dissolved in 100 ml of benzene and 2g of optically active support were added to the solution. The support was crystalline P(S)MP obtained from a monomer having optical purity of 89.6% and extracted for 48 h with boiling decaline.

To the above suspesnion, 200 ml of CH$_3$OH were slowly added and

then the solvents were distilled. The supported polymer thus obtained suspended in methanol was added to a thermostatted chromatographic column containing 18g of P(S)MP as previously described.[3]
After successive elution with ethylacetate, diethylether-ethyl-
acetate (1:1), diethylether and benzene, the content of the chro-
matrographic column was extracted by boiling diethylether benzene, methylcyclohexane and then toluene. The results are reported
in Table 6.

Table 6. Optical Activity of the Fractions Obtained in the
Fractionation of Sample A.

Fractionation Method

Elution chromatography[a,b]	Boiling Solvents Extraction[c]	%	$[\alpha]_D^{20}$ [e]
Ethylacetate	-	6.92	+ 3.7
Ethylacetate	-	12.61	- 3.2
Ethylacetate-Diethylether (1:1)	-	14.79 1.53	- 13.5 - 2.4
Diethylether	-	4.66	- 12.5
Benzene	-	12.97	- 19.1
Benzene[d]	-	9.26	- 4.3
-	Diethylether	4.66	- 1.6
-	Benzene	20.33	+ 19.9
-	Benzene	6.41	+ 17.0
-	Methylcyclohexane	4.66	+110.5
-	Toluene	1.17	+ 31.3

[a] Fractions of 20 ml were collected. [b] at 25°C if not otherwise
stated. [c] for 24 h. [d] at 60°. [e] in methylcyclohexane (C=1-5 mg/ml)

Fractionation of Samples C and Z

The samples C and Z supported on P(S)4MP as described for sample A (170 mg and 190 mg on 2 g of support) suspended in methanol were cautiously dropped into a cavity prepared in the middle of the support (18g) that had been placed in the cellulose vessel of the Kumagawa extractor. The polymers were then extracted with a series of solvents for periods of 6 to 100 hours, as shown in Tables 3 and 4.

Table 7. Optical Activity and Intrinsic Viscosity of the Fractions Obtained in the Fractionation of Sample C by Boiling Solvents Extraction.

Solvents	Extraction Time h	%	$[\alpha]_D^{20}$ a	$[\eta]$ b dl/g
Dibutylether	25	1.0	− 44	−
	75	3.38	− 17	−
Pentane	12	12.75	− 38	3.75
	25	16.65	− 34	3.15
	50	2.19	− 57	2.40
Hexane	18	13.64	− 14	0.75
	25	10.68	− 17	0.50
	75	6.46	− 72	3.45
Benzene	25	5.63	+ 49	3.0
	25	6.76	+ 35	2.40
	25	3.32	+ 37	3.0
	50	1.9	+ 53	−
	75	1.0	+ 78	−
Methylcyclohexane	25	14.65	+118	−

a in methylcyclohexane (c=1-5 mg/ml). b in methylcyclohexane at 60°.

Stereoelective Polymerization of Racemic 4MH

The experiments were carried out as previously described [14]. As an example, experiment 8 of Table 4 is described.

0.286 g of a catalyst prepared by milling $TiCl_4$(−)MA(Ti,9.65%)with anhydrous $MgCl_2$(Ti,3.9%) were suspended in 50 ml of n-heptane. To

the suspension a solution of 2 ml Al$(iC_4H_9)_3$ freshly distilled and
1.208 g of (-)MA$(n_D^{25}=1.5182$, $[\alpha]_{546}^{25}-95.5)$ in 20 ml of 4MH was added at room temperature. After 22 hours the polymerization was

Table 8. Optical Activity of the Fractions Obtained in the
          Fractionation of Sample z by Boiling Solvents Extraction.

Solvent	Extraction Time h	%	$[\alpha]_D^{20}$ [a]
Diethylether	25	4.6	-45
	50	26.5	-80
Pentane	6	11.7	-223
	12	4.4	-156
	30	1.6	-117
	100	0.9	-98
Hexane	6	8.8	-66
	12	4.0	-92
	30	1.0	+ 4
	100	1.5	+31
Heptane	6	1.7	+28
	12	1.5	+24
	75	0.9	+110
Benzene	75	6.6	+153
Cyclohexane	75	21.0	+233
	100	1.6	+206
Methylcyclohexane	100	1.7	+159

[a] in methylcyclohexane (c=1-5 mg/ml)

interrupted by adding 5 ml of CH_3OH. Solvent, not-polymerized monomer and methanol were evaporated in vacuum, collected at -180°C
and finally distilled with a Perkin Elmer 251 Autoanularstill.

The recovered monomer had $[\alpha]_{365}^{25}$ - 0.253, corresponding to an optical purity of 2.3%. The polymer was purified with a mixture of
diethylether, CH_3OH and conc. HCl and finally washed with CH_3OH
and dried under vacuum at 80°C. In this way 9.4 g of polymer were
obtained, corresponding to a conversion of 65%.

The polymer was successively extracted with boiling solvents, giving 9% of acetone soluble polymer , 4.5% of ethylacetate soluble polymer ($[\alpha]_{365}^{25}$+15.4), 28% of diethylether soluble polymer ($[\alpha]_{365}^{25}$-14.50); 5% of diisopropylether soluble fraction ($[\alpha]_{365}^{25}$-8.8) 52.8% of benzene soluble polymer ($[\alpha]_{365}^{25}$-30.83, M.p.217.5°C; $[\eta]$ = 4.78 ; methylcyclohexane 60°).

ACKNOWLEDGEMENT

The authors thank Montedison Company, Milano, Italy, for partial support of this work.

REFERENCES

1. U. Giannini, A.Cassata, P.Longi, R.Mazzocchi, Belg.Pat.785332 (1972).
2. P.Pino, R.Mülhaupt, Angew.Chem.(Int.Ed.), 19,857 (1980).
3. P.Pino, G.Montagnoli, F.Ciardelli, E.Benedetti, Makromol.Chem. 93, 158 (1966).
4. P.Pino, F.Ciardelli, G.P.Lorenzi, Makromol.Chem., 70,182 (1964).
5. G.Montagnoli, D. Pini, A.Lucherini, F.Ciardelli, P.Pino, Macromolecules 2, 684 (1969).
6. A.Oschwald, Dissertation E.T.H. Zurich (1974).
7. F.Ciardelli, G.Montagnoli, D.Pini, O.Pieroni, C.Carlini, E.Benedetti, Makromol.Chem. 147, 53 (1971).
8. P.Pino, F.Ciardelli, G.Montagnoli, O.Pieroni, J.Polymer Sci.(B) 5, 307 (1967).
9. P.Pino, F.Ciardelli, G.P.Lorenzi, J.Am.Chem.Soc.85, 3888 (1963).
10. H.Ringger, P.Pino unpublished results; H.Ringger Dissertation E.T.H. Zurich (1976).
11. C.Carlini, F.Ciardelli, La Chimica e l'Industria 63,486 (1981).
12. C.Carlini, H.Bano, E.Chiellini, J.Polymer Sci.,(9-1) 10,2803 (1972).
13. R.Mülhaupt, Dissertation E.T.H. Zurich (1981).
14. P.Pino, F.Ciardelli, G.P.Lorenzi, Makrom.Chem.70, 188 (1964).
15. P.Pino, G.Fochi, U.Giannini unpublished results.
16. P.Pino, G.Fochi, O.Piccolo, U.Giannini unpublished results.
17. P.Pino, R.Mülhaupt, B.Rotzinger, manuscript in preparation.
18. J.Brandrup,E.H.Immergut,Polymer Handbook,page IV9,J.Wiley N.Y. (1979).
19. P.L.Luisi, G.Montagnoli, M.Zandomeneghi, Gazz.Chim.Ital.97, 222 (1967).
20. P.Neuenschwander, P.Pino, Makromol.Chem. 18, 737 (1980.
21. P.Pino, Adv. Polym. Sci., 4, 393 (1965).

SUPPORTED CATALYSTS FOR POLYPROPYLENE: ALUMINUM ALKYL-ESTER CHEMISTRY

Arthur W. Langer, Terry J. Burkhardt and John J. Steger

Corporate Research Laboratories
Exxon Research and Engineering Company
Linden, New Jersey 07036

SYNOPSIS

The reactions of aluminum alkyl cocatalysts with the ethyl benzoate (EB) component of $MgCl_2$ supported $TiCl_4$ catalysts have a strong influence on both activity and stereospecificity. NMR investigations show that $AlEt_3$ alkylates ethyl benzoate by the following reaction: $PhCO_2Et + 3\ AlEt_3 \longrightarrow 1/2\ (Et_2AlOEt)_2 + Et_3Al \cdot Et_2AlOCEt_2Ph$. The rate of ethyl benzoate consumption decreases with lower $Al/PhCO_2Et$ ratios. Atmospheric pressure polymerization studies reveal that the interaction of ethyl benzoate with the catalyst is responsible for achieving high isotacticity as measured by heptane insolubles (% HI). The aluminum alkoxide products from the alkylation reaction increase HI indirectly by complexing $AlEt_3$, lowering the free $[AlEt_3]$ and the $AlEt_3/PhCO_2Et$ ratio. However, lower free $[AlEt_3]$ also reduces polymerization rate. The introduction of steric bulk into the aluminum alkyl component minimizes the ester alkylation reaction while maintaining catalyst activity. Two types of cocatalysts have been found which give significantly better catalyst performance than the $AlEt_3$ cocatalyst: (1) sterically hindered trialkyl aluminum cocatalysts, such as $s-Bu_2AlEt$ and $t-Bu_2AlEt$ and (2) certain aluminum dialkyl and diaryl amides such as $Et_2Al-2,2,6,6$-tetramethylpiperidide and Et_2AlNPh_2.

Ethyl benzoate improves HI by inactivating the nonstereo-specific polymerization sites to a greater degree than the stereospecific site. This has led to the finding that $2,2,6,6$-tetra-methylpiperidine is highly selective in complexing the nonstereo-specific sites with minimal interaction toward stereospecific sites.

225

INTRODUCTION

With respect to the stereospecific polymerization of propylene, Pino and Mulhaupt[1] recently provided an excellent review of the historical background on Ziegler catalysis and the major contributions in the development of industrial catalysts and processes. Proceeding from Ziegler's[2] broad discovery of catalysts based on alkyl metals and transition metal salts, one finds that the most useful one based on $TiCl_4$ made very low crystallinity polypropylene (40% HI) and that Natta's[3] improved crystalline $TiCl_3$ produced higher crystallinity products (90% HI) but at very low rate. The delta-$TiCl_3 \cdot 1/3$ $AlCl_3$ catalyst[4], which was subsequently commercialized as $TiCl_3AA$, made high crystallinity polypropylene (93% HI) at practical rates. During the 1960's a large amount of work was done to improve the catalyst by adding a wide assortment of Lewis bases which increased polypropylene crystallinity to about 95% HI, but this generally caused some loss of activity. More recently, $TiCl_3$ catalysts with low aluminum content have been developed in which the aluminum residue has been extracted as an ether complex.[5-8] This resulted in a 3-5 fold increase in activity at about 95% HI.

Supported catalysts have been successfully used for ethylene polymerization for many years but these have generally utilized high surface area metal oxide supports which are non-stereospecific. Titanium catalysts supported on $MgCl_2$ have the potential of being several times more active than the low aluminum $TiCl_3$ catalysts at the same stereospecificity. Polypropylene catalysts are stereospecific by virtue of the asymmetric sites which are produced by reaction with aluminum alkyl at the edges of chloride layer structures such as alpha, gamma or delta $TiCl_3$. Thus it is not surprising that $MgCl_2$, which has the cubic closest packing of chlorides similar to that of gamma $TiCl_3$, should be the support of choice. Tornqvist[9] has shown how the $TiCl_4$ would fill in the chloride vacancies at the edges of a $TiCl_3$ structure to stabilize the reactive positions and this applies as well to $MgCl_2$. In order to generate a sufficient number of edge sites for the titanium, it is necessary to activate the $MgCl_2$ by grinding and/or treatment with Lewis bases.[10-13] A typical catalyst made in this fashion with ethyl benzoate as the Lewis base component was used in this study (a) to examine the reactions between the cocatalyst and the base, (b) to elucidate the mechanism by which the base improves stereospecificity, and (c) to prepare improved catalyst compositions based on the findings.

RESULTS AND DISCUSSION

A. Aluminum triethyl-ethyl benzoate chemistry. We concentrated our initial studies on the $AlEt_3$/ethyl benzoate cocatalyst chemistry to determine how this dynamic system influences catalyst performance. Equimolar mixtures of $AlEt_3$ and ethyl benzoate are known to form an observable 1:1 complex in which the aluminum alkyl is coordinated to the ethyl benzoate carbonyl oxygen. However, at higher $AlEt_3$/ester ratios this complex is alkylated giving aluminum alkoxides which yield primarily $PhCEt_2OH$ and ethanol after hydrolysis.[14] Since we were concerned with the aluminum alkyl compounds generated by the alkylation reaction during the polymerization, we examined this system by ^{1}H NMR.

Figure 1 shows the reaction mixture at an $AlEt_3$/ester ratio of 2/1 with 46% ethyl benzoate remaining. The three quartets labeled A, B and C are particularly informative. They have been assigned as A = OCH_2 group of the 1:1 $AlEt_3$-ethyl benzoate complex (see Figure 2), B = OCH_2 group of $Et_2AlOCH_2CH_3$ and C = CCH_2 group of the hemialkoxide, $Et_5Al_2OC(CH_2CH_3)_2Ph$. The latter assignment is supported by comparison with the NMR obtained from a sample prepared by reaction of a 2:1 mole ratio of $AlEt_3$ with propiophenone. (Figure 3).

A mechanism which accounts for the formation of these products is shown in Scheme I. The initially formed 1:1 $AlEt_3$-ethyl benzoate complex is alkylated by a second mole of $AlEt_3$ to give II via a 2:1 intermediate complex or transition state. Loss of Et_2AlOEt from II gives the 1:1 $AlEt_3$-propiophenone complex which rapidly reacts with a third mole of $AlEt_3$ to give $Et_5Al_2OCEt_2Ph$. Although II and III have not been observed, traces of propiophenone in the analysis of this reaction after hydrolysis support the intermediacy of III. The mechanism subsequent to the formation of $AlEt_3$:propiophenone complex is the same as that reported for $AlMe_3$ alkylation of benzophenone.[15] The overall reaction results in consumption of two moles of $AlEt_3$ per mole of ester with a third mole of $AlEt_3$ complexed as the hemialkoxide.

$$PhCO_2Et + AlEt_3 \longrightarrow Ph\overset{\overset{O:AlEt_3}{\|}}{C}OEt \quad I$$

$$I + AlEt_3 \longrightarrow \begin{bmatrix} Et_3Al:\overset{|}{O}\text{-}AlEt_2 \\ Ph\overset{|}{C}OEt \\ \overset{|}{Et} \end{bmatrix} \quad II$$

$$II \longrightarrow Et_2AlOEt + \begin{bmatrix} \overset{O:AlEt_3}{\|} \\ PhCEt \end{bmatrix} \quad III$$

$$III + AlEt_3 \longrightarrow \begin{matrix} Et_3Al:OAlEt_2 \\ Ph\overset{|}{C}Et \\ \overset{|}{Et} \end{matrix}$$

Scheme I

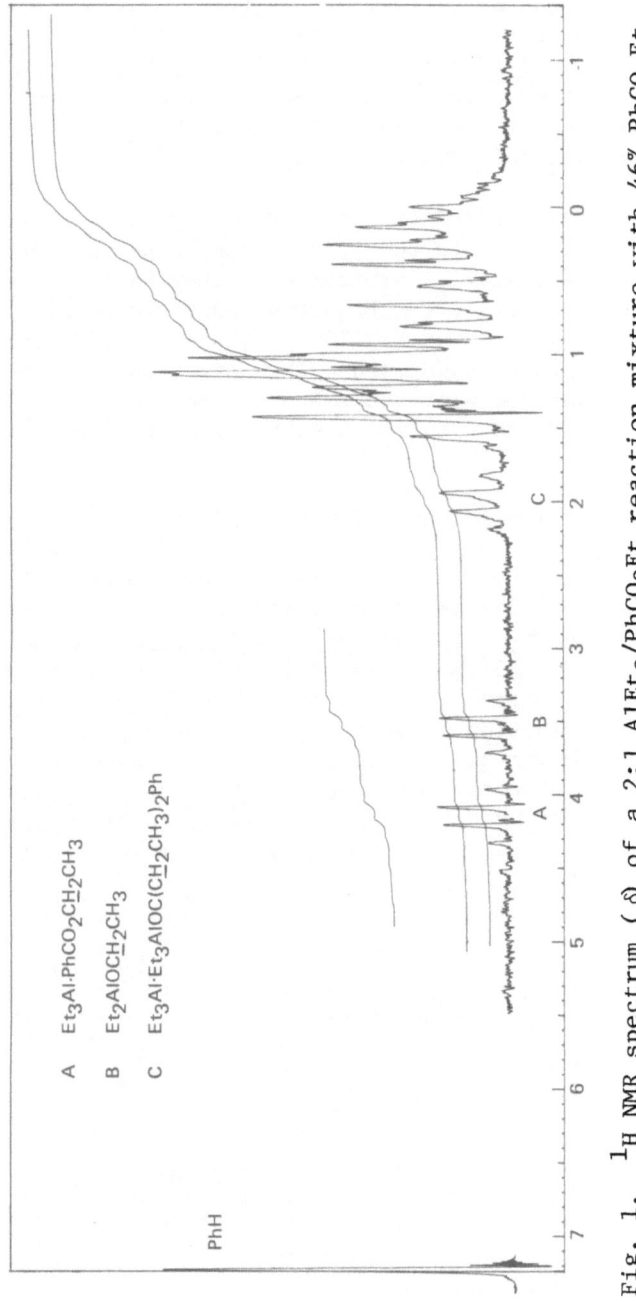

A   Et$_3$Al·PhCO$_2$C$\underline{H}_2$CH$_3$

B   Et$_2$AlOC$\underline{H}_2$CH$_3$

C   Et$_3$Al·Et$_3$AlOC(C$\underline{H}_2$CH$_3$)$_2$Ph

PhH

Fig. 1. ^{1}H NMR spectrum (δ) of a 2:1 AlEt$_3$/PhCO$_2$Et reaction mixture with 46% PhCO$_2$Et remaining.

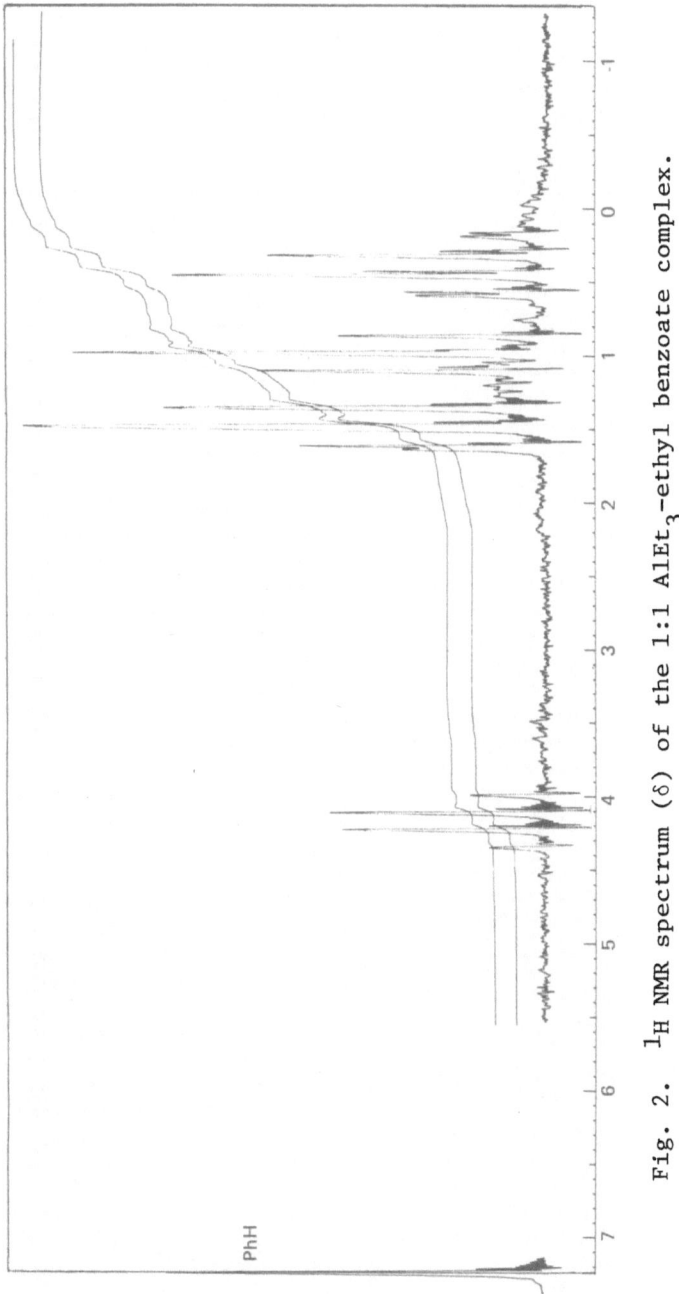

Fig. 2. ^{1}H NMR spectrum (δ) of the 1:1 AlEt$_3$-ethyl benzoate complex.

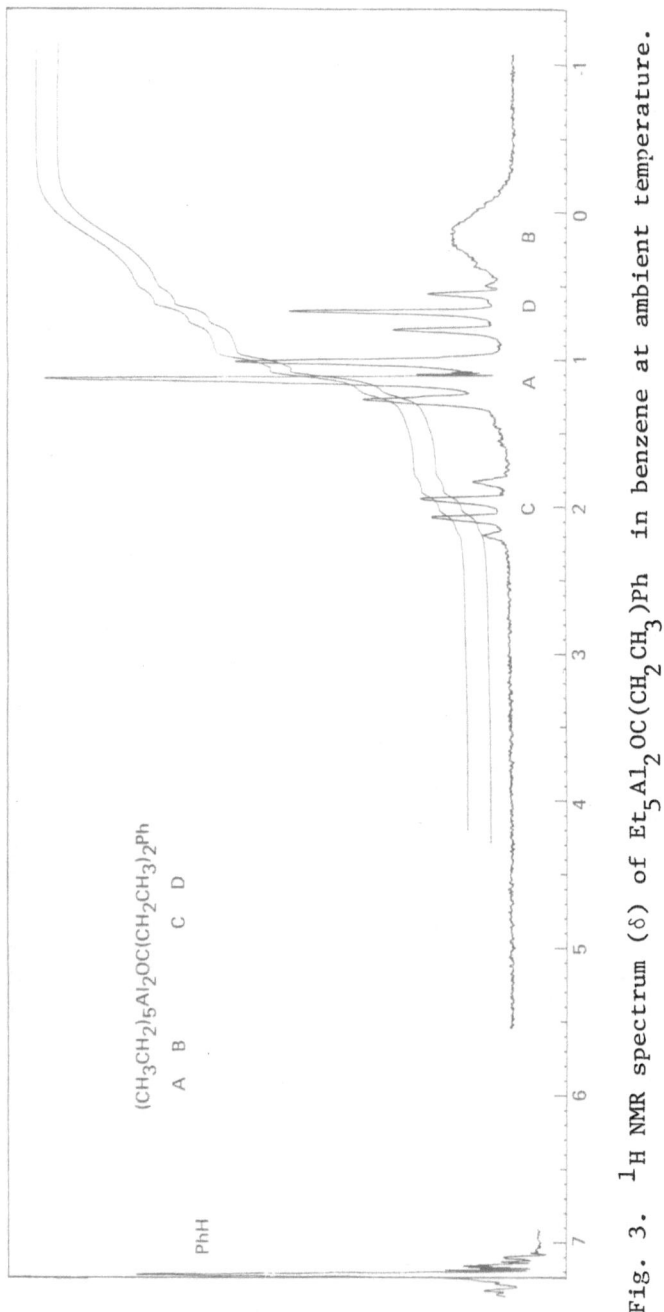

Fig. 3. ^{1}H NMR spectrum (δ) of $Et_5Al_2OC(CH_2CH_3)Ph$ in benzene at ambient temperature.

Variable temperature NMR spectra were used to understand the properties of the aluminum alkoxide products, their interaction with $AlEt_3$, and their influence on the catalyst performance.   Both Et_2AlOEt and Et_2AlOn-Bu have been reported to be dimeric.[16] Consistent with the dimeric structure is the unchanged NMR spectrum of Et_2AlOn-Bu over the temperature range $+30°C$ to $-40°C$.   The variable temperature NMR of equimolar mixtures of Et_2AlOn-Bu and $AlEt_3$ (Figure 4) again shows no change in the $AlCH_2$ group attributed to Et_2AlOn-Bu.   On the other hand the $AlCH_2$ quartet of $AlEt_3$ broadens with decreasing temperature reaching a coalescence point at ca. $-40°C$. A bridge-terminal exchange of Et groups in the $AlEt_3$ dimer accounts for this coalescence[17].

This observation demonstrates that there is not a significant amount of mixed dimer (hemialkoxide) formed via the equilibrium reaction of a 1:1 mole ratio of Et_2AlOn-Bu and $AlEt_3$ dimers.   This has also been recognized for Me_2AlOEt and $AlMe_3$ mixtures.[18]

Examination of $Et_5Al_2OCEt_2Ph$ by variable temperature NMR reveals the absence of any free $AlEt_3$ dimer (Figure 5).   The changes in the spectrum as the temperature is decreased are from bridge terminal exchange of ethyl groups within the hemialkoxide.

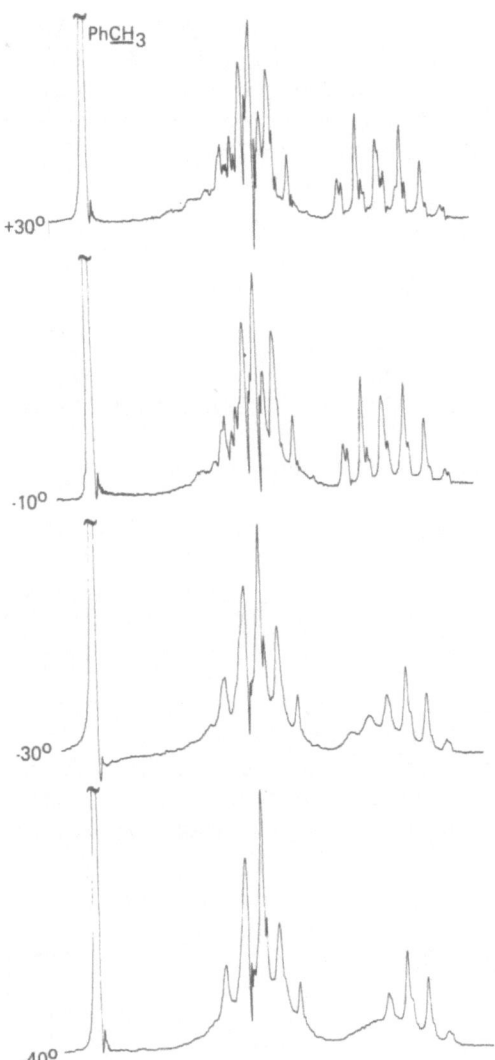

Fig. 4. Variable temperature ^{1}H NMR spectra of an equimolar
mixture of Et_2AlOn-Bu and $AlEt_3$ in toluene.

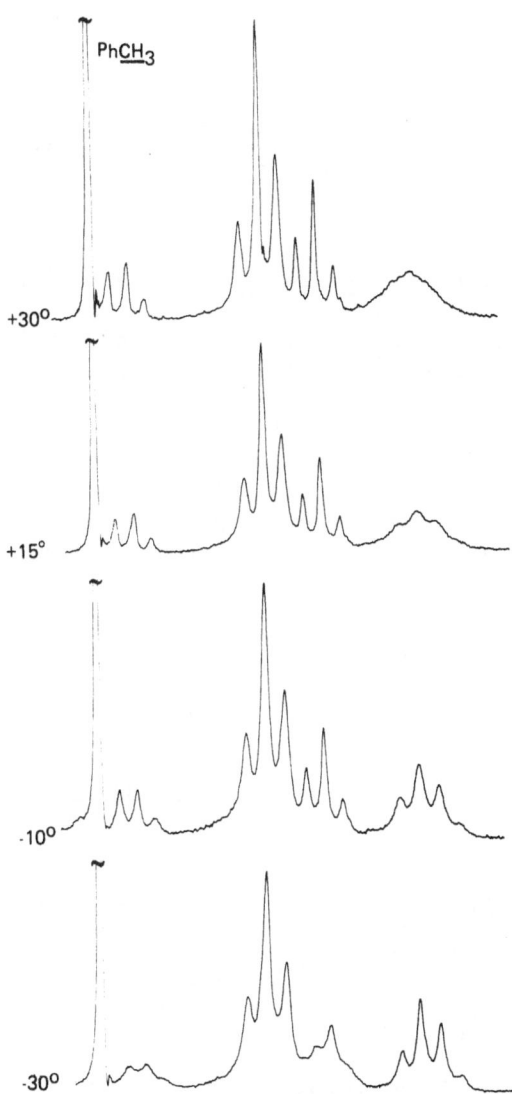

Fig. 5. Variable temperature ^{1}H NMR spectra of $Et_5Al_2OC(CH_2CH_3)_2Ph$ in toluene.

The +30°C coalescence temperature is 25°C higher than that found for $Me_5Al_2OCMe_2Ph$.[18] Presumably the steric bulk of the aluminum alkoxide group destabilizes the aluminum alkoxide homodimer and facilitates the formation of the mixed dimer (hemialkoxide) with $AlEt_3$.

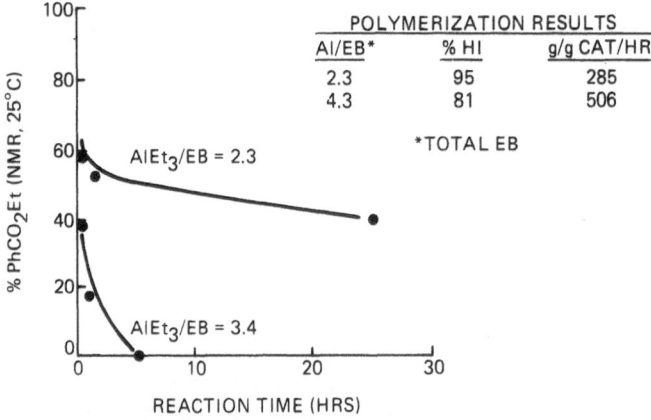

$$(Et_2AlOCEt_2Ph)_2 + (AlEt_3)_2 \rightleftharpoons 2\ Et_2Al \underset{Et}{\overset{\overset{\displaystyle Et}{\overset{\displaystyle PhCEt}{\underset{\displaystyle |}{\underset{\displaystyle O}{}}}}{\diagup\diagdown}} AlEt_2$$

    B. Alkylation reaction rates. Figure 6 illustrates the marked effect that the $AlEt_3$/ester ratio has on the consumption rate of ethyl benzoate. Using ^{1}H NMR and ethyl benzoate concentrations of approximately 0.75$\underline{M}$, one can easily monitor the ethyl benzoate loss by integrating the OCH_2 groups of the ester and $Et_2AlOCH_2CH_3$ product. At a 1:1 stoichiometry, no alkylation of ethyl benzoate occurs. However, at Al/ester mole ratios from > 1 to < 3 reaction is rapid initially and then slows dramatically. At $AlEt_3$/ester > 3 the reaction proceeds until all of the ester is consumed. This behavior is consistent with our mechanistic scheme

AI/EB*	% HI	g/g CAT/HR
2.3	95	285
4.3	81	506

POLYMERIZATION RESULTS

*TOTAL EB

Fig. 6. Effect of $AlEt_3$/ethyl benzoate (EB) ratio on consumption of ethyl benzoate as a function of time (^{1}H NMR). Also shown are the polymerization results carried out at two $AlEt_3$/EB ratios.

which necessitates three moles of $AlEt_3$ to consume one mole of
ethyl benzoate, and is analogous to the $AlMe_3$/benzophenone system
studied by Ashby.[15] They found that the alkylation rate increased
dramatically when the $AlMe_3$/ketone ratio was increased from 1:1 to
less than 2:1, but after 50% of the ketone is consumed the reaction
rate slows. The reaction goes to completion when $AlMe_3$/ketone$\geq$2.
The similar rate observed for both systems is due to a reduction
in free AlR_3 not only from the alkylation reaction, but also from
mixed dimer formation with the aluminum alkoxide products.

The reaction rates of several hindered AlR_3 compounds[19] are
substantially slower than $AlEt_3$ (Figure 7). The formation of the
1:1 complex is indicated by its characteristic color. We believe
the rate decrease is due to increased steric hindrance in the
formation of a 2:1 Al/ester transition state (or complex).
Similarly complete suppression of the ethyl benzoate alkylation can
be achieved by substituting the aluminum with bulky $-NPh_2$ or

−N⟨ ⟩ groups.

C. Polymerization rates. Figure 6 shows the polymerization
results at $AlEt_3$/ethyl benzoate ratios of 2.3 and 4.3 (ethyl
benzoate concentrations of 8.7×10^{-4} $\underline{M}$, respectively) based on the
sum of ethyl benzoate in the catalyst and cocatalyst components.
As the ratio is increased, which accelerates the ethyl benzoate

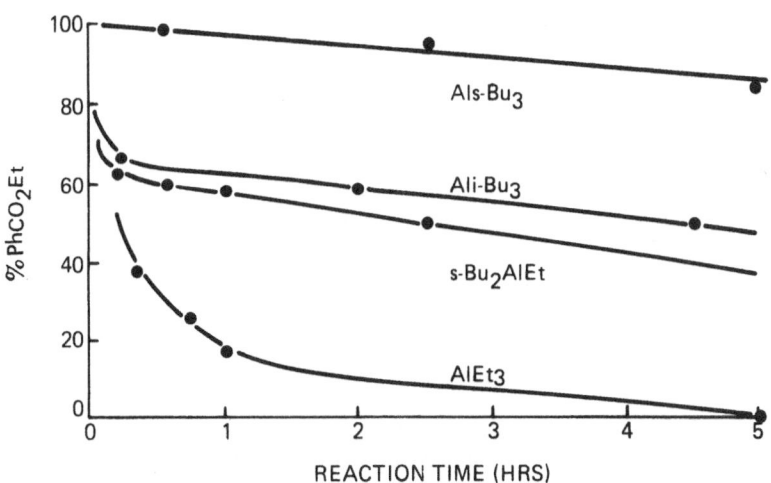

Fig. 7.   Ethyl benzoate consumption as a function of time
for several hindered AlR_3 compounds (^{1}H NMR).

consumption, the % HI decreases from 95% to 82% while the
polymerization rate increases from 285 g/g/hr to 506 g/g/hr. These
results suggest a correlation between alkylation rate, HI, and
polymerization rate. However, the correlation does not distinguish
whether ethyl benzoate or the reaction products are responsible
for changes in the polymerization performance.

Figure 8 shows the effect of adding $Et_2AlOCEt_2Ph$ and
$Et_2AlOn-Bu$ (similar to the Et_2AlOEt alkylation product) to several
polymerization runs described in Table 1. Both increase HI, but
their influence is significantly smaller than ethyl benzoate when
compared on an equimolar basis. This demonstrates that ethyl
benzoate is largely responsible for increasing the HI and decreasing
the polymerization rate. In accordance with the $AlEt_3$-aluminum
alkoxide equilibrium established by low temperature NMR, the
aluminum alkoxides increase HI indirectly by complexing free $AlEt_3$
thus slowing the $AlEt_3$ consumption of ethyl benzoate. The
hindered $Et_2AlOCEt_2Ph$ is more effective than the unhindered
$Et_2AlOn-Bu$. The decline in free $AlEt_3$ concentration is also
manifest in the decreased polymerization rate (Table 1).

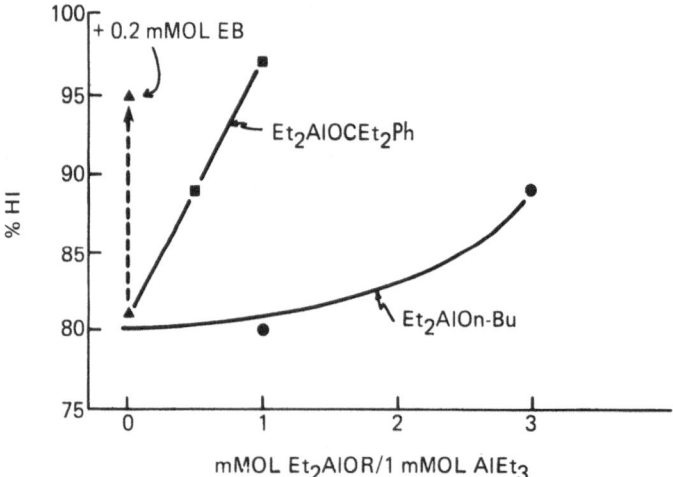

Fig. 8.   Effect of ethyl benzoate (EB), $Et_2AlOn-Bu$, or
           $Et_2AlOCEt_2Ph$ on polypropylene heptane insolubility
           (% HI, see Table 1).

Table 1. Effect of Et_2AlOR on Catalyst Performance[a]

$AlEt_3$ (mmol)	$Et_2AlOn-Bu$ (mmol)	$Et_2AlOCEt_2Ph$ (mmol)	Ethyl Benzoate (mmol)	Rate g/g/ cat/hr	HI (%)	MW $(x10^{-3})$
1.0	--	--	--	506	81	238
1.0	1.0	--	--	488	80	210
1.0	3.0	--	--	454	89	270
1.0	--	0.5	--	446	89	292
1.0	--	1.0	--	66	97	508
1.0	--	--	0.2	285	95	421
1.0	1.0	--	0.2	173	96	411
1.0	--	1.0	0.2	11	98	588

[a]0.2g catalyst Al.

In one approach to circumvent the ethyl benzoate alkylation reaction, several hindered esters were tested in the polymerization system. The $AlEt_3$ alkylation rates of hindered esters[20] and hindered ketones[21] are known to be slower than the unhindered parent compounds. In fact we have found that at NMR concentration the alkylation rate of ⬡-$CO_2CH_2CH_3$ is very slow compared to that

of ethyl benzoate (only 22% loss of ester in 22 hours). However, contrary to expectations, the hindered esters are ineffective in improving the HI (Table 2). To rationalize this result an ester-catalyst interaction must be involved for HI improvement to be achieved. As shown in the ester series $PhCO_2Et$, $PhCO_2n-Bu$, $PhCO_2i-Bu$ and $PhCO_2t-Bu$, this interaction decreases as the ester becomes more hindered.

As an alternate approach to decreasing ethyl benzoate alkylation, several hindered trialkylaluminum compounds have been tested as cocatalysts. As shown in Table 3, a clear trend is evident. The increasingly hindered AlR_3 compounds improve the HI without substantial loss in polymerization activity. In light of the relative reaction rates shown in Figure 7, the HI improvement can be attributed to decreased ethyl benzoate alkylation. This approach has been extended to certain hindered diethyl aluminum amides which do not alkylate ethyl benzoate. The unhindered amides Et_2AlNEt_2 and $Et_2AlN(cyclohexyl)_2$ do not activate the catalyst presumably because of their stable dimeric structure. However,

because the dimeric structures of Et_2AlNPh_2 and Et_2AlN⬡ are

Table 2.   Hindered Esters as Third Components

Catalyst[a]	AlEt$_3$ (mmol)	Ester (mmol)	Rate (g/g cat/hr)	HI (%)	MW ($\times 10^{-3}$)
B1	0.4	0.12 PhCO$_2$Et	310	95	378
B1	0.4	0.12 PhCO$_2$n–Bu	336	91	319
B1	0.4	0.12 PhCO$_2$i–Bu	406	87	302
B1	0.4	0.12 PhCO$_2$t–Bu	401	73	233
A1	1.0	0.2 PhCO$_2$Et	285	95	421
A1	1.0	0.2 —PhCO$_2$Et	378	89	303

[a]0.08g catalyst B1; 0.2g catalyst A1 (see page 243).

Table 3.   Effect of Hindered AlR$_3$'s on Catalyst Performance

AlR$_3$[a]	Rate (g/g cat/hr)	HI (%)	MW ($\times 10^{-3}$)
AlEt$_3$	506	82	238
Ali–Bu$_3$	520	88	315
s–Bu$_2$AlEt	501	89	307
t–Bu$_2$AlEt	489	90	313
s–Bu$_3$Al[b]	461	92	357

[a]1.0 mmole AlR$_3$; 0.2g catalyst A1.
[b]0.2g catalyst A2.

sufficiently destabilized by electronic weakening and steric
hindrance, respectively, the monomeric aluminum amide is available
for catalyst activation. Table 4 summarizes the polymerization
results for these aluminum amide systems. The alkyl aluminum
diaryl amides have improved HI/rate blance when compared with other
cocatalyst systems we have studied.

   D. Role of ethyl benzoate in controlling stereospecificity.
From the preceding studies it is clear that preservation of the
ethyl benzoate is the key to maintaining catalyst stereospecificity.
To understand how the ethyl benzoate operates to improve the polymer
HI, experiments using Et_2AlNPh_2 cocatalyst were carried out at low
ratios of ethyl benzoate to titanium. The polymerization rate
was then divided into two components, the rate at 100% HI and the
rate at 0% Hi in order to observe the base effects on the isospecific
and nonspecific sites separately. The results are shown in Table 5
and Figure 9. Clearly as the amount of ethyl benzoate is increased,
both rates decline. However, on a percentage basis the rate at 0%
HI declines more rapidly than the rate at 100% HI. In fact at
ester/Ti = 0.6 the rate at 0% HI has dropped 71% while the rate
at 100% HI has declined only 33%. These results can be explained
by the ethyl benzoate poisoning all the polymerization sites, but
being somewhat more selective toward the nonspecific sites.

Table 4. Effect of Alkyl Aluminum Amide Cocatalysts.

Cocatalyst (mmol)	Rate (g/g cat/hr)	HI (%)	MW (x 10^{-3})
2.0 $AlEt_3$[a]	419	66	151
2.0 Et_2AlNPh_2[a]	308	95	398
2.0 $Et_2AlN(o-tolyl)_2$[a]	324	95	351
1.0 Et_2AlNEt_2[b]	0	--	--
2.0 $Et_2AlN(cyclohexyl)_2$[b]	<5	--	--
2.0 Et_2AlN [a]	70	99	585

[a]0.2g catalyst A2.
[b]0.2g catlayst Al.

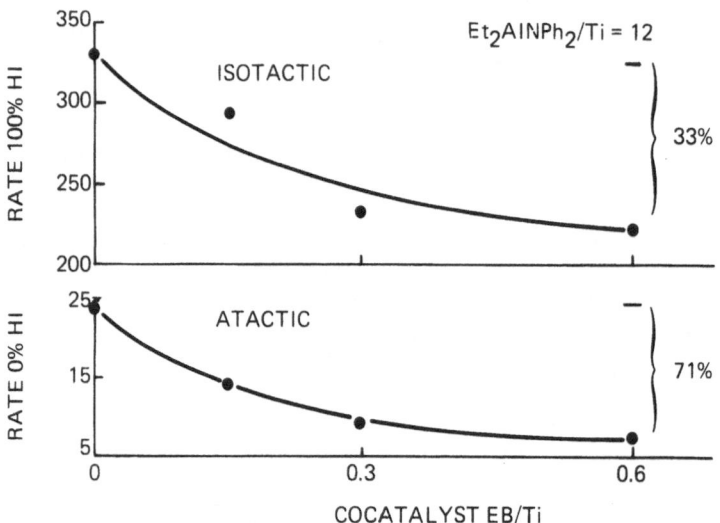

Fig. 9.   Effect of ethyl benzoate (EB) addition on the 100%
          HI and 0% HI polymerization rates (g polymer/g cat/hr.)

Table 5.   Comparison of Ethyl Benzoate and 2,2,6,6-Tetramethyl-
           piperidine Selectivities[a]

PhCO$_2$Et[b]/Ti	Rate (g/g cat/hr)	HI (%)	MW (x10^{-3})	Rate @ 100% HI (g/g cat hr)	Rate @ 0% HI (g/g cat/hr)
0.0	354	93.3	368	330	24
0.15	306	95.5	372	292	14
0.30	241	96.1	414	232	9
0.60	228	96.8	449	221	7

HN⬡[b]/Ti

0.08	308	96.7	436	298	10
0.30	316	98.1	497	310	6
0.60	295	97.9	521	289	6

[a]1.6 mmol Et$_2$AlNPh$_2$; 0.16g catalyst A3.
[b]Added as cocatalyst component.

Bases have been widely used with all types of Ziegler-type catalyst systems to improve steric control. However, they are also very useful as probes to assist in the continuing quest for a fuller understanding of the catalyst mechanism.[22] In this regard we have studied a wide range of bases[23] and have found that 2,2,6,6-tetramethylpiperidine is effective in poisoning only nonspecific sites. Figure 10 illustrates the remarkable selectivity of this hindered amine. Addition of small amounts (0.08 amine/Ti) causes a much sharper drop in the 0% HI rate than does the same amount of ethyl benzoate. The best result is obtained at 0.30 amine/Ti ratio. The rate of 0% HI has dropped by 75% while the 100% HI rate declined only 6%. Greater amounts of amine only result in further loss in the 100% HI rate.

The implication of this selectivity with regard to site structure and further catalyst improvements are still under active investigation.

CONCLUSIONS

The ethyl benzoate Lewis base present in $MgCl_2$ supported $TiCl_4$ catalyst system has been shown to be largely responsible for high stereospecificity in propylene polymerization. Low stereospecificity is obtained when it is present in insufficient amounts

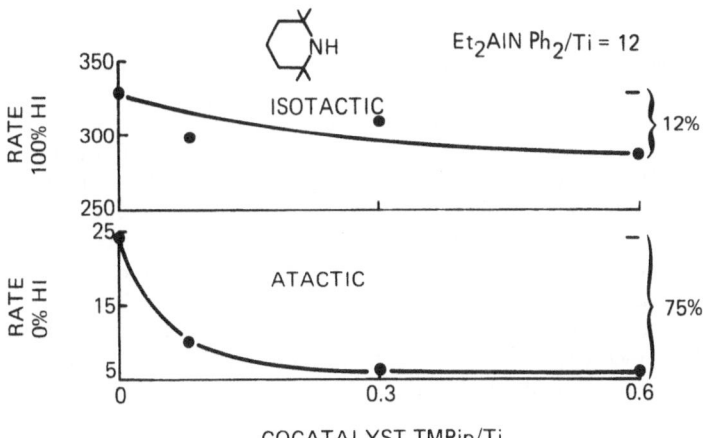

Fig. 10.   Effect of 2,2,6,6-tetramethylpiperidine (TMPip)
           addition on 100% HI and 0% HI polymerization rates
           (g polymer/g cat/hr).

relative to titanium and when it is depleted by an alkylation
reaction involving the aluminum trialkyl cocatalyst. By using ^{1}H
NMR studies the rate of alkylation has been shown to be dependent
upon the aluminum alkyl/ester ratio and the reactivity of the
aluminum alkyl compound. The most rapid alkylation pathway occurs
by attack of an aluminum trialkyl on a 1:1 complex of aluminum
trialkyl and ethyl benzoate. This pathway can be minimized by
two bulky secondary and tertiary alkyl groups which sterically
hinder formation of the 2:1 transition state complex. An
electronically weakened or sterically hindered amine group
attached to aluminum is even more effective and renders the
aluminum alkyl virtually non-alkylating towards ethyl benzoate.
Polymerization studies have demonstrated that these aluminum
compounds still retain their ability to alkylate and activate
the titanium catalyst. Such cocatalyst systems greatly enhance
the stereospecificity of the supported catalysts.

The primary way in which ethyl benzoate increases selectivity
to isotactic polymer is by inactivating the nonstereospecific
polymerization sites more extensively than the stereospecific sites.
As a consequence, increasing the ratio of ethyl benzoate to titanium
decreases total catalyst activity while increasing the percentage
of isotactic polymer. Other Lewis bases can function in the same
manner, but due to different electronic and steric factors they would
be expected to exhibit different selectivities than ethyl benzoate
for the nonstereospecific sites. 2,2,6,6-tetramethylpiperidine
was shown to be highly selective in complexing nonstereospecific
sites with minimal loss of activity from the stereospecific sites.

EXPERIMENTAL SECTION

General Data. All manipulations involving air and moisture
sensitive compounds were carried out in N_2 swept glove boxes.
n-Heptane was dried to < 0.5 ppm H_2O by N_2 sparging. Benzene
and toluene were dried with CaH_2. Infrared spectra were obtained
in the indicated solvent on Beckman Instruments IR8 or Perkin Elmer
Corporation 567 spectrophotometers and are reported in cm^{-1}. ^{1}H
NMR spectra were recorded on Varian Associates A-60 or EM-360A
instruments. Unless otherwise noted, chemical shifts are reported
in δ with solvent references corrected to TMS = 0 by using the
following: C_6H_6 = 7.24, C_6D_6 = 7.15, dioxane = 3.56. Splitting
patterns are designated as s(singlet), d(doublet), t(triplet),
q(quartet), m(multiplet) or br(broadened peak). All coupling
constants are reported in Hertz. Melting and boiling points are
uncorrected.

All commercial aluminum alkyls were obtained from Texas
Alkyls as neat liquids and used as received. Et_2AlOn-Bu,
Et_2AlNEt_2 and $Et_2AlN(cyclohexyl)_2$ were prepared by literature

procedures.[24] Ethyl benzoate was purchased from Matheson, Coleman
and Bell and dried over CaH_2.

Catalyst Synthesis Anhydrous $MgCl_2$ was obtained by HCl
sparging of molten $MgCl_2$ (Alfa, technical grade #102076) at 90°C
in a silica reactor.

Catalysts Al, A2 and A3 were prepared by placing $MgCl_2$
(0.4 mol), ethyl benzoate (0.08 mol), and $TiCl_4$ (0.08 mol) in a
dry 1 ℓ stainless steel ball mill containing 135 stainless steel
balls of 1.6 cm diameter. The ball mill was rotated at 110 rpm for
4 days. The light yellow contents of the jar were loaded into a
500 ml round bottom flask, $TiCl_4$ (200 ml) was added and heated to
80°C and stirred for 4 hours. The slurry was filtered hot through
a medium frit filter and the solids were washed five times with
50°C heptane. The solids were rotoevaporated at 50°C to constant
weight.

Catalyst B1 was prepared by placing $MgCl_2$ (0.2 mol), ethyl
benzoate (0.04 mol) and $TiCl_4$ (0.04 mol) in a dry Shatter Box
(Spec Industries, 6.5 inch diameter) and ground for 1 hour. The
yellow solids were recovered and charged to a 250 ml flask. $TiCl_4$
(81 ml) was added, the temperature raised to 138°C and held for
four hours with stirring. The catalyst was recovered as in A.

The elemental analysis for these catalyst preparations is
shown in Table 6.

Table 6.   Elemental Analaysis

Catalyst	Mg[a] (%)	Ti[a] (%)	Cl[b] (%)	C (%)	H (%)
Al	17.9	3.16	60.6	12.7	1.28
A2	16.6	3.43	60.4	13.0	1.12
A3	17.8	3.94	61.0	13.0	1.23
B1	21.2	2.26	66.7	5.9	0.56

[a]Atomic absorption.
[b]By $AgNO_3$ titration.

Polymerization procedure. Atmospheric pressure polymerizations were carried out in a 1 liter baffled glass reactor using
pure n-heptane diluent dried to <0.5 ppm water. Polymerization
grade propylene was further purified at 950 kPa by passing it
through a bed of reduced copper oxide catalyst at 220°C to remove
oxygen and then through 3A molecular sieves at 25°C to remove water.
Typically, the propylene feed to the reactor analyzed less than 1
ppm water and oxygen. All operations were carried out in a dry
nitrogen atmosphere.

A solution of the alkyl metal cocatalyst in 480 ml n-heptane
was charged to a dropping funnel in a glove box, attached to the
reactor with nitrogen flowing and then charged to the reactor. The
catalyst was weighed into a tube fitted with a stopcock and a septum
cap which was also attached to the reactor with nitrogen flowing.
The reactor temperature was then increased to 65°C while saturating
the solution with propylene through a dip tube. A syringe containing 20 ml n-heptane was used to rinse the catalyst into the reactor
while excess propylene was passing through the reactor at 1.01 atm.
Temperature was controlled at 65°C while propylene was fed at a rate
to maintain an exit rate of 200-800 cc/min.

After one hour, propylene was stopped, and the reactor contents
were poured into 1 liter isopropanol, stirred at least one hour,
filtered, and the polymer was vacuum dried overnight at 70°C. The
filtrate was stripped and vacuum dried at 50°C to recover soluble
product from which the weights of catalyst, cocatalyst and nonvola-
tile additives were subtracted to obtain the yield of soluble polymer. A standard boiling heptane extraction was carried out on the
solid polymer, and the result was combined with the yield of alco-
hol-soluble polymer to obtain the percent heptane insolubles (%HI)
on total product. Molecular weight was determined from the solution viscosity in decahydronaphthalene at 130°C using the Kinsinger
equation (J. Phys. Chem. 63, 2002, (1959).

$Et_3Al.Et_2AlOCEt_2Ph$. Triethylaluminum (2.283 g, 20 mmol) was
dissolved in benzene (20 ml) and cooled to 5°C. Propiophenone
(1.342 g, 10 mmol) dissolved in benzene (5 ml) was slowly added to
give an orange, homogeneous solution. This was refluxed ten minutes and evaporated to constant weight. A colorless liquid weigh-
ing 3.6 g (99%) was obtained. NMR (C_6D_6): 7.38 (m, 5H), 2.1 (q,
4H) 1.2 (t, 15H), 0.75 (t, 6H), 0.28 (br, 10H).

$(t-Bu)_2AlEt$. Neat $EtAlCl_2$ (25.39 g, 200 mmol) was dissolved
in n-heptane (160 ml) and cooled to -15°C. After addition of t-
BuLi (222 ml, 1.8 M, 400 mmol), the solution was allowed to come
to room temperature overnight. The solution was heated to 80°C for
one hour, filtered hot and washed with heptane (50 ml). The filtrate was then stripped (4 hours RT, 2 hours 50°C, 1 hr 80°C). A

yellow liquid (25.37 g) was obtained. A portion of this liquid
(~ 16 g) was distilled in a 50 ml micro distillation apparatus.
The distillate (10.7 g) contained < 10 ppm Cl. NMR (55°C, C_6D_6):
1.11 (t), 0.97 (s), 0.25 (q)

s-Bu_2AlEt. A procedure analogous to the preparation of
t-Bu_2AlEt was used. NMR (dioxane): 1.8-0.5 (m, 19H), 0.2 to
-0.6 (m,4H).

s-Bu_3Al. To a 12 wt % solution of s-BuLi (82 g, 0.15 mol) was
added $MeAlCl_2$ (5.65 g, 0.05 mol) dissolved in heptane (95 ml). A
white precipitate slowly formed. After stirring at 35°C overnight
the reaction mixture was heated to 90°C for 8 hours, stirred
overnight at 35°C and filtered. Rotoevaporation gave a viscous
yellow oil (7.15 g). NMR (dioxane, PhH ref) was consistent with
s-Bu_3AlMe$^-$ Li$^+$: 1.5-0.6 (m, 24H), -0.26 (br, 3H), -1.44 (br, s
3H). Vacuum distillation gave 2.3 g (23% of a clear, water white
liquid, boiling at 45-60°C (< 0.1 mm Hg), NMR (dioxane): 2.9 -
1.85 (m, 24H), 1.18 (m, 3H).

AlR_3 Alkylation of ethyl benzoate. The following illustrates
a typical procedure. Ethyl benzoate (0.451 g, 3 mmol) was dissolved in benzene (2 ml) and cooled to 5°C. AlR_3 (10.2 mmol) was
slowly added to give an orange solution. Benzene (0.5 ml) was used
to rinse the AlR_3 vial. The solution was quickly transferred
to an NMR tube and kept at -78°C until the ^{1}H NMR was run at
ambient temperature. Repeated integrations of the CH_2 quartet
of the ethyl benzoate and the CH_2 quartet of the $R_2AlOCH_2CH_3$
produced was used to determine the quantity of EB remaining.

Ethyl 2,6-dimethylbenzoate.[25] 2,6-Dimethylbenzoic acid (4.8
g, 32 mmol) was dissolved in 100% H_2SO_4 (30 ml), pured into cold,
absolute EtOH (200 ml) and stirred for 75 minutes. Water (350 ml)
and Et_2O were added, the two phases separated and the aqueous
layer was again extracted with Et_2O. The two ether layers were
combined, washed with H_2O (5x) and saturated K_2CO_3 solution (4x),
and then dried over K_2CO_3. Filtration and evaporation gave 1.6 g
of crude ester. Vacuum distillation of 80°C (0.7 mm Hg) gave
the pure ester. IR (PhH): 1718, 1261, 1246, 1104, 1070, 1016,
765. NMR (CDCl , TMS) 7.2-6.9 (m, 3H), 4.40 (q, J = 7, 2H),
2.32 (s, 6H), 1.40 (t, J = 7, 2H).

$AlEt_3$ alkylation of ethyl 2,6-dimethylbenzoate. The ester
(0.267 g, 1.5 mmol) was dissolved in 1 ml C_6H_6 and cooled to
5°C. The $AlEt_3$ (0.856 g, 7.5 mmol) was slowly added. One quarter
ml C_6H_6 was used to rinse the $AlEt_3$ vial. The reaction mixture
was transferred to an NMR tube and cooled to -78°C until the ^{1}H
NMR was run at ambient (30°C) temperature. IR analysis indicated
1:1 complex formation (ν_{CO} = 1642, $\Delta\nu$ = 76 cm^{-1}). After 22 hours
NMR integration indicated 78% unreacted ester.

Di-2-Tolylamine. To a 250 ml two neck flask equipped with N_2
inlet overhead stirrer, Dean Stark Trap and condenser was added
o-toluidine (25 g, 0.24 mol), o-iodotoluene (43.6 g, 0.20 mol),
K_2CO_3 (27.6 g, 0.20 mol, powdered), copper powder (12.6 g, 0.20
mol) and o-dichlorobenzene (30 g). This mixture was heated to reflux for 3.5 hours. Evidence of H_2O indicated that the reaction
had taken place. Approximately 100 ml xylene was added to facilitate
removal of the H_2O. Reflux was continued for 2.5 hours and the
xylene and water were collected and removed. The reaction mixture
was filtered hot and washed with hot o-dichlorobenzene (35 ml).

The crude reaction mixture was vacuum distilled to give 10.82 g
(25%) of an orange liquid boiling at 115°-119°C (0.02 mm) which
solidified on cooling. Recrystallization from pentane gave 5.7 g
(14%) of light yellow crystals, mp 51-56°C (lit.[26] 48-50°C).
IR (nujol): 3435, 2950, 2860, 1470, 1300, 747, 712. NMR (C_6D_6,
TMS); 6.9 (bs, 8H), 4.75 (br s, 1H), 2.0 (s, 6H).

$Et_2AlN(o-tolyl)_2$. n-Butyl lithium (12.0 ml, 1.6 $\underline{M}$, 19.2 mmol)
was placed in a 4 oz. bottle and cooled to 0-5°C. To this was
slowly added di-2-tolylamine (3.79 g, 19.2 mmol) dissolved in heptane (24 ml). A light yellow precipitate formed with gas evolution. The reaction was warmed to ambient temperature and stirred
one hour. To this was added Et_2AlCl (2.315 g, 19.2 mmol) dissolved
in heptane (10 ml). After stirring 64 hours, the white slurry was
vacuum filtered, and the solids washed with heptane (5 ml, 4x).
The filtrate was vacuum stripped to give 4.8 g (89%) of a white
solid: mp 59-66°C; NMR (C_6D_6, 65°C); 6.9 (s), 2.0 (s, 6H) 1.0
(t,6H) 0.25 (q, 4H).

Et_2AlN⬡ . To a hexane solution of n-BuLi (50 ml, 1.6 $\underline{M}$,

0.08 mol, Aldrich Chemical Co.) at 0°C was slowly added 2,2,6,6-
tetramethylpiperidine (11.30 g, 0.08 mol). This mixture was slowly
warmed to 12°C at which time gas evolution occurred. Further warm-
ing to ambient temperature gave a milky suspension. After one hour
Et_2AlCl (9.65 g, 0.08 mol) dissolved in heptane (10 ml) was added
and stirred 96 hours at ambient temperature. The white suspension
was vacuum filtered and washed with heptane (80 ml). The filtrate
was rotoevaporated to remove solvent. The residue was vacuum distilled to give 12.2 g (68%) of a water-white liquid. bp 55-58°C
(< 0.1 mm Hg); NMR (C_6H_6): 1.9-1.0 (m, 22H), 0.24 (q, 4H).

Et_2AlN⬡ Alkylation of ethyl benzoate. Ethyl benzoate

(0.075 g, 0.5 nmol) was dissolved in C_6H_6 (1 ml) and cooled to 5°C.
The aluminum amide (0.564 g, 2.5 mmol) was slowly added with 0.25
ml C_6H_6. The resulting orange solution was placed in an NMR tube
and cooled to -78°C until the spectrum was run. No alkylation was

detected after 24 hours.   IR analysis of the initial reaction
mixture indicated 1:1 complex formation (ν_{CO} = 1645 cm^{-1}).

Et_2AlNPh_2. Diphenylamine (16.92 g. 0.1 mol) dissolved in
toluene (30 ml) was slowly added to $AlEt_3$ (11.42 g, 0.1 mol
dissolved in 20 ml toluene. Gas evolution occurred during the
addition. The reaction mixture was heated to 90°C for 1 hr.
Evaporation of the solvent gave 24.8 g (98%) white solid. NMR
(dioxane):   6.9 (m, 10H), 0.9 (t, J = 8, 6H), =0.1 (q, J = 8, 4H).

Et_2AlNPh_2 alkylation of ethyl benzoate. The aluminum amide
(0.633 g, 2.5 mmol) was dissolved in C_6H_6 (1.5 ml) by heating to
$\sim$60°C. Immediately upon cooling this was added to ethyl benzoate
(0.075 g, 0.5 mmol) giving an orange solution which was transferred
to an NMR tube. After 24 hours, no change in the spectrum could
be detected.

ACKNOWLEDGMENTS

The authors are indebted to Karl Geissel, Walter Funk, Douglas
Barist and John Mangus for their skillful experimental assistance.

REFERENCES

1. P. Pino, R. Mulhaupt, Angew, Chem. Int. Ed. Engl. 19, 857 (1980).
2. K. Ziegler, Belg. Pat. 533,362. K. Ziegler, E. Holzkamp,
        H. Breil, H. Martin, Angew, Chem. 67, 541 (1955).
3. G. Natta, J. Polymer Sci. 16, 143 (1955).
4. E. Tornqvist, C. W. Seelbach, A. W. Langer, Jr., (Exxon),
        U.S. Pat. 3,128,252 (1964). E. Tornqvist, A. W. Langer, Jr.,
        (Exxon), U.S. Pat. 3,032,510 (1962).
5. Brit. Pat. 895,595 (1962) (Hoechst). K. K. G. Rust, A. G. M.
        Gumboldt, K. F. Horndler, S. Sommer, E. Heitzer (Hoechst,
        assigned to Hercules), U.S. Pat. 3,058,970 (1962).
6. J. P. Hermans, P. Henrioulle (Solvay), DBP 2,213,086 (1972),
        U.S. Pat, 4,210,738 (1980) and 4,210,735 (1980).
7. M. Yokoyama, A. Yamada, S. Okosi, T. Katou, S. Yoshida,
        (Mitsubishi Petrochem.), U.S. Pat, 4,151,111 (1979).
8. H. Ueno, N. Inaba, T. Makishima, K. Watanabe, S. Wada (Exxon),
        U.S. Pat. 4,182,691 (1980).
9. E. Tornqvist, Ann. N.Y. Acad. Sci, 155, 447 (1969).
10. W. A. Hewett, E. C. Shokal (Shell), U.S. Pat, 3,238,146 (1966).
11. A. Mayr, P. Galli, E. Susa, G. DiDrusco, E. Cischetti
        (Montecatini), Brit. Pat. 1,286,867 (1969).
12. N. Kashiwa, T. Tokuzumi, O. Fujimura, H. Fujimura (Mitsui
        Petrochem.), U.S. Pat. 3,642,746 (1972).
13. P. Longi, U. Giannini, A. Cassata (Montecatini), Brit. Pat.
        1,335,887 (1973).
14. T. Mole and E. A. Jeffrey, Organoaluminum Compounds, Elsevier
        Publishing Company (1972), p. 302.

15.   E. C. Ashby, J. Laemmle, and H. M. Neumann, J. Amer. Chem. Soc., $\underline{90}$, 5179 (1968).
16.   E. G. Hoffmann, Ann. Chem. $\underline{629}$, 104 (1960).
17.   J. Smidt, M. P. Groenewege and H. de Vries, Recueil $\underline{81}$, 729 (1962).
18.   E. A Jeffrey and T. Mole, Aust. J. Chem. $\underline{23}$, 715 (1970).
19.   For reactions of $Ali-Bu_3$ and $AlEt_3$ with ethyl benzoate see also D. Adenhaim and J. L. Namy, Tet Letters, 3011 (1972).
20.   Y. Baba, Bull. Chem. Soc., Japan $\underline{41}$, 1022 (1968).
21.   S. Pasynkiewicz and E. Sliwa, J. Organometal. Chem. $\underline{3}$, 121 (1965).
22.   A. W. Langer, Jr., Ann. N.Y. Acad. Sci. $\underline{295}$, 110 (1977).
23.   A. W. Langer, T. J. Burkhardt, J. J. Steger, unpublished results.
24.   T. Mole, Australian J. Chem. $\underline{19}$, 373 (1966); G. E. Coates and J. Graham, J. Chem. Soc., 233 (1963).
25.   M. S. Newman, J. Amer. Chem. Soc. $\underline{63}$, 2431 (1941).
26.   C. L. Frye, G. A. Vincent, G. L. Hauschildt. J. Amer. Chem. Soc. $\underline{88}$, 2727 (1966).

"LIVING" COORDINATION POLYMERIZATION OF PROPYLENE AND

ITS APPLICATION TO BLOCK COPOLYMER SYNTHESIS

Yoshiharu Doe and Satoshi Ueki

Department of Chemical Engineering, Tokyo
Institute of Technology, Ookayama, Meguro-ku,
Tokyo 152, Japan

Tominaga Keii

Department of Chemistry, Science University
of Tokyo, Kagurazaka, Shinjuku-ku
Tokyo 162, Japan

SYNOPSIS

A soluble $V(acac)_3/Al(C_2H_5)_2Cl$ system was found to polymerize propylene to give a syndiotactic "living" polypropylene having narrow molecular weight distribution ($\overline{M_w}/\overline{M_n}$=1.05-1.20) at temperatures below -65°C. The coupling reaction between iodine-terminated living polypropylene and monofunctional living polystyrene anion was examined at 50°C and shown to result in the quantitative formation of monodisperse AB block copolymer of propylene and styrene. In addition, sequential polymerization of ethylene and propylene was attempted in the presence of the soluble $V(acac)_3/Al(C_2H_5)_2Cl$ system to synthesize block copolymers.

INTRODUCTION

An ideal living polymerization takes place when propagation of all chains is started simultaneously and chain transfer and termination are absent[1,2].   These limiting conditions have been achieved in the anionic polymerization of vinyl monomers or dienes with organic derivatives of alkali metal as initiators, which gives "living" polymers with a narrow molecular weight distribution[3-7].   The finding of the unique "living" character of chain ends in the anionic polymerization system led to the successful preparation of well-characterized block copolymer, relatively free from homopolymer impurities[3,8-11].   The most important application of this type of block copolymerization involves synthesis of styrene-butadiene block copolymers of AB and ABA structures with desired chain length.

In the field of coordination polymerization, many patents and publications claim the synthesis of block copolymers of ethylene and propylene using heterogeneous Ziegler-Natta catalysts[12-16].   However, well-characterized block copolymers comparable to styrene-butadiene block copolymers have not been successfully synthesized from olefins with the help of Ziegler type catalysts, since the life time of growing chains was finite.

Recently, we found that a soluble catalyst system of $V(acac)_3$ (acac = acetyl acetonate) with $Al(C_2H_5)_2Cl$ initiates a living polymerization of propylene to afford monodisperse polymers ($\bar{M}w/\bar{M}n$ = 1.05-1.20) at temperatures below -65°C[17,18]. The polymerization of propylene with this soluble catalyst has been proved to be a coordination polymerization[19,20].   The purpose of this study is the synthesis of a new tailored block copolymer of AB structure using this living coordination catalyst. The block copolymer synthesis has been attempted on the basis of two methods : (1) sequential coordination polymerization of ethylene and propylene, and (2) coupling reaction between iodine-terminated living polypropylene and monofunctional living polystyrene anion, as represented by equation (1).

$$\sim\!\!\sim\!\!\sim CH_2\text{-}\overset{-}{C}H\ \overset{+}{Li}\ +\ I\text{-}\overset{\overset{\displaystyle CH_3}{|}}{C}H\text{-}CH_2\!\!\sim\!\!\sim\!\!\sim$$

$$\longrightarrow\ \sim\!\!\sim\!\!\sim CH_2\text{-}CH\text{-}\overset{\overset{\displaystyle CH_3}{|}}{C}H\text{-}CH_2\!\!\sim\!\!\sim\!\!\sim\ +\ LiI\qquad (\ 1\ )$$

Using the latter method, we have succeeded in the preparation of well-defined diblock copolymers of propylene and styrene.

EXPERIMENTAL

Materials

V(acac)$_3$ (Alfa Division Ventron Co.), Al(C$_2$H$_5$)$_2$Cl (Japan Aluminium Alkyl Co.) and CH$_3$(CH$_2$)$_3$Li (Wako pure Chemicals) were used without further purification. Propylene (purity 99.8 % ; the impurity was propane) or ethylene (purity 99.9 % ; the impurity was ethane) was used after passing it through columns of sodium hydroxide and phosphorus pentoxide. Styrene was washed with aqueous alkaline solution, dried with LiAlH$_4$, and distilled under reduced pressure prior to use. Nitrogen (extra pure grade) was used after passing it through a column of reduced copper metal. A reagent grade of toluene was dried on 4A molecular sieve and saturated with dry nitrogen prior to use.

Polymer Synthesis

Polypropylene   A three-necked glass flask equipped with a magnetic stirrer was used as a reactor. Propylene was condensed into toluene in a reactor kept at the polymerization temperature (-78°C, -65°C or -48°C). The amount of toluene used was adjusted to be 100 ml as the total volume of the solution. Then, given amounts of Al(C$_2$H$_5$)$_2$Cl and V(acac)$_3$, in that order, were charged under a nitrogen stream. The polymerization was timed from the addition of the V(acac)$_3$ component, and quenched at a given time by adding 100 ml of an ethanol solution of hydrochloric acid or by adding 22 ml of toluene solution (0.5 mol/l) of iodine. The polymers produced were washed several times with 500 ml of ethanol and dried in vacuo at room temperature.

Polyethylene   The procedures of ethylene polymerization were the same as those for polypropylene synthesis, excepting that the polymerization was carried out at -78°C under a constant pressure (76 Torr) of ethylene.

Polystyrene   Monofunctional living polystyrene anion was prepared at 50°C in toluene using n-butyllithium as initiator along the method described by Hsieh[21].

Diblock Copolymer Synthesis

Ethylene-Propylene Block Copolymer   Attempts were made to
prepare diblock copolymers of ethylene and propylene by using
the sequential polymerization technique.   A sequential
polymerization was tried at -78°C by admitting ethylene monomer
into a living polymerization system of propylene with the
$V(acac)_3/Al(C_2H_5)_2Cl$ catalyst.   Using propylene as a second
monomer, the polymerization was repeated as follows.   After
the polymerization of ethylene was carried out at -78°C for
0.5 h under a pressure of 60 torr in the presence of $V(acac)_3$
and $Al(C_2H_5)_2Cl$, ethylene monomer unreacted in a reactor was
pumped out.   Then, propylene monomer (830 mmol) was admitted
into the polymerization solution, and the polymerization was
further continued for 6 h at -78°C.   The resulting polymers
were examined by ^{13}C NMR and GPC analyses.

Propylene-Styrene Diblock Copolymer   The coupling reactions
between iodine-terminated living polypropylene and monofunctional
living polystyrene anion have been carried out at 50°C in toluene.
In these reactions almost equivalent numbers of living polystyrene
and polypropylene chains [N], which are calculated from the
relation [N] = polymer weight/$\bar{Mn}$, were used.   A given amount
of orange solution of monofunctional living polystyrene was
added gradually into the toluene solution of iodine-terminated
polypropylene at 50°C under dry nitrogen atomosphere.   The
solution mixture was kept at the temperature for 20 h.   The
product was isolated by precipitation from ethanol and dried
for GPC analysis.

Gel-Permeation Chromatogram

Gel-permeation chromatograms of polymers were recorded
using a Waters Model-200 GPC equipped with a differential
refractometer and five column set with porosities 10^7, 10^6, 10^5,
10^4 and 10^3 Å at 135°C.   The solvent was o-dichlorobenzene
and the flow rate was 1.0 ml/min.   A 1.0 ml aliquot of 0.1
w/v % o-dichlorobenzene solution of polymer was injected.   A
molecular weight calibration curve was obtained on the basis of
the universal calibration[22,23] with ten standard polystyrene
samples of narrow distribution of molecular weights 2,100 to
2,610,000.   From GPC curves, the number-average and weight-
average molecular weights ($\bar{Mn}$ and $\bar{Mw}$) were calculated by
standard procedures.   The correction of GPC peak spreading
(zone spreading) for the polydispersity parameters ($\bar{Mw}/\bar{Mn}$) was
made by the method of Hamielec and Ray[24,25].

^{13}C NMR Spectra ^{13}C NMR Spectra

^{13}C NMR spectra of polymers were recorded at 140°C in
o-dichlorobenzene (20-30 w/v %), using a JEOL PS-100 spectrometer
with the PFT-100 Fourier transform system operating at 25.14 MHz.

RESULTS AND DISCUSSION

"Living" Coordination Polymerization of Propylene

The yields and the molecular weights of polypropylenes
produced in the course of polymerization at -78, -65 and -48°C
with the soluble $V(acac)_3/Al(C_2H_5)_2Cl$ system are given in
Table I.   All of the polymerizations were carried out to
conversion of propylene monomer below several percent.   As
Fig. 1a shows, the yield of polymer is proportional to the
polymerization time at respective temperature, which indicates
that the propagation centers are formed completely just after
starting polymerization and active during the polymerization
without any deactivation.   Figs. 1b and 1c show the number-
average molecular weight $\bar{M}n$ of polymers and the number of polymer
chains per vanadium atom, [N], calculated from the relation
[N] = polymer yield/$\bar{M}n$ during the polymerization.   The $\bar{M}n$
of polymers produced at -78°C and -65°C are proportional to the
polymerization time.   As a result, the number of polymer chains
[N] remains almost constant during the course of polymerization,
indicating that the polymerization of propylene takes place
without chain transfer.   On the contrary, in the polymerization
at -48°C the number of polymer chains [N] increases with the
polymerization time, indicative of the existence of chain
transfer reactions.   From the above results, we have arrived
at the conclusion that the soluble $V(acac)_3/Al(C_2H_5)_2Cl$ system
initiates a "living" polymerization of propylene at temperatures
below -65°C[17].
Fig. 2 shows the GPC elution (molecular weight distribution)
curves of polypropylenes obtained in different polymerization
time at -78°C.   The unimodal peak of the curves shifted toward
low elution counts (high molecular weights) with an increase in

Table I.   Results of propylene polymerization at different
temperatures with the soluble $V(acac)_3/Al(C_2H_5)_2Cl$ system[a].

Run	Temp. (°C)	Time (h)	Yield (g)	Molecular Weight		$[N]^c$ (mol/mol-V)
				$10^{-4} \cdot \bar{Mn}$	$\bar{Mw}/\bar{Mn}^b$	
P- 1	-78	1.0	0.11	0.604	1.11	0.036
P- 2	-78	1.5	0.14	0.919	1.09	0.030
P- 3	-78	2.0	0.25	1.25	1.17	0.040
P- 4	-78	2.0	0.27	1.20	1.13	0.045
P- 5	-78	3.0	0.38	1.50	1.10	0.051
P- 6	-78	3.0	0.35	1.60	1.14	0.044
P- 7	-78	4.0	0.51	2.50	1.17	0.041
P- 8	-78	5.0	0.75	3.52	1.12	0.043
P- 9	-78	5.0	0.70	3.18	1.15	0 044
P-10	-78	6.0	0.75	3.75	1.12	0.040
P-11	-78	8.0	1.07	4.42	1.14	0.048
P-12	-78	10.0	1.18	5,16	1.16	0.046
P-13	-78	15.0	1.76	9.13	1.18	0.039
P-14	-65	1.5	0.23	1.46	1.07	0.032
P-15	-65	2.0	0.56	3.08	1.09	0.036
P-16	-65	3.0	0.62	4.13	1.14	0.030
P-17	-65	4.0	1.02	5.43	1.13	0.038
P-18	-65	5.0	1.17	6.04	1.17	0.039
P-19	-48	0.33	0.15	1.40	1.37	0.021
P-20	-48	0.67	0.43	2.81	1.45	0.031
P-21	-48	1.33	0.82	3.22	1.40	0.051
P-22	-48	2.0	1.82	4.47	1.37	0.081
P-23	-48	2.0	1.87	4.50	1.37	0.083

[a]Polymerization conditions ; propylene = 830 mmol, $V(acac)_3$ =
0.5 mmol, $Al(C_2H_5)_2Cl$ = 5.0 mmol and toluene solution = 100 ml.
[b]Corrected GPC peak spreading.
[c]Number of polymer chains produced per vanadium atom.

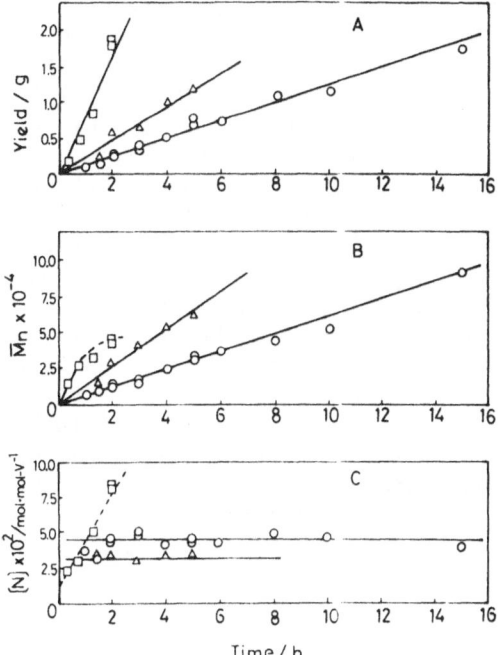

Fig. 1.   Time dependence of yields and M̄n of the resulting
polymers and of the number of polymer chains produced per
vanadium atom [N] in polymerization of propylene with the
soluble $V(acac)_3/Al(C_2H_5)_2Cl$ system at −78, −65 and −48°C;
(○) at −78°C; (△) at −65°C; (□) at −48°C.
Polymerization conditions are given in Table I.

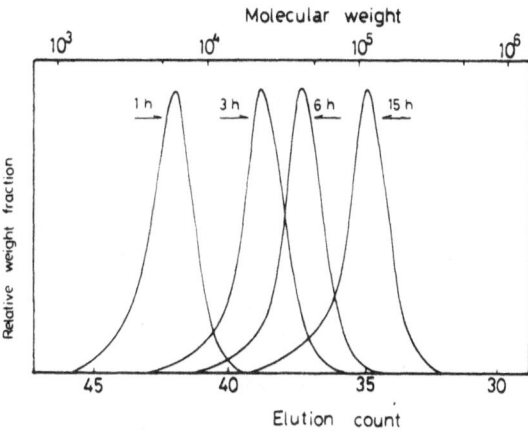

Fig. 2.   GPC elution (molecular weight distribution)
curves of polypropylenes obtained after different
polymerization times at −78°C in the presence of the
soluble $V(acac)_3/Al(C_2H_5)_2Cl$ system.

polymerization time.   The values of polydispersity, $\overline{Mw}/\overline{Mn}$,
corrected for GPC peak spreading are listed in Table I.   The
values are 1.13± 0.05 in the polymerizations at -78 and -65°C,
being close to the theoretical value ($\overline{Mw}/\overline{Mn}$ = 1.0) for an ideal
living polymerization.   On the other hand, in the polymerization
at -48°C the values were as much as 1.41±0.04, as anticipated
in the polymerization involving chain transfer reactions.
        The stereoregularities of polypropylenes obtained at -78,
-65 and -48°C were determined from the methyl carbon resonance
in the ^{13}C NMR spectra[26,27].   The results are shown in Table
II.   The soluble V(acac)$_3$/Al(C$_2$H$_5$)$_2$Cl system gives a syndiotactic
form of polypropylene, as reported by Natta et al[28].
        The kinetics of the "living" polymerization of propylene
with the soluble V(acac)$_3$/Al(C$_2$H$_5$)$_2$Cl catalyst system have been
studied by us[17].   The catalyst exhibiting "living" character
has proven quite helpful in the precise determination of the
number of propagation chains, since the chain ends remain alive.
The dependence of number of propagation chains [N*] on the
concentration of Al(C$_2$H$_5$)$_2$Cl, [A], is expressed by

$$[N*] = \alpha \left(\frac{K_A[A]}{1 + K_A[A]}\right) \qquad\qquad (2)$$

The values of the constants, α and K_A, at -78°C were 0.080 (mol/
mol of V) and 26 (1/mol of Al), respectively.   The polymerization
rate is represented by the following equation:

Table II.   Stereoregularity of polypropylene obtained at
different temperatures with the soluble V(acac)$_3$/Al(C$_2$H$_5$)$_2$Cl
system.

| | Polymerization | | Stereoregularity | | | |
| | | | Triad Fractions[a] | | | Dyad[b] |
Run	Temp.(°C)	Time(h)	[rr]	[rm]	[rr]	[r]
P-5	-78	3.0	0.652	0.324	0.024	0.814
P-16	-65	3.0	0.638	0.328	0.034	0.802
P-22	-48	2.0	0.635	0.330	0.035	0.800

[a] Determined from the methyl region of ^{13}C NMR spectra.
[b] Calculated from triad fraction.

Table III.   Values of k_s and K_M and their activation energies.

Temp. (°C)	k_s (h^{-1})	E_{k_s} (kcal/mol)	K_M (1/mol)	E_{K_M} (kcal/mol)
-78^a	$(1.9\pm0.7)\times10^2$		0.37 ± 0.01	
-65^a	$(4.2\pm1.9)\times10^2$	8.1 ± 3.4	0.35 ± 0.03	-0.81 ± 0.20
-48	$(3.2\pm1.3)\times10^3$		0.28 ± 0.02	

^{a}From reference 17.

$$Rate = k_s \left(\frac{K_M[M]}{1 + K_M[M]}\right) [N^*] \qquad (3)$$

Where, [M] is the concentration of propylene monomer.   The values of the constants k_s and K_M determined at −78, −65 and −48 °C are summarized in Table III, together with their activation energies.   Equation (3) can be interpreted in terms of a coordination polymerization mechanism in which the polymerization of propylene occurs by a sequence of two successive reactions : (1) propylene monomer coordination to active vanadium and (2) a subsequent insertion of the coordinated monomer into a living polymer chain attached to the metal, V^*-P_n (eq. 4).

$$V^*-P_n + C_3H_6 \underset{K_M}{\overset{C_3H_6}{\rightleftharpoons}} \bar{V}^*-P_n \overset{k_s}{\longrightarrow} V^*-P_{n+1} \qquad (4)$$

Here, the constants K_M and k_s represent the equilibrium constant for the propylene monomer coordination and the rate constant for the insertion, respectively.

Sequential Polymerization of Ethylene and Propylene

A block copolymer of AB structure is commonly prepared by the sequential addition of monomer B to a monofunctional living polymer A.   We have attempted to prepare AB block copolymers of ethylene and propylene using the sequential polymerization technique.   Sequential polymerizations of ethylene and propylene were carried out at -78°C with different orders of monomer addition in the presence of soluble $V(acac)_3/Al(C_2H_5)_2Cl$ catalyst, and the resulting polymers were analysed by ^{13}C NMR and GPC.   We found that none of ^{13}C NMR and GPC data showed any evidences for the formation of a definite AB block copolymer of ethylene and propylene, free from homopolymer impurities. The resulting polymer always had a broad molecular weight distribution ($\bar{M}w/\bar{M}n$ = 1.6-2.5) and contained a random copolymer of ethylene and propylene.   The formation of these polymers is attributed to the following reasons:

 (1) The removal of unreacted monomer from a toluene solution of monofunctional living polymer A was incomplete.

 (2) Chain transfer reactions took place at -78°C in the polymerization of ethylene.   As shown in Table IV, the number of polymer chains [N] increased during the polymerization of ethylene at -78°C, suggesting that chain transfer reactions take place at -78°C in the presence of ethylene   In fact, the molecular weight distributions of polyethylenes produced are broad in comparison with those of polypropylenes (see Fig. 3). The polydispersity parameter, $\bar{M}w/\bar{M}n$, of polyethylenes increased with the polymerization time and approached 2.0 in accordance with a most probable distribution of chain length.

Table IV.   Results of ethylene polymerization at -78°C with the soluble $V(acac)_3/Al(C_2H_5)_2Cl$ system[a].

Run	Time(h)	Yield(g)	Molecular Weight		[N][c]
			$10^{-5} \cdot \bar{M}n$	$\bar{M}w/\bar{M}n$[b]	(mol/mol-V)
E-1	0.5	0.05	0.964	1.75	0.026
E-2	0.67	0.07	1.13	1.80	0.031
E-3	1 0	0.18	1.45	2.03	0.061
E-4	1.5	0.30	2.77	2.06	0.054
E-5	2.0	0.38	3.11	2.02	0.061
E-6	3.0	0.57	3.27	2.00	0.087

[a] Polymerization conditions ; ethylene = 76 Torr, $V(acac)_3$ = 0.02 mmol, $Al(C_2H_5)_2Cl$ = 5.0 mmol and toluene solution = 100 ml.

[b] Corrected GPC peak spreading.

[c] Number of polymer chains produced per vanadium atom.

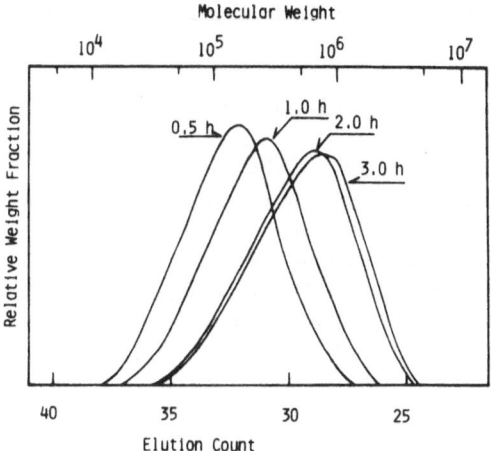

Fig. 3.   GPC elution (molecular weight distribution) curves of polyethylenes obtained after different polymerization times at -78°C in the presence of the soluble V(acac)$_3$/Al(C$_2$H$_5$)$_2$Cl system.

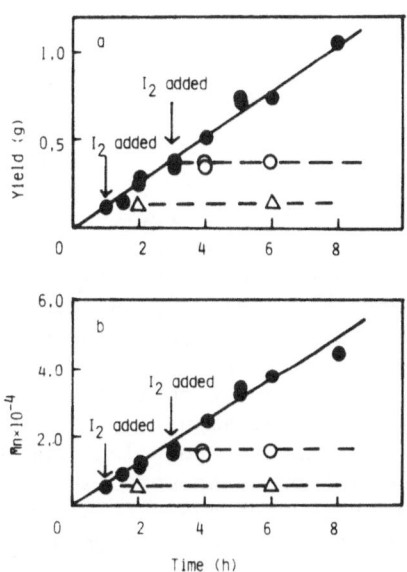

Fig. 4.   Changes in polymer yield and $\bar{M}n$ induced by the addition of I$_2$ during the polymerization of propylene at -78°C.   Polymerization conditions; propylene =830 mmol, V(acac)$_3$ = 0.5 mmol, Al(C$_2$H$_5$)$_2$Cl = 5.0 mmol, toluene solution = 100 ml and I$_2$ = 11 mmol; (●) without I$_2$; (○) and (△) with I$_2$ of 11 mmol.

Preparation of Propylene-Styrene Diblock Copolymers

        We propose here a novel method to prepare tailored AB
block copolymers of propylene and styrene.   The method is
based on the coupling reaction between monofunctional living
polystyrene anion and iodine-terminated living polypropylene,
as depicted in the introductory section.   The reaction is an
application of the Wurtz condensation reaction[29].
        The iodine-terminated living polypropylenes were prepared
as follows.   The living polymerization of propylene was done
at $-78°C$ in the presence of the soluble $V(acac)_3/Al(C_2H_5)_2Cl$
system, followed by introducing 22 ml of toluene solution
(0.5 mol/l) of I_2 at the prescribed time.   As shown in Fig. 4,
I_2 is a very effective inhibitor for the polymerization.
The resulting polymers were washed several times with 500 ml of
ethanol to remove free iodine and dried in vacuo at room
temperature.   The monodispersity ($\bar{M}w/\bar{M}n$ = 1.07-1.15) of the
polymers was confirmed by GPC analysis.   The reaction between
iodine and a living polypropylene chain attached to vanadium
may be written as

$$\sim\sim CH_2-\overset{\overset{\displaystyle CH_3}{|}}{CH}-V^* + I_2 \longrightarrow \sim\sim CH_2-\overset{\overset{\displaystyle CH_3}{|}}{CH}-I + V-I \quad (5)$$

Equation (5) is rationalized by the previous findings: (1) the
syndiotactic polymerization of propylene with soluble vanadium-
based catalysts occurs via secondary insertion of propylene
into vanadium-carbon bond[30-33], and (2) I_2 reacts quantitatively
with titanium-polymer bond to yield an iodine-polymer bond[34,35].
        The coupling reaction between the iodine-terminated living
polypropylene and a living polystyrene anion was examined at
50°C in toluene solution.   The amounts and molecular weights
of component polymers used for the coupling reaction are given
in Table V.   The reaction products were isolated by
precipitation from ethanol, dried and made up to a 0.1 w/v %
solution in o-dichlorobenzene for GPC analysis.   The GPC
elution curves of the products are shown in Figs.5 and 6
respectively, together with those of component polymers.   The
GPC curves of polypropylene and polystyrene measured by the
differential refractometer appear in opposite directions.
The relative response of polypropylene to polystyrene in the
differential refractometer was determined as 1.2 w/w.
        In Fig. 5, the GPC peak of product shifts toward higher
molecular weights compared with those of the component polymers,

Table V.   The amounts and the molecular weights of component polymers used for the coupling reactions at 50°C.

Run	Component polymer	Weight (g)	Molecular Weight		Number of polymer chains (mmol)
			$10^{-4} \cdot \bar{M}n$	$10^{-4} \cdot \bar{M}n$	
B-1	PP[a]	0.60	1.48	1.69	0.041
	PS[b]	0.46	1.13	1.44	0.041
B-2	PP[a]	0.18	0.636	0.706	0.027
	PS[b]	0.92	3.68	4.56	0.025

[a] Iodine-terminated polypropylene.

[b] Monofunctional living polystyrene.

and appears at 35000 molecular weight based on polystyrene calibration.   The molecular weight of 35000 is just the sum of the molecular weights at the peaks of component polymers, 13000 and 22000, indicating that the iodine-terminated polypropylene and the monofunctional polystyryllithium have coupled to yield a new AB block copolymer.   In order to evaluate the efficiency of block copolymer formation, the molecular weight of iodine-terminated polypropylene was chosen to be as low as possible in order to obtain a significant change in the retention times between the polypropylene and the product after equimolar mixing of living polymer chains.   Fig. 6 shows the result obtained with the iodine-terminated polypropylene of 6300 in $\bar{M}n$.   The GPC peak of the polypropylene component could not be detected in the curve of the product in Fig. 6. This result strongly shows that the block copolymer formation of AB structure is quantitative.

The method we demonstrate here may be applicable to many other coupling reactions between iodine-terminated living polypropylene and monofunctional or difunctional living polymers, yielding well-defined and monodisperse block copolymers of AB and ABA types containing a polypropylene segment.

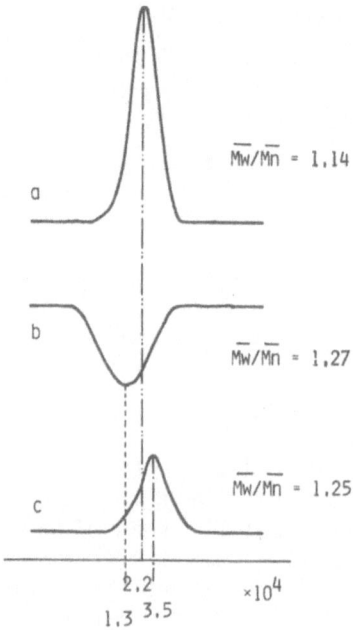

$\overline{Mw}/\overline{Mn} = 1.14$

$\overline{Mw}/\overline{Mn} = 1.27$

$\overline{Mw}/\overline{Mn} = 1.25$

2.2

1.3 3.5

×10⁴

Molecular Weight (polystyrene equivalent)

Fig. 5.   GPC curves of (a) iodine-terminated living
polypropylene; (b) ethanol-terminated monofunctional
living polystyrene; and (c) the product of the coupling
reaction (run B-1).

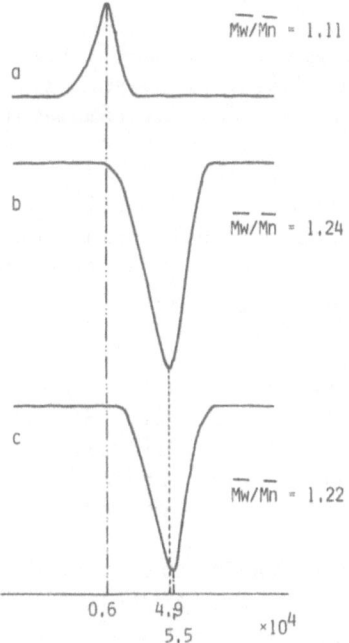

Fig. 6.   GPC curves of (a) iodine-terminated living polypropylene; (b) ethanol-terminated monofunctional living polystyrene; and (c) the product of the coupling reaction (run B-2).

REFERENCES

1. P. J. Flory, J. Am. Chem. Soc., 62, 1561(1940).
2. L. Gold, J. Chem. Phys., 28, 91(1958).
3. M. Szwarc, Nature(London), 178, 1168(1956); M. Szwarc, M. Levy and R. Milkovich, J. Am. Chem. Soc., 78, 2656(1956).
4. M. Szwarc, "Carbanions, Living Polymers and Electron Trancfer Processes", Interscience, New York, 1968.
5. H. Hostalka, R. V. Figini and G. V. Schulz, Makromol. Chem., 71, 198(1964).
6. H. L. Hsieh and O. F. McKinney, J. Polym. Sci. B, 4, 843(1966).
7. H. Hirota, M. Nakayama and N. Ise, J. Chem. Soc., Faraday trans. I, 68, 58(1972).
8. M. Szwarc and A. Rembaum, J. Polym. Sci., 22, 189(1962).
9. M. Morton and F. R. Ells, J. Polym. Sci., 61, 721(1962).
10. L. J. Fetters and M. Morton, Macromolecules, 2, 453(1969).
11. M. Morton, R. F. Kammereck and L. J. Fetters, Macromolecules, 4, 11(1971).
12. G. Natta and I. Pasquon, Adv. Catal., 11, 1(1959).
13. G. Bier and G. Lehman, "Copolymerization", G. E. Ham, Ed. Interscience, New York, 1964, Chapter 4, Section B, p149.
14. G. Bier, Angew. Chem., 73, 186(1961).
15. E. G. Kontos, E. K. Esterbrook and R. D. Gilbert, J. Polym. Sci., 61, 69(1962).
16. P. Prabhu, A. Schindler, M. H. Theil and R. D. Gilbert, J. Polym. Sci , Polym. Lett. Ed., 18, 389(1980); J. Polym. Sci., Polym. Chem. Ed., 19, 523(1981).
17. Y. Doi, S. Ueki and T. Keii, Macromolecules, 12, 814(1979).
18. Y. Doi, S. Ueki and T. Keii, Makromol. Chem., 180, 1359 (1979).
19. Y. Doi, S. Ueki, S. Nagahara and T. Keii, "IUPAC International Symposium on Macromolecules", at Florence (1980).
20. Y. Doi, S. Ueki and T. Keii, Polymer, 21, 1352(1980).
21. H. Hsieh, J. Polym. Sci. A, 3, 153,163,173(1965).
22. Z. Grubisic, P. Rempp and H. Benoit, J. Polym. Sci. B, 5, 753(1967).
23. T. Ogawa, S. Tanaka and S. Hoshino, Kobunshi Kagaku, 29, 6(1972).
24. A. E. Hamielec and W. H. Ray, J. Appl. Polym. Sci., 13, 1319(1969).
25. A. E. Hamielec, J. Appl. Polym. Sci., 14, 1519(1970).
26. L. F. Johnson, F. Heatly and F. A. Bovey, Macromolecules, 3, 175(1970).

27. Y. Doi and T. Asakura, Makromol. Chem., $\underline{176}$, 507(1975).
28. G. Natta, I. Pasquon and A. Zambelli, J. Am. Chem. Soc.,
     $\underline{84}$, 1488(1962).
29. A. Wurtz, Ann. Chim. Phys., $\underline{44}$, 275(1855); Ann., $\underline{96}$,
     364(1855).
30. T. Suzuki and Y. Takegami, Bull. Chem. Soc. Jpn., $\underline{43}$, 1484
     (1970).
31. A. Zambelli, C. Tosi and C. Sacchi, Macromolecules,
     $\underline{5}$, 649(1972).
32. A. Zambelli, P. Lacattelli and G. Bajo, Macromolecules,
     $\underline{12}$, 154(1979).
33. Y. Doi, Macromolecules, $\underline{12}$, 248(1979).
34. J. C. W. Chien, J. Am. Chem. Soc., $\underline{81}$, 86(1959).
35. H. Schnecko, K. A. Jung and L. Grosse, Makromol. Chem.,
     $\underline{148}$, 67(1971).

MECHANISM OF ZIEGLER-NATTA POLYMERIZATION

ON THE BASIS OF DATA ON THE NUMBER OF ACTIVE

CENTERS AND THEIR REACTIVITY

V.A. Zakharov, G.D. Bukatov, and
Yu.I. Yermakov

Institute of Catalysis
Novosibirsk, 630090
USSR

INTRODUCTION

Quantitative data on the number of active centers are very informative for the formulation of concepts on mechanism of catalytic polymerization and for interpretation of polymerization kinetics. Systematic study of the effect of the catalyst composition and the polymerization conditions on the number of active centers and their reactivity (rate constants of different steps of overall polymerization process) is important for understanding: i) the role of various components of catalytic system in the formation of active centers; ii) the plausible composition of active centers; iii) the mechanism of separate polymerization steps; iv) the possible reasons for the increased activity of supported catalysts in comparison with conventional Ziegler-Natta catalysts.

A survey of the different techniques of the determination of C_p and K_p*) for catalytic polymerization of olefins is given in reviews (1,2). The technique based

---

*) The list of abbreviations is given at the end of the paper.

on direct determination of the number of metal-polymer
bonds by their interaction with radioactive quenching
agents is preferable. Tritiated alcohols ROT were widely
used as radioactive quenching agents. However, during
polymerization by two-component (Ziegler-Natta-type)
catalysts, alcohol interacts not only with active centers
containing the transition metal-polymer bonds, but decomposes also inactive aluminium-polymer bonds which are
formed during chain transfer reactions with the cocatalyst. Usually the number of inactive aluminium-polymer
bonds is 10-100 times higher than the number of active
centers (4). To apply tritiated alcohols for the Cp determination is possible only for the case when the
number of inactive Al-polymer bonds is comparable with
the number of active centers. A probalble example is
the polymerization of 4-methyl-pentene-1 by VCl_3 +
+ AlR_xCl_{3-x} catalyst (5). A more universal technique
of Cp determination is based on the application of selective quenching agents interacting only with the
active metal-polymer bonds. Carbon monoxide and carbon
dioxide labelled by carbon-14 are examples of such
quenching agents (6-8). Using these radioactive compounds
the systematic data on the effect of catalyst composition and the polymerization conditions on Cp and Kp
were obtained when the monomers were ethylene and propylene.

Here on the basis of data on the number of active
centers and their reactivity (rate constant of chain
propagation and transfer reactions) some aspects of the
polymerization mechanism by bulk and supported Ziegler-
Natta type catalysts are discussed.

OLEFIN POLYMERIZATION BY
CATALYTIC SYSTEMS CONTAINING
BULK CHLORIDES OF TITANIUM
AND VANADIUM

Effect of Catalyst Composition
on C_p and K_p

The number of active centers varies drastically
with the condition of preparation and the composition
of titanium chloride catalysts (Table 1). The lowest
Cp value is observed for one-component $TiCl_2$ catalyst.
In this catalyst active centers are apparently formed
by oxidative addition of olefin to the small number

of coordinatively unsaturated surface titanium ions (10).
Catalysts containing α-TiCl$_3$ prepared by interaction
of TiCl$_4$ with hydrogen also contain a low concentration
of active centers. The value of Cp does not exceed one
per cent of the number of titanium ions adjacent to the
surface. This low number of centers is in agreement with
the Cossee and Arlman concepts (11,12) and the Rodrigues
et al. experimental data (13) on the localization of
active sites on side planes and outlets of spiral dis-
locations of TiCl$_3$ crystallites.

For δ-TiCl$_3$·0.3 AlCl$_3$ samples activated by dry
grinding, the number of propagation centers is signifi--
cantly higher and reaches 0.8×10^{-2} mol/mol Ti. This
value is close to the number of surface titanium ions
calculated from BET measurements ($\sim$11 m^2/g). In polymerization, however, the catalyst surface increases due
to the catalyst disintegration by growing polymer (14-
16). A correlation is found (17,18) between the activity
and the size of primary crystallites defined by the
X-ray method. Catalyst disintegration into the primary
crystallites has been analyzed (19). Apparently, the
"working" surface of the δ-TiCl$_3$·0.3 AlCl$_3$ sample can
attain 70-80 m^2/g after disintegration of the sample
with the "initial" (BET) surface area of 11 m^2/g (19).
In this catalyst system the maximum number of propagation centers corresponds to 10-15% of the total number
of surface titanium ions. In δ-TiCl$_3$ the surface of
side planes can comprise a significant portion of the
overall surface area. So the high surface concentration
of active sites is consistent with the concept that the
propagation centers are localized on side planes of crystallites of TiCl$_3$.

Reactivity of active centers in the propagation
(Kp values, see Table 1) was constant with variation of
the cocatalyst composition and did not change in the
presence of AlCl$_3$ in titanium trichloride. These data
are consistent with the concept of monometallic active
centers whose reactivity is dependent only on the type
of transition metal compounds (an increase of K$_p$ was
observed when VCl$_3$ was used instead of TiCl$_3$.

Effect of Polymerization
Parameters on C$_p$ and K$_p$

Steady-state number of propagation centers was not
influenced by the concentration of monomer. A rise of
C$_p$ was observed with increase of temperature (20). Due

Table 1. Data on the Maximum Activity,
         Number of Propagation Centers and
         Propagation Rate Constants during
         Polymerization of Propylene on
         Titanium and Vanadium Chlorides

Catalyst	R max $\dfrac{\text{g polymer}}{\text{mmol } M \cdot h \cdot atm}$	$C_p \times 10^2$ $\dfrac{\text{mol}}{\text{mol } M}$	K_p $\dfrac{l}{\text{mol} \cdot s}$
$TiCl_2$	0.005	0.00013	76
$TiCl_2 + AlEt_2Cl$	0.004	0.00008	94
$\alpha - TiCl_3 + AlEt_2Cl$	0.11	0.003	71
$\delta - TiCl_3 \cdot O,3AlCl_3 + AlEt_2Cl$	7.7	0.17	90
$\delta - TiCl_3 \cdot O,3AlCl_3 + AlEt_3$	28	0.58	100
$\delta - TiCl_3 \cdot O.3AlCl_3 + Al(i-Bu)_3$	36	0.80	90
$VCl_3 + Al(i-Bu)_3$	7.0	0.02	370

(Polymerization temperature 70°C; $M=Ti$, V;
the data are given for the fraction insoluble
in boiling ether (in the case of $TiCl_3$ and
VCl_3) and in boiling n-heptane (in the case
of $TiCl_2$).

to this effect the overall activation energy is higher
than the activation energy of the propagation reaction
(see Table 2). The difference between E_{ov} and E_p is dependent on the catalytic system. This difference decreases in the row (20) $AlEt_2Cl > AlEt_3 > Al(i-Bu)_3$. For polymerization at relatively low temperatures ($\leqslant 30°C$) a
decrease of C_p was observed (21) when the concentration
of $AlEt_3$ was increased from 0.015 to 0.3 mole/l. The
change of the number of active centers with variation
of the reaction temperature or cocatalyst concentration
is reversible (20,21). These data may be explained by
the reversible adsorption of organoaluminium cocatalyst
on the monometallic active center:

Table 2. Activation Energies[a] for Olefin
Polymerization on Titanium and
Vanadium Chlorides

Catalyst	Monomer	E_{ov} kcal/mol	E_p kcal/mol
δ-TiCl$_3$ 0.3AlCl$_3$+AlEt$_3$	C_2H_4	6.0	3.0
δ-TiCl$_3$ 0.3AlCl$_3$+AlEt$_2$Cl	C_3H_6	11.0	5.5
VCl$_3$+Al(i-Bu)$_3$	C_2H_4	3.7	0.6

a) E_{ov} and E_p are calculated from Arrhenius
plots for steady-state polymerization
rate and propagation rate constant respectively (20).

(1)

The equilibrium constant for adsorption of AlR$_3$ on
active centers (reaction 1) was calculated on the base
of the change of C_p with temperature (21). This constant
was shown to be 3.9×10^{-10} exp (-18500/RT) l/mol. The
values of (E_{ov}-E_p) for olefin polymerization by TiCl$_3$+
+AlR$_3$ catalyst may change from 0 to 10 kcal/mol depending on polymerization parameters (21).

Mechanism of Propagation Reaction

Polymerization of olefins in the presence of
transition metal compounds is characterized by low
activation energies of the propagation step. These activation energies are not higher than E_p for radical polymerization (Table 3). The concerted mechanism of the
propagation step (see 24-26) provides the high degree
of the compensation of energy of ruptured bonds by the
energy of the newly formed bonds. For such a mechanism

Table 3. Propagation Rate Constants for
         Olefin Polymerization on Various
         Active Centers

Catalyst	Active center	Monomer	K_P 1/mol c
$TiCl_3 + AlEt_3$	Ti – C	C_2H_4	8.0×10^5 exp $(-3000/RT)$
$VCl_3 + Al(i\text{-}Bu)_3$	V – C	C_2H_4	1.8×10^5 exp $(-600/RT)$
$TiCl_3 + AlEt_2Cl$	Ti – C	C_3H_6	3.0×10^5 exp $(-5,500/RT)$
$AlEt_3$ [a]	Al – C	C_3H_6	2×10^3 exp $(-12,000/RT)$
Radical initiator [b]	$\sim$C	C_2H_4	6.7×10^6 exp $(-5000/RT)$

a) data from (22)
b) data from (23)

the precoordination of olefin on the metal ion is
necessary:

$$L_xM\text{-}P + C{=}C \underset{K_{-1}}{\overset{K_1}{\rightleftarrows}} L_xM\text{-}P \overset{\overset{\displaystyle C{=}C}{\downarrow}}{\ } \overset{K_2}{\longrightarrow} L_xM\text{-}\underset{|}{\overset{|}{C}}\text{-}\underset{|}{\overset{|}{C}}\text{-}P \quad (2)$$

The insertion of ethylene into the aluminium-alkyl bond
also proceeds via the step of preliminary olefin coordination (27), but is characterized by a higher activation
energy (Table 3). For both types of active centers (containing transition or non-transition metal ions) the
activation energy of olefin coordination is determined
by the energy loss for the rearrangement in the coordination structure of the active center. The activation
energy of the insertion step is due to the energy loss
at rupture of the double bond of an olefin (25-27).

     For the two-step mechanism of propagation the ex-
perimental K_P value is determined by the following

expression:

$$K_P = K_1 \cdot K_2 / (K_2 + K_{-1} + K_1 \cdot C_M) \tag{I}$$

A first order of propagation in respect to the monomer concentration is observed. It may correspond to the following cases:

i) propagation rate is determined by the rate of monomer coordination

$$K_2 \gg (K_{-1} + K_1 C_M) \tag{II}$$

$$K_P = K_1$$

ii) propagation rate is determined by the concentration of the coordinated olefin and by the rate of insertion

$$K_{-1} \gg (K_2 + K_1 C_M) \tag{III}$$

$$K_P = K_2 \cdot K_1 / K_{-1}$$

According to the Natta et al. (28) the rate-determining step of the propagation reaction of styrene and its derivatives by titanium trichloride is the monomer coordination ($K_P = K_1$).

In some papers (28,30) the overall activation energy of olefin polymerization was considered to be an activation energy of the propagation step. These values (10-14 kcal/mol) were considered to be too high if they corresponded to the monomer coordination as a rate-determining stage. However, E_P (1-6 kcal/mol) is rather lower than E_{ov} (see above).

The concept that the olefin coordination is a rate-determining step is in accordance with the values of frequency factors, calculated from the experimental data on K_P and E_P. The comparison of the experimental values of A_P and calculated frequency factors for different steps is given in Table 4. The calculated values of frequency factors for olefin coordination agree well with the experimental values of A_P. For olefin insertion A_i values are expected to be $10^4 - 10^5$ lower than calculated frequency factors for the simple unimolecular reaction of olefin desorption. The relation $A_2 > A_{-1}$ of the frequency factors of the two unimolecular reactions accounts for the possible relation of steric effects at

Table 4. Activation Energies and Frequency
Factors Corresponding to Different
Cases of the 1st Order Polymerization
Kinetics

	Meaning of E_P [a]	Meaning of A_P
$K_P=K_1$	$E_P=E_1$	$A_P=A_1=(0.3-3)\cdot 10^5$ 1/mol·s [b]
$K_P=K_2\cdot K_1/K_{-1}$	$E_P=E_2-q$	$A_P=A_1\cdot A_2/A_{-1}\ll A_1$ [c]
Experimental data [d]	1-6 kcal/mol	$(2-8)\cdot 10^5$ 1/mol·s

a) E_1,E_2,q – activation energies of adsorption, insertion steps and adsorption heat
   respectively.
b) Estimated according to the activated
   complex method for ethylene adsorption [9]
c) The expression follows from the assumption that $A_2 \gg A_{-1}$.
d) See Table 3.

formation of a complex four-centered intermediate during
insertion reaction compared with simple olefin desorption (20).

Chain Transfer Reactions

        In olefin polymerization by Ziegler-Natta catalysts
chain transfer reaction with the participation of cocatalyst, monomer and hydrogen occurs. Spontaneous transfer
due to β-hydrogen shift reaction also proceeds (28,30,
33). The ratios of the transfer rate constants to the
Kp may be calculated from data on the effect of the
transfer agent concentration on molecular weight of
polymer (29,33,34). For chain transfer with organoaluminum cocatalyst the K(t,AlR$_3$)Kp values may be also calculated from the data on the number of metal-polymer
bonds determined by the quenching of polymerization
with radioactive alcohol (1,4,5). The absolute values
of rate constants for the transfer reactions in (4,35)
were calculated using the data on Kp determined with
the application of [14]CO as a selective quenching agent.

        Tables 5 and 6 give the data on transfer reaction

Table 5. Rate Constants for the Chain Transfer
Reations with Organoaluminum
Cocatalyst in Polymerization on
δ-TiCl$_3\cdot$0.3AlCl$_3$

Cocatalyst Monomer[a)	K(t,AlEt$_3$) $1^{0.5}$/mol$^{0.5}$s	K(t,AlEt$_3$)/K$_D^{-0.5}$ [b) or K(t,Al(i-Bu)$_3$), 1/mol s
AlEt$_3$ } C$_2$H$_4$	2.3	120
Al(i-Bu)$_3$	-	8.7
AlEt$_3$ } C$_3$H$_6$	0.045	3.3
Al(i-Bu)$_3$	-	0.2

a) Temperature of polymerization 80°C for
   C$_2$H$_4$ and 70°C for C$_3$H$_6$;
b) K$_D$ - equilibrium constant of $\left[\text{AlEt}_3\right]_2$
   dissociation.

rate constants for olefin polymerization of titanium
trichloride.

The values of K(t) for the transfer with cocatalyst
depend on the type of organoaluminum compound and monomer (Table 5).

$$K(t,\text{AlEt}_3,C_2H_4) > K(t,\text{Al}(i\text{-Bu})_3,C_2H_4) >$$
$$K(t,\text{AlEt}_3,C_3H_6) > K(t,\text{Al}(i\text{-Bu})_3\ C_3H_6)$$

This series may be due to the increase of the steric
hinderance for the adsorption of organoaluminum cocatalyst when AlEt$_3$ is changed for Al(i-Bu)$_3$ and C$_2$H$_4$ is
changed for C$_3$H$_6$.

Rate constants for the transfer with monomer do not
depend on the type of cocatalyst but are determined by
the type of the monomer (35). For propylene polymerization K(t) values are much lower (about 20 times) than
for ethylene polymerization (Table 6). For TiCl$_3$ based

Table 6. Rate Constants of Chain Transfer
Reactions for Polymerization of
Olefins on δ-TiCl$_3$ 0.3AlCl$_3$+AlEt$_3$
(Polymerization temperature is
80°C for C$_2$H$_4$ and 70°C for C$_3$H$_6$)

K(t,x)	Monomer	
	C$_2$H$_4$	C$_3$H$_6$
K(t,AlEt$_3$),l$^{0.5}$/mol$^{0.5}$. s	2.3	0.045
K(t,M)/ω ,l/mol·s	0.084	0.004
K(t,H$_2$)/ω ,l$^{0.5}$/mol$^{0.5}$. s	2.3	0.59
$\tau \cdot \omega$, s [a]	25	160

a) Mean time of the polymer chain growth,
calculated for the monomer concentra-
tion 0.5 mol/l from eq. $\tau = P_W / \omega K_p \cdot C_M$
(P_W - steady-state mean-viscosity degree
of polymerization independent of the
time of polymerization, $\omega = M_W / M_n$).

polymerization catalysts the chain transfer with monomer
is dominant at 70-80°C in the absence of H$_2$ when the
monomer concentration is high enough ($>$0.5 mol/l).
The role of the chain transfer with AlEt$_3$ increases
with decrease of temperature; at 50°C this transfer
reaction becomes predominant for ethylene polymerization (4). Rate constants of transfer reactions with
H$_2$ for ethylene and propylene polymerization do not
differ to such an extent as propagation rate constants
for these monomers. As a result the regulation of molecular weight by an introduction of H$_2$ is more facile
for propylene than for ethylene polymerization.

The data on the mean time of polymer chain growth
(Table 6) appears to contradict the experimental fact
that the mean-viscosity molecular weight of polymer
increased with polymerization time during 30-60 min
(when polymerization temperature is 70-80°C). But the
change of molecular weight during the first stage of
polymerization may be due to the transfer reaction
with impurities present in the reaction mixture or
with the soluble products of the catalyst component

interaction. The occurrence of these supplementary transfer reactions have to change the molecular weight distribution with time (36). However, at the very first
stage of polymerization (<15 s) the formation of some
polymer chains with high molecular mass, close to that
one at steady-state condition was observed.

SUPPORTED TWO-COMPONENT (ZIEGLER
TYPE) OLEFIN POLYMERIZATION CATALYSTS

Composition of the Supported Ziegler Type Catalysts

A great number of supported catalysts, containing
titanium chlorides are known. These catalysts differ
in the composition of support and in the procedure of
their preparation. For catalysts prepared by supporting
$TiCl_4$ on SiO_2 it was shown (37,38) that during catalyst
interaction with organoaluminum cocatalyst the destruction of the surface oxychloride type complexes
$\geqslant Si-O-TiCl_3$ proceeds with the formation of the dispersed phase of $TiCl_3$. This phase plays the role of a real
polymerization catalyst.

The most active catalysts are prepared when various
magnesium compounds are used either as supports or as
modifying agents.

In this paper catalysts prepared from titanium
chloride and compounds of magnesium are designated as
Ti-Mg catalysts. Critical consideration of all known
techniques of the preparation of Ti-Mg polymerization
catalytic systems suggests that in all cases magnesium
halogenides are the real component determining the high
activity of these catalysts (38). In some cases magnesium halogenides are formed "in situ" during the catalyst
preparation, for example, during interaction of MgO,
magnesium alkoxides or organomagnesium compounds with
$TiCl_4$. The strong binding of $TiCl_4$ with anhydrous $MgCl_2$
and the formation of highly active catalysts are achieved when the dispersed samples of $MgCl_2$ with nearly
X-ray amorphous structure are used as support (39).
It was supposed (39) that titanium chloride interacts
with the surface defects of the crystallites of $MgCl_2$
or with the X-ray amorphous $MgCl_2$, mixed crystals being
formed.

When supported catalysts are treated with $AlEt_3$, $Ti(IV)$ is reduced. According to ESR data (40) isolated ions of Ti(III) in chloride environment and alkylated Ti(III) ions are formed. But the considerable amount of titanium ions is not revealed by ESR signals due to the formation of the aggregates of Ti(III) ions. Probably active centers may be formed both from isolated and aggregated ions of Ti(III). So in supported Ti-Mg polymerization catalysts $MgCl_2$ plays the same role as the $TiCl_3$ phase in bulk catalysts in respect to localization of the active centers on the surface defects of the cryatalline structure of metal chlorides.

Unsolvated alkyl halogenides of magnesium may be recommended as the initial component for the preparation of highly active Ti-Mg catalysts of various composition (41). With the use of these compounds different catalysts may be prepared, for example, i) $TiCl_4/MgCl_2 \cdot (P)$ supported catalysts (P is the polymeric organic residue included in $MgCl_2$)(42); ii) $TiCl_3/MgCl_2$ bulk catalysts (43); iii) $TiCl_2/MgCl_2$ catalysts, active without addition of organoaluminum cocatalyst (44). In what follows the data on the number of active centers and the rate constants of separate steps of polymerization by Ti-Mg catalysts will be discussed.

Ethylene Polymerization on Ti-Mg Catalysts

Different techniques were used (36,45-48) to obtain the data on Cp and Kp values for ethylene polymerization by Ti-Mg catalysts. Table 7 gives the results of the determination of the number of active centers and propagation rate constants with the use of ^{14}CO as selective quenching agent. The change in the composition of Ti-Mg catalysts practically has no effect on the Kp values. Propagation rate constants for polymerization on Ti-Mg catalysts were the same as for bulk $TiCl_3$. So the use of $MgCl_2$ as a support or activating component for polymerization catalysts does not effect the reactivity of active centers in propagation step. The drastic increase in the activity of Ti-Mg catalysts in ethylene polymerization is due only to the rise in the number of active centers. For supported catalysts with activity at the level 60-80 kg of C_2H_4/g Ti x h x atm, the number of active centers is 0.04-0.06 mol per g atom of Ti. But activity of Ti-Mg catalysts may be considerably increased by the proper choice of the preparation

Table 7. Number of Active Centers and Propagation Rate Constants for Ethylene Polymerization on Catalysts of Various Composition (80°C)

Catalyst	Content of Ti, wt %	C_M, mol/l	Activity, $\dfrac{kg\ PE}{g\ Ti \times h}$	Yield, $\dfrac{kg\ PE}{g\ Ti}$	$C_P \times 10^2$ mol/mol Ti	$K_P \times 10^{-4}$ l/mol s
δ-TiCl$_3$ · 0.3AlCl$_3$+AlEt$_3$	25	0.15	1.9	1	0.052	1.2
TiCl$_2$/MgCl$_2$ [a]	15	0.15	8.8	1.15	0.25	1.1
	15	0.15	13	2.0	0.30	1.4
TiCl$_3$/MgCl$_2$+AlEt$_3$	0.9	0.15	60	12	1.6	1.2
	3.7	0.15	18	4	0.52	1.1
TiCl$_4$/MgCl$_2$(P)+AlEt$_3$	2.4	0.3	75	5.4	1.2	1.0
	0.5	0.15	140	72	4.4	1.0
	0.27	0.6	440	240	3.2	1.1
	0.27	0.3	320	50	3.7	1.4
	0.27	0.3	4.7 [b]	200	0.054	1.4

a) The catalyst is active without an organoaluminum cocatalyst; b) CH$_3$OH was introduced into the reactor at the rate 220 kg PE/g^T h. After the rate decrease up to 4.7 kg PE/g Ti·h polymerization was stopped by 14CO. C_P and K_P were calculated from the number of radioactive tags and the corresponding value of the polymerization rate.

technique and modifying additives. In Table 8 the data
on the maximum number of active centers for catalyst
composition and polymerization parameters providing
maximum activity are given. For supported Ti-Mg catalysts it becomes possible to achieve the high degree
(>10 per cent) of the use of transition metal in the
formation of the active centers.

A specific feature of the ethylene polymerization
by Ti-Mg catalysts is the formation of polymer with
lower molecular weight in comparison with polyethylene,
prepared using conventional $TiCl_3$ catalysts (49). The
regulation of molecular weight by hydrogen is also more
facile for Ti-Mg catalysts. The reason for this is
higher values of the rate constants for transfer reactions with monomer and H_2 for ethylene polymerization
on Ti-Mg catalysts (Table 9) in comparison with polymerization on $TiCl_3$ (Table 6). Especially great differences between two catalysts were observed for the chain
transfer with hydrogen. This reaction is first order
in respect to the H_2 concentration for polymerization
on Ti-Mg catalyst, while for polymerization on $TiCl_3$
the reaction order is 0.5 (see Fig. 1). Rate constants
for transfer with $AlEt_3$ are close for both cases
(Tables 6 and 9). So the change of the active center
composition by the introduction of the magnesium ions
in the second coordination sphere of titanium ions may
result in the change of the reactivity of active centers
in chain transfer reactions.

Propylene Polymerization on Ti-Mg Catalysts

The rate of propylene polymerization may be considerably increased when supported or bulk Ti-Mg catalysts are used, but the polymer with low stereoregularity is formed (50,51). The increase of the stereospecificity of highly active catalysts may be achieved by the
introduction of electron donor compounds in the composition of Ti-Mg catalyst and in polymerization mixture
(52). By the variation of the type of organoaluminum
cocatalyst and the content and concentration of electron
donor modifiers it is possible to produce polypropylene
with isotacticity from 45 to 95% (50).

Higher activity of the supported Ti-Mg catalysts
in propylene polymerization is due both to the increase
of the number of active centers and to the evident

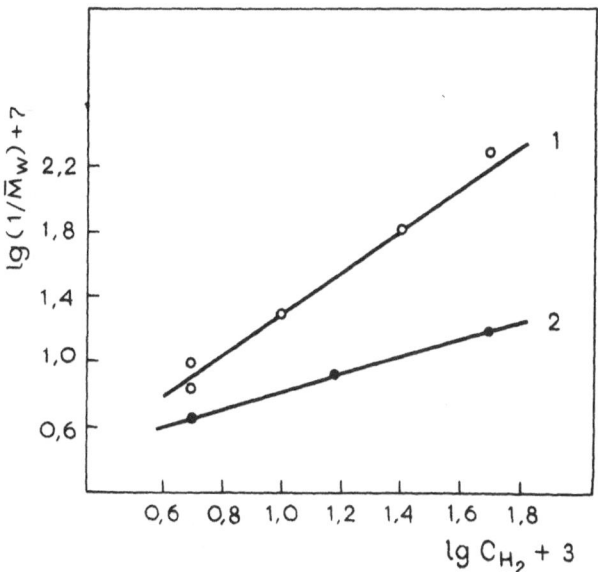

Fig. 1. Dependence of the polyethylene molecular weight
on the hydrogen concentration at ethylene polymerization on $TiCl_4/MgCl_2(P)+AlEt_3$ (curve 1)
and $\delta-TiCl_3 \cdot 0.3AlCl_3+AlEt_3$ (curve 2).
Temperature is 80°C.

increase of the propagation rate constant (53)(Table
10) in comparison with the polymerization by $TiCl_3$.
Higher values of K_P were found both with the one-component Ti-Mg catalyst active without organoaluminum
compounds and with the two-component supported catalyst.
Addition of the stereoregulating modifier (ethyl-benzoate) to Ti-Mg catalysts practically does not change
the K_P value. Recently the data were published (54)
on K_P values for propylene polymerization in the presence of $TiCl_4$-(ethyl-benzoate)-$MgCl_2$ catalyst. These
data are based on the measurement of the mean-number
molecular weight of polymer produced during a short
reaction time (5s). The values of K_P (500-1000 1/mol s)
are quite compatible with the results obtained with
the use of the ^{14}CO quenching technique.

Table 8. Maximum Activity and the Maximum
          Number of Active Centers for Ethylene
          Polymerization on Catalysts of
          Various Composition

Catalyst	R max $\dfrac{g\ PE}{g\ Ti \cdot h \cdot atm}$	C_P b) mol/mol Ti
δ-TiCl$_3 \cdot$0.3AlCl$_3$+AlEt$_3$	1400	0.001
TiCl$_3$/SiO$_2$+AlR$_3$	16000	0.012
TiCl$_2$/ MgCl$_2$a)	25000	0.018
TiCl$_3$ / MgCl$_2$a)+AlEt$_3$	87000	0.063
TiCl$_4$/MgCl$_2$ (P,AlCl$_3$)+AlEt$_3$	520000	0.38

a) For these systems with unsteady-state
   kinetics the values of the initial maximum
   activity are given; in other cases the data
   on steady-state activity are presented.
b) For calculation the value K_P=1.1$\cdot$10^4 1/mol s
   was used.

Table 9. Rate Constants for the Chain
          Transfer Reactions at Ethylene
          Polymerization on TiCl$_4$/MgCl$_2$(P)+
          +AlEt$_3$ (80°C)

$K(t,M)/\omega$ 1/mol$\cdot$s	$K(t,AlEt_3)$ $1^{0.5}$/mol$^{0.5}$s	$K(t,H_2)/\omega$ 1/mol$\cdot$s
0.22	1.70	7

Propagation rate constants for isotactic and atactic addition for propylene polymerization on Ti-Mg catalysts show close values (Table 11). This means that
the ratio between isotactic and atactic fractions in a
polymer is determined mainly by the relative numbers

Table 10. Number of Propagation Centers and Propagation Rate Constants for Propylene Polymerization on Catalysts of Various Composition (Temperature 70°C; C_M = 1 mol/l)

Catalyst	Quench-ing agent	Rate[a] $\dfrac{\text{g } C_3H_6}{\text{g Ti·h}}$	Yield, g/gTi	$C_P^{a)} \cdot 10^2$ $\dfrac{\text{mol}}{\text{mol Ti}}$	$K_P^{a)} \cdot 10^{-2}$ $\dfrac{\text{l}}{\text{mol·s}}$
δ-TiCl₃·0.3AlCl₃+AlEt₃	14CO	600	150	0.16	1.1
TiCl₂·nMgCl₂	14CO	180	45	0.012	4.7
	14CO₂	96	15	0.0045	6.9
TiCl₄/MgCl₂(P)+AlEt₃	14CO	12000	18500	0.52	7.4
	14CO	21600	10000	0.86	8.1
b) TiCl₄·EB/MgCl₂(P)+AlEt₃	14CO	7200	770	0.27	8.7
	14CO	26400	760	0.68	12.5

a) Data are presented for the fraction insoluble in boiling heptane;

b) EB - ethylbenzoate.

Table 11. Number of Active Centers and
Propagation Rate Constants for
Atactic[a] and Isotactic[b] Fractions
of Propylene (C_M = 1 mol/l;
Temperature 70°C)

Catalyst	$TiCl_2$ / $MgCl_2$			$TiCl_4$/$MgCl_2$(P)+AlEt$_3$	
Polymerization rate, g C_3H_6/gTi·h	760	520	340	27000	48000
C_p,i·10^3,mol/mol Ti	0.14	0.14	0.06	5.2	8.6
C_p,a·10^3,mol/mol Ti	0.16	0.32	0.16	4.1	6.3
K_p,i·10^{-2},1/mol·s	4.7	2.7	5.6	7.4	8.1
K_p,a·10^{-2},1/mol·s	6.8	2.6	2.7	6.2	7.3
Quenching agent	^{14}CO	^{14}CO$_2$	^{14}CO$_2$	^{14}CO	^{14}CO

a) Fraction soluble in boiling ether;
b) Fraction insoluble in boiling heptane.

of stereospecific and non-stereospecific active centers.

Table 12 gives the data on the maximum activity measured in the authors' laboratory for propylene polymerization by catalysts of different composition and corresponding numbers of active centers in these catalysts. The number of active centers for ethylene and propylene polymerization are close when identical samples of Ti-Mg catalyst were used. For example, the number of propylene polymerization active centers (including atactic ones) in $TiCl_4$/$MgCl_2$+AlEt$_3$ catalyst (see Table 12) is 0.06 mol/mol Ti. The same catalyst sample in ethylene polymerization shows the activity 70 kg C_2H_4/g Ti x h x atm, so the number of active centers (taking into account that K_p=1.1x10^4 1/mol s) is 0.055 mol/mol Ti.

The increase of K_p for propylene polymerization on Ti-Mg catalysts is hardly to be explained by the change of the electronic state of titanium ions in the active centers because this change has to influence also the K_p values for active centers at ethylene polymerization. It is possible to suppose that the titanium ion in active

Table 12. Maximum Number of Active Centers
for Propylene Polymerization on
Catalysts of Different Composition[a]

Catalyst	R max $\dfrac{\text{kg } C_3H_6}{\text{g Ti·h·atm}}$	C_P max $\cdot 10^2$ mol/mol Ti	$K_P \cdot 10^{-2}$ 1/mol· s
δ -TiCl$_3 \cdot$ 0.3 AlCl$_3$	0.63	0.65	0.9
TiCl$_4$/MgCl$_2$·(P)	35	4.2	8.0
TiCl$_4 \cdot$ EB/MgCl$_2$·(P)	28	2.8	10.0

a) Data are presented for the polymer fraction insoluble in boiling ether.

centers of Ti-Mg catalysts is more accessible for propylene coordination due to the change in the steric environment of the chlorine atoms in comparison with
active centers on the surface of TiCl$_3$. This results
in lowering the activation energy of the coordination
step for propylene but not for ethylene. This explanation of the change of K_P values for propylene polymerization is possible if the rate determining step of propagation on Ti-Mg catalysts is the olefin coordination as
it was proposed for polymerization on TiCl$_3$.

Aknowledgement

        The authors aknowledge the contribution of Mr.
S. Shepelev (the determination of the number of active
centers for VCl$_3$ and Ti-Mg catalysts), Mr. S. Sergeev
(the preparation of Ti-Mg catalysts for propylene polymerization) and Mr. S. Makhtarulin (the preparation of
Ti-Mg catalysts for ethylene polymerization).

LIST OF ABBREVIATIONS

A - frequency factor.

E - activation energy (E_{ov} - for over-all process, E_P -
for propagation).

C_{H_2} – concentration of hydrogen.

C_M – concentration of monomer.
C_P – concentration of active centers.
$C_{P,a}$ – concentration of atactic active centers.
$C_{P,i}$ – concentration of isotactic active centers.
$C_{P,max}$ – concentration of active centers, corresponding
     to the maximum observed polymerization rate.
K_P – propagation rate constant ($K_{P,a}$ – for atactic addi-
     tion; $K_{P,i}$ – for isotactic addition).
$K(t,x,M)$ – rate constant of the chain transfer reaction
     with agent X for polymerization of the monomer M,
     e.g. $K(t, AlEt_3, C_2H_4)$ – rate constant of the
     transfer reaction with $AlEt_3$ from polymerization of
     C_2H_4.
P – polymer residue in a catalyst.
PE – polyethylene.

REFERENCES

1. H. Schnecko, W. Kern, Chem.-Ztg. Chem. Appar., 94,
     229 (1970).
2. Yu.I. Yermakov and V.A. Zakharov, Uspek. Khim., 41,
     377 (1972).
3. G.F. Feldman, E. Perry, J. Polym. Sci., 46, 217
     (1960).
4. G.D. Bukatov, N.B. Chumaevskii, V.A. Zakharov, G.I.
     Kuznetsova and Yu.I. Yermakov, Macromol. Chem.,
     178, 953 (1977).
5. P.J.T. Tait, in "Coordination Polymerization" ed.
     by J.C.W. Chien, Acad. Press, N.Y.-San Francisco-
     London, 1975, p. 155.
6. Yu.I. Yermakov, V.A. Zakharov and G.D. Bukatov,
     Proc. V Int. Congr. Catal., Amsterdam-N.Y., 1973,
     v. 1, p. 339.
7. V.A. Zakharov, G.D. Bukatov, N.B. Chumaevskii and
     Yu.I. Yermakov, React. Kinet. Catal. Lett., 1,
     247 (1974).
8. Yu.I. Yermakov and V.A. Zakharov, in "Coordination
     Polymerization" ed. by J.C.W. Chien, Acad. Press,
     N.Y.-San Francisco-London, 1975, p. 91.
9. G.D. Bukatov, V.A. Zakharov and Yu.I. Yermakov,
     Kinetika i Kataliz, 16, 645 (1975).
10. V.A. Zakharov, G.D. Bukatov, V.K. Dudchenko and
     Yu.I. Yermakov, Kinetika i Kataliz, 16, 417
     (1975).
11. E.J. Arlman, P. Cossee, J. Catal., 3, 89, 99 (1964).
12. E.J. Arlman, J. Catal., 5, 178 (1966).
13. L.A.M. Rodriques and H.M. van Looy, J. Polym. Sci.,

A1, 4, 1971 (1966).

14. V.M. Buls, T.L. Higgins, J. Polym. Sci., A1, 8, 1025 (1970).
15. W.R. Carradine and H.F. Rase, J. Appl. Polym. Sci., 15, 889 (1971).
16. A. Munoz-Escalona and A. Parada, Polymer., 20, 474 (1979).
17. Z. Wm. Wilshinsky, R.W. Looney and G.M. Tornqvist, J. Catal., 28, 351 (1973).
18. V.A. Zakharov, P.A. Zhdan, E.E. Vermel and S.G. Artamonova, Kinetika i Kataliz, 16, 1184 (1975).
19. N.B. Chumaevskii, V.A. Zakharov, G.D. Bukatov and Yu.I. Yermakov, Makromol. Chem., 177, 747 (1976).
20. V.A. Zakharov, N.B. Chumaevskii, G.D. Bukatov and Yu.I. Yermakov, Macromol. Chem., 177, 763 (1976).
21. V.A. Zakharov, G.D. Bukatov, N.B. Chumaevskii and Yu.I. Yermakov, Macromol. Chem., 178, 967 (1977).
22. P.E.M. Allen, A.E. Byers, Trans. Farad. Soc., 67, 1718 (1971).
23. Kh.S. Bagdasarjan, Theory of Radical Polymerization, Nauka, M., 1966.
24. P. Cossee, J. Catal., 3, 80 (1964).
25. V.I. Avdeev, I.I. Zakharov, V.A. Zakharov, G.D. Bukatov and Yu.I. Yermakov, J. of Structural Chemistry, 18, 525 (1977).
26. O. Novaro, E. Blaisten-Barojas, E. Clementi, G. Giunchi and M.E. Ruizvizcaya, J. Chem. Phys., 68, 2337 (1978).
27. I.I. Zakharov and V.A. Zakharov, React. Kinet. Catal. Lett., 14, 169 (1980).
28. G. Natta, F. Danusso and D. Sianesi, Macromol. Chem., 30, 236 (1959).
29. N.M. Chirkov, Proc. IUPAC Int. Symp. on Macromolecular Chemistry, Budapest, Akad. Kiado, 1971, p. 365.
30. T. Keii, Kinetics of Ziegler-Natta Polymerization, Kodansha, Tokyo, 1972.
31. W.L. Smith and T. Wartik, J. Inorg. Nucl. Chem., 29, 629 (1967).
32. A.T. Clocks and K.W. Egger, J. Chem. Soc., Faraday Trans. 1, 68, 423 (1972).
33. G. Natta and I. Pasquon, Adv. Catal., 11, 1 (1959).
34. L.P. Ivanov, Yu.I. Yermakov and A.I. Gelbshtein, Vysokomol. Soed., A 9, 2422 (1967).
35. V.A. Zakharov, N.B. Chumaevskii, Z.K. Bukatova, G.D. Bukatov and Yu.I. Yermakov, React. Kinet. Catal. Lett., 5, 429 (1976).
36. L.L. Böhm, Polymer, 19, 562 (1978).

37. N.G. Maksimov, E.G. Kushnareva, V.A. Zakharov, V.F. Anufrienko, P.A. Zhdan and Yu.I. Yermakov, Kinetika i Kataliz, 15, 738 (1974).
38. V.A. Zakharov, Kinetika i Kataliz, 21, 892 (1980).
39. S.I. Makhtarulin, E.M. Moroz and V.A. Zakharov, React. Kinet. Catal. Lett., 9, 269 (1978).
40. S.I. Makhtarulin, V.A. Zakharov, E.M. Moroz and N.G. Maksimov, in: Catalysts containing supported complexes, Novosibirsk, Institute of Catalysis, 1, 1980, p. 205.
41. V.A. Zakharov and Yu.I. Yermakov, in: Catalysts and Catalytic Processes, Institute of Catalysis, 1977, p. 135.
42. V.A. Zakharov, S.I. Makhtarulin, Yu.I. Yermakov and V.E. Nikitin, USSR Patent, N 726702; E.E. Vermal, V.A. Zakharov, S.I. Makhtarulin and V.E. Nikitin, USSR Patent, N 701698, Bull. izobr. N 45, 1979.
43. S.I. Makhtarulin, V.A. Zakharov and V.E. Nikitin, USSR Patent, N 667232, Bull. izobr., N 22, 1979.
44. S.I. Makhtarulin, V.A. Zakharov and Yu.I. Yermakov, USSR Patent, N 677187, Bull. izobr., n 1, 1981.
45. V.A. Zakharov, N.B. Chumaevskii and S.I. Makhtarulin, React. Kinet. Catal. Lett., 2, 329 (1975).
46. A.A. Baulin, V.N. Sokolov and A.A. Semenova, Vysokomol. Soed., 17A, 46 (1976).
47. D. G. Boucher, I.W. Parsons and R.N. Haward, Macromol. Chem., 175, 3461 (1974).
48. H. Meyer and V.H. Reichert, Angew. Makromol. Chem., 57, 211 (1977).
49. V.A. Zakharov, Z.K. Bukatova, S.I. Makhtarulin, N.B. Chumaevskii and Yu.I. Yermakov, Vysokomol. Soed., 21A, 466 (1979).
50. E.E. Vermel, V.A. Zakharov and Z.K. Bukatova, Vysokomol. Soed., 22A, 22 (1980).
51. K. Gardner, I.W. Parsons and R.N. Haward, J. Polym. Sci., Polym. Chem. Ed., 16, 1683 (1978).
52. V. Giannini, A. Cassata, P. Songi and R. Mazzochi, BRD Patents, 2230 728, 2230 752.
53. S.N. Sheplev, G.D. Bukatov, V.A. Zakharov and Yu.I. Yermakov, Kinetika i Kataliz, 22, 258 (1981).
54. E. Suzuki, M. Tamura, Y. Doi and T. Keii, Macromol. Chem., 180, 2235 (1979).

SOME ASPECTS OF COORDINATION HOMO- AND COPOLYMERIZATION OF HIGH α-OLEFINS, CYCLIC AND ALLENE HYDROCARBONS

B. A. Krentsel, V. I. Kleiner, E. A. Mushina
and L. L. Stotskaya

The Topchiev Institute of Petrochemical
Synthesis Academy of Sciences of the U.S.S.R.
Moscow, U.S.S.R.

INTRODUCTION

The discovery of the principles and methods of stereospecific polymerization made with the help of complex metallic catalytic systems has in essence caused
a revolution in macromolecular chemistry which has meant
that many of the established notions have had to be re-
considered and has given rise to new works and ideas
in the polymerization of the carbochain and hetero-
atomic monomers.

In particular, these approaches include the concept of donor-acceptor interactions which plays an
important part in the polymerization of many monomers,
both in the presence of compounds of transition metals
and under the influence of other initiating systems.

On the other hand, of significant theoretical and
practical interest are investigations of the possibi-
lities of coordination polymerization of high α-olefins and their analogs complicated by the side processes of isomerization of these monomers.

π -allyl complexes have greatly assisted the understanding of the coordination and stereoregulation
mechanism of the polymerization of unsaturated hydrocarbons. These complexes have not only made it possible
to obtain stereoregular polymers of dienes, but to

bring about the polymerization of allene hydrocarbons.

Of course, in one article it is impossible to consider all the aspects of co-ordination polymerization on
which olefin polymerization laboratory at the Topchiev
Institute of Petrochemical Synthesis of the U.S.S.R.
Academy of Sciences has been concentrating. Therefore
we are going to confine ourselves to considering just
those aspects with which the most important experimental
results are connected, in our opinion, and which are of
general interest for the further development of ideas
on co-ordination polymerization.

THE PECULIARITIES OF IONIC CO-ORDINATION HOMO -AND COPOLYMERIZATION OF HIGH α-OLEFINS

The polymerization of high olefins is characterized by
many peculiarities, in particular, by isomerization
transformations, high catalyst consumption, etc. Without
knowing these peculiarities it is impossible to create
effective methods of obtaining new types of polyolefins
which would possess higher thermal stability and satisfactory mechanical properties.

For a few years now we have been investigating general regularities and peculiarities of homo- and copolymerization of vinyl cyclohexane /VCH/, 4-methyl pen-
tene-1 /MP/ and 3-methyl butene-1 /MB/l.

It has been found that during the polymerization of
VCH in the presence of various complex organometallic
catalysts (Table 1) as well as VCH polymerization at
temperatures of 40-80°C a number of competitive reactions occur, among them isomerization.

Using low-temperature gas-liquid chromatography it
has been shown that VCH isomerization does not stop at
the stage of double bond migration into the ß - position with the formation of ethylidene cyclohexane. The
double bond migrates farther into the ring, resulting
in the formation of a mixture of VCH endocyclic isomers
(1-ethyl-, 3-ethyl- and 4-ethyl cyclohexenes). Besides
exo- and endocyclic isomers, ethyl cyclohexane has been
found among the by- products of VCH transformation. In
general, the process of VCH polymerization is described
by the following scheme:

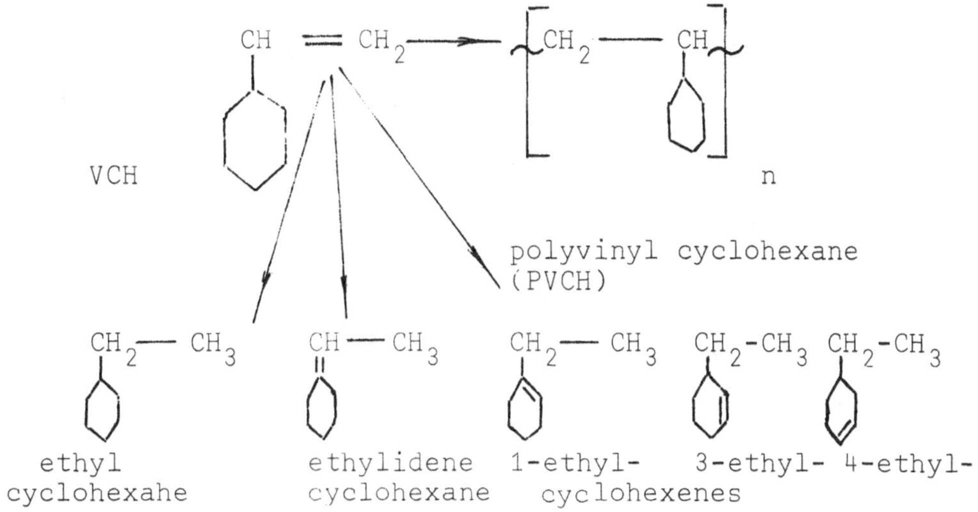

polyvinyl cyclohexane (PVCH)

ethyl cyclohexahe   ethylidene cyclohexane   1-ethyl- cyclohexenes   3-ethyl- 4-ethyl-

        The formation of ethyl cyclohexane may be attri-
buted to the following reactions:
1) the catalytic hydrogenation of VCH by hydrogen evolved
upon the formation of a catalytic complex;
2) the decomposition on treating the sample under chromatography with alcohol of primary and secondary ethylcyclo-
hexyl organo- aluminium compounds or their hydrides formed during the VCH interaction with organo-aluminium
compounds. Alkyl-aluminium hydrides that react with VCH
are always present in organo-aluminium compounds at temperatures above 40°C at which the VCH polymerization
is performed;
3) the disproportionation of organo-titanium compounds

        Table 1 presents comparative data on, the activities
of various catalysts in VCH polymerization reactions,
its isomerization and transformation into ethyl
cyclohexane. None of the catalytic systems under the
conditions studied gives complete conversion of the
monomer into a polymer. It is thought that this is
conditioned by the formation of low reactive VCH isomers,
which deactivate the catalyst upon adsorption at the
active sites of propagation. At the same time, the above
deactivation does not influence the yield of by-products
which testifies to the different nature of the active
sites at which the reaction of the VCH polymerization
and isomerization and monomer transformation into ethyl
cyclohexane take place. The yield of by-products of
the VCH transformation grows as the temperature, reaction
duration, Al/Ti ratio and catalyst concentration, are
increased.

1.

$$CH=CH_2 \qquad\qquad CH_2 \text{---} CH_3$$

$$+ \quad H_2 \longrightarrow$$

2.

$$CH=CH_2 \quad + \quad HAl(i - C_4H_9)_2$$

$$(i - C_4H_9)_2Al-CH_2-CH_2 \qquad (i-C_4H_9)_2AlCHCH_3$$

ROH

$$CH_3\text{---}CH_2$$

$$+ \quad i - C_4H_{10} + Al(OR)_3$$

3.

$$CH=CH_2 \quad + \quad HTiCl_3 \rightleftharpoons H\text{---}CH_2\text{---}CH\text{---}TiCl_3$$

$$CH_3 \text{---} CH_2 \quad + \quad CH_2 = CH \quad + \quad TiCl_3$$

The replacement of $TiCl_4$ by $TiCl_3$ with the same organo-aluminium component increases the polymer yield by 20%, that is, apparently, conditioned by the lower isomerizing activity of catalysts containing TiCl3. Thus, it can be seen from Table 1 that the yield of VCH isomers on the $TiCl_3$ + Al(iso-C_4H_9)$_3$ system is five times less than that on the $TiCl_4$ + Al(iso-C_4H_9)$_3$ system.

At the same time, it can be seen that during the reaction of VCH with titanium salts only $TiCl_3$ causes isomerization. On the basis of these data the conclusion was drawn that VCH isomerization passes through the stage of the preliminary formation of a VCH π-complex with a transition metal and subsequent double bond migration according to the π-allyl or σ-alkyl mechanism.

Table 1.   VCH transformations into ethylcyclohexane,
           isomers with an internal double bond and
           PVCH under the influence of various catalysts

Catalyst	Yields of transformation products, mass %		
	Ethyl cyclo-hexane	Isomers with an internal double bond	PVCH
1. $TiCl_3/Al(i-C_4H_9)_3$	9.5	5.4	70
2. $TiCl_3/Al(C_2H_5)_3$	10.2	9.7	55
3. $TiCl_4/Al(i-C_4H_9)_3$	8.4	28.8	50
4. $TiCl_4/Al(C_2H_5)_2Cl$	0.0	0.0	10
5. $TiCl_4/Al(i-C_4H_9)_2Cl$	2.0	2.0	1
6. α-$TiCl_3$	0.0	10.3	0
7. $TiCl_4$, $Al(C_2H_5)_2Cl$	0.0	0.0	0
8. $Al(C_2H_5)_3$	3.0	0.0	0
9. $Al(i-C_4H_9)_3$	28.6	3.5	0

$80^{\circ}C$; 4 hr; $[VCH]o = 2$ mole/l; $[Ti]o = 0.0537$ mole/l;
Al/Ti= 3; solvent - n-heptane.

Bearing in mind the well-known fact that the transition metal is reduced during the interaction of organo-
aluminium compounds with titanium salts, it can be
considered reasonable that the proposed mechanism of VCH
isomerization is also valid for double bond migration
in the monomer at its polymerization in the presence of
the catalytic complex.

The suggested mechanism for the various reactions
of VCH transformation is in agreement with the order
of these reactions with respect to the monomer
as determined from kinetic data. The order of the
polymerization reaction and the overall reaction of
the VCH isomerization into isomers with an internal
double bond is close to unity, and the reaction of
the VCH transformation into ethyl cyclohexane has a
zero order with respect to the monomer.

The above data were taken as the basis for
elaborating the method of VCH polymerization in a
condensed propane medium on the catalytic $TiCl_3$ + AlR_3

system, allowing the reduction of the side reactions
of monomer transformation to a minimum and making it
possible to obtain PVCH with an 80-85% yield instead
of the 50% known before[2]. The use of condensed propane
as a diluent ensures an increase in the polymer yield,
simplifies the process of diluent recirculation and
makes it possible to regenerate the unreacted VCH.
All this is conditioned by a decrease in the rate
constant of VCH isomerization into isomers with an
internal double bond.

Such are the peculiarities of the VCH polymerization on the complex organometallic catalysts, which
are to be taken into account when PVCH synthesis
takes place and which are also typical of MB and MP
polymerization.

During the MB polymerization in the presence of
various catalytic systems (Table 2) the polymerization
process is accompanied by MB isomerization into
2-methyl butene-2 and 2-methyl butene-1. Systems
containing LiC_4H_9 or $Al(C_2H_5)_3$ possess the maximum
isomerization activity.

According to the data on the temperature dependence
of the initial isomerization rate it was found that
the activation energy of MB isomerization into 2-methyl
butene-1 and 2-methyl butene-2 in the presence of
the catalytic $TiCl_4$ + $Al(iso-C_4H_9)_3$ system is equal
to 88.0 $\pm$ 3.3kJ/mole (in the temperature range of
20-80°C). For the $TiCl_4$ + $Al(C_2H_5)_3$ system the activation
energy of isomerization decreases as the temperature
rise and makes up 84.0 $\pm$ 6.3 kJ/mole at 15-40°C and
22.0 $\pm$ 6.3 kJ/mole at 60-80°C. These results and the
presence of $TiCl_2$ in the solid component of the
catalytic complex testify to the fact that MB isomerization mainly takes place at active sites containing
Ti^{+2}, according to the π-allyl or δ-alkyl mechanism.

Just as in VCH polymerization, in the case of
MB it is also impossible to achieve a 100% monomer
conversion into a polymer. However, in MB polymerization on organometallic catalysts the yield of poly-3-
methyl butene-1 (PMB) does not exceed 3-8 g/g Ti
(for PVCH the corresponding value is over 100 g/g Ti)
and hardly depends on the nature of the catalytic
system. The polymerization activity of catalytic systems determined from the values of the initial MB

Table 2. MB transformation under the influence
of various catalysts

Catalyst	Yields of the transformation products, mass %		
	2-methyl butene-2	2-methyl butene-1	Polymer
$TiCl_4/LiC_4H_9$	32	0.67	12
$TiCl_4/Al(C_2H_5)_3$	18	0.22	20
$TiCl_4/Al(i-C_4H_9)_3$	16	0.11	28
$TiCl_3/Al(C_2H_5)_3$	13	0.10	20
$TiCl_3/Al(C_2H_5)_2Cl$	0	0.0	12
α-$TiCl_3$	0.1	0.01	0
$TiCl_4$, YCl_3, $Al(C_2H_5)_3$	0	0	0
$Al(i-C_4H_9)_3$, LiC_4H_9			
$Al(C_2H_5)_2Cl$			
$TiCl_3/Al(i-C_4H_9)_3$	2.6	0.08	22

80°C; 6hr; $[MB]_o$ = 1.88 mole/l, $[Ti]_o$ = 0.057 mole/l;
Al/Ti = 2; solvent - n-heptane.

polymerization rates corresponds to the following
series: $Al(iso-C_4H_9)_3$ + $TiCl_4 >$ $Al(C_2H_5)_3$ + $TiCl_4 \approx$
$Al(iso-C_4H_9)_3$ + $TiCl_3 >$ $Al(C_2H_5)_2Cl$ + $TiCl_4$. The MB
polymerization reaction is of the first order with
respect to the monomer. The rate dependence of MB
polymerization on the catalyst concentration is
described by the law of the first order for the $TiCl_4$ +
$Al(iso-C_4H_9)_3$ system and of the second order for the
$TiCl_4$ + $Al(C_2H_5)_3$ and $TiCl_3$ + $Al(iso-C_4H_9)_3$ systems.

Irrespective of the catalyst concentration, the
rate of MB polymerization decreases with time, and
the yield of PMB reaches its maximum value in a fairly
short period of time, after which it remains practically
unchanged. It should be emphasized that the MB polymerization rate does not drop at the expense of monomer
consumption to form the polymer: in a number of cases
the reaction mixture contains up to 80-90% of unreacted
MB. The peculiarities of MB polymerization observed
are conditioned, on the one hand, by the adsorption
of 2-methyl butene-1 and 2-methyl butene-2, formed

from MB as a result of the competitive side isomerization reaction, on catalytic active sites of polymerization and on the other hand, by PMB adsorption
on the catalytic sites which leads to the diffusion
retardation of MB polymerization. The latter fact is
of specific significance for PMB - a polymer that
is practically insoluble and unswelling in the known
solvents of polyolefins.

Unlike MB, MP polymerization on the complex organometallic catalysts results in the formation of
poly-4-methyl pentene-1 (PMP) with a relatively high
yield reaching hundreds of grams of the polymer per
gram of the transition metal. However, in this
case the isomerization of MP into 4-methyl pentene-2
and 2-methyl pentene-2 occurs simultaneously. As can
be seen from Table 3, according to their activities
in the isomerization reaction the catalysts may be
placed in the following series: $TiCl_4 + Al(iso-C_4H_9)_3 >$
$TiCl_3 + Al(iso-C_4H_9)_3 > VCl_4 + Al(iso-C_4H_9)_3 > VCl_3$
$+ Al(iso-C_4H_9)_3$. The yield of isomers is in direct
proportion to the catalyst concentration. The mechanism
of MP isomerization is similar to MB and VCH isomerization.

Table 3   Dependence of the yield of 4-methyl
           pentene-2 on the catalyst concentra-
           tion during MP polymerization

Catalyst	The molar ratio of MP to 4-methyl pentene-2	
	$[M]^a_o/[MP]_o = 0.02$	$[M]_o/[MP]_o = 0.07$
$TiCl_3/Al(i-C_4H_9)_3$	1 : 0.18	1 : 1.10
$TiCl_4/Al(i-C_4H_9)_3$	1 : 0.23	1 : 4.80
$VCL_3/Al(i-C_4H_9)_3$	1 : 0.10	1 : 0.25
$VCl_4/Al(i-C_4H_9)_3$	1 : 0.10	1 : 0.33

[a] $[M]_o$ - transition metal of the catalyst; $80°C$;
4 hr; $[MP]_o$ = 2 mole/l; Al/M = 3; solvent - n-heptane

The study of the influence of β-isomers on the process of MP polymerization showed that the introduction of 4-methyl pentene-2 and 2-methyl pentene-2 into the reaction mixture decreased the PMP yield somewhat but did not affect the molecular mass and the stereoregularity of the polymer. This testifies to the existence of two types of active sites where isomerization and polymerization of MP proceed. The presence of β-isomers in the reaction medium results in partial deactivation of the polymerization active sites, in a decrease in their concentration, and, as a consequence, to a decrease in the MP polymerization rate. The weak deactivation influence of β-isomers formed during MP polymerization on the polymerization catalytic activity and the possibility of obtaining PMP with a high yield are conditioned by a considerable difference in the rate constants of ionic co-ordination polymerization and isomerization of MP.

Summing up all this, it should be emphasized that the side reactions of isomerization have a strong effect on ionic co-ordination polymerization of the olefins with a relatively low reactivity (VCH, MB) and have hardly any effect on the polymerization of monomers whose polymerization rate constants are much higher than those of isomerization rate (MP).

The isomerization of high α-olefins into isomers with an internal double bond apparently proceeds on the active sites containing Ti^{+2}. This is why the complex organometallic catalysts on the basis of $TiCl_4$ or $TiCl_3$ and $Al(C_2H_5)_2Cl$ in which the content of Ti^{+2} sites is low because of the weak reducing capacity of diethylaluminium chloride, do cause very little isomerization of VCH, MB and MP.

The polymerization reactions of VCH, MB and MP in the presence of complex organometallic catalysts that induce no isomerization of monomers are characterized by the regularities common to α-olefin polymerization on Ziegler-Natta catalysts.

Thus, the MP polymerization on the $TiCl_3 + Al(C_2H_5)_2Cl$ system reveals an induction period whose duration decreases with the rise in temperature and makes up 13 and 3 min at 30 and 45°C respectively. MP polymerization at 60° begins without an induction period. On the basis of these data the values of the initiation rate constants at 30 and 45°C were determined, equalling $1.07 \cdot 10^{-3}$ and $4.6 \cdot 10^{-3}$ l/mole·s respectively.

Within the temperature range investigated one could observe an increase in the molecular weight of PMP in the process of polymerization. This justifies the supposition that MP polymerization leads to the formation of "living" macrochains of PMP.

The formation of "living" macrochains of PMP during MP polymerization in the presence of the catalytic $TiCl_3$ + $Al(C_2H_5)_2Cl$ system and the similar behaviour of propylene upon polymerization on this catalytic system make it possible to obtain block copolymers of MP with propylene by successive polymerization of the above monomers on the "living" macrochains of both polypropylene and PMP.

Block copolymers of VCH with propylene containing up to 15 mole % of VCH links were obtained similarly by the reaction of successive VCH additions on the "living" polypropylene macrochains. The elementary rate constant of the VCH block copolymerization (at 30°C) which is equal to 0.002 l/mole·s was determined from the values of the effective rate constant of VCH block-copolymerization with the "living" polypropylene macro-chains ($0.13 \cdot 10^{-3}$ l/mole $TiCl_3$·s).

Besides the successive polymerization of MP and VCH with propylene to obtain blockcopolymers the copolymerization of MP, VCH and MB with each other and with other olefins was investigated (Table 4).

Table 4   Constants of ionic coordination copolymerization of VCH (M_I) with various monomers

Comonomer	r_1	r_2	$r_1 \cdot r_2$
Propylene	0.049	80	3.9
4-Vinyl cyclo-hexene-1	0.21	3.78	0.78
MP	0.41	3.14	1.29
Styrene	0.18	2.12	0.38
MB	0.98	1.02	1

An anylysis of the data obtained on ionic co-ordination copolymerization of VCH with various monomers in the presence of catalytic systems based on titanium salts and aluminium alkyls permits us to arrange these monomers according to their reactivity in the following series: propylene $>$ 4-vinyl cyclo-hexene-1 $>$ MP $>$ styrene $>$ MB $\approx$ VCH.

The distribution of monomer links in copolymers of VCH with various comonomers depends on the nature of the substituent in the comonomer. Copolymers of VCH with ethylene and propylene are products with block distribution of ethylene and propylene links $r_1 \cdot r_2 > 1$). Apparently, the tendency to form elements of block structure is also typical of MP copolymers with propylene ($r_1 = 0.31 \pm 0.05$; $r_2 = 6.44 \pm 0.50$; $r_1 \cdot r_2 = 2$). This supposition is substantiated by the data of IR-spectroscopy, X-ray examination and DTA of copolymers. VCH copolymerization with 4-vinyl cyclo-hexene-1, MP and MB leads to copolymers with statistical distribution of the links (for all pairs $r_1 \cdot r_2 \approx 1$). VCH copolymers with styrene exhibit a tendency to monomer alteration ($r_1 \cdot r_2 < 1$).

The experimental results considered characterize the peculiarities of co-ordination polymerization of high α-olefins and must be taken into account when effective methods to obtain polymers on their basis are to be worked out.

COORDINATION POLYMERIZATION OF ALLENE HYDROCARBONS

The allene, the simplest representative of the class of allene hydrocarbons by its physical properties occupies an intermediate position between ethylene and acetylene.

The two double bonds, and hence the methylene groups of the allene are in reciprocally perpendicular planes. This stereochemical picture follows from the description of the allene by atomic orbitals. In this model the central atom forms two collinear sp-σ-bonds with the terminal sp^2-hybridized carbon atoms. The two electrons at the central carbon atom occupy p-orbitals situated at a right angle to one another. π-bonds are formed by overlapping the p-orbitals of the central atom and the p-orbitals of the terminal carbon atoms.

Such a structure of cumulative Diene-Allene leads
to certain peculiarities in the polymerization reaction
under the influence of transition metals complexes
and so it deserves special consideration.

Previously we investigated polymerization of allene
and its derivatives under the influence of bis-π-al-
lylnickel halide complexes. The influence of the halide
nature on the reactivity of the complex was shown.
Thus, the relative allene polymerization rate increases
in the series $Cl < Br < I$. Such ligands as triphenylphos-
phine and bipyridyl inhibit the reaction. The main
kinetic regularities of the allene and 1,1-dimethyl-
allene polymerization were determined on bis-π-al-
lylnickel bromide. The IR-spectra testify to the
following structures of the obtained polymers:[3,4]

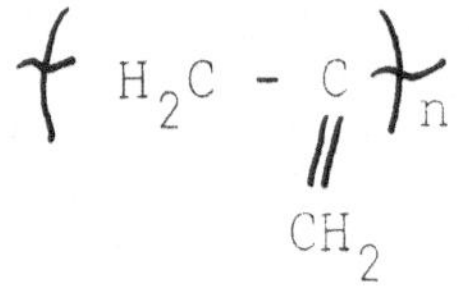

$$\left. \left\{ H_2C - \underset{\underset{CH_2}{\|}}{C} \right\}_n \right.$$

polyallene

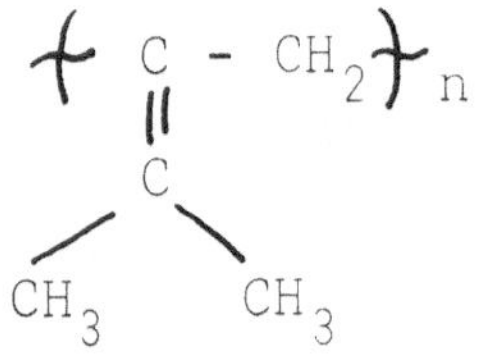

$$\left\{ \underset{\underset{\underset{CH_3 \quad CH_3}{\diagup \diagdown}}{C}}{\underset{\|}{C}} - CH_2 \right\}_n$$

polydimethylallene

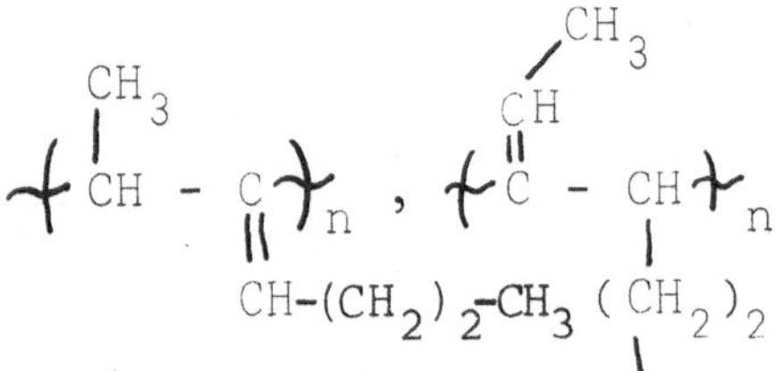

$$\left\{ \underset{\underset{CH-(CH_2)_2-CH_3}{\|}}{\underset{CH_3}{\overset{|}{CH}} - C} \right\}_n , \quad \left\{ \underset{\underset{(CH_2)_2}{|}}{\overset{\overset{CH_3}{\diagup}}{\underset{\|}{CH}}} - CH \right\}_n$$

polyheptadiene - 2,3

$$\left\{ \underset{C}{C} = C - C \right\}_n \qquad \underset{CH_3}{\overset{|}{}}$$

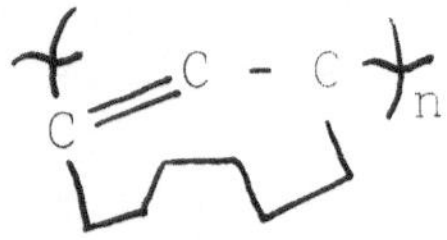

polycyclononadiene

One of the most important results of the present
work was the establishment by experiments that the
polymerization of allene hydrocarbons under the influence of bis-π-allylnickelhalides proceeds according
to the type of "living" chains:
the polymer chain resumes its propagation when a new
portion of the monomer is added. The intrinsic viscosity
of polydimethylallene increases with time, corresponding
to the increase in the polymer molecular weight determined
by the osmometry method from 30.000 (η int. = 0.025)
to 67,000 (η int. = 0.184).

To elucidate the nature of the active sites in the allene hydrocarbons polymerization under the influence of bis-$\widetilde{\pi}$-allylnickelhalide, we investigated adducts, formed in mixing benzene solutions of dimethylallene and bis-$\widetilde{\pi}$-allylnickelbromide.

We showed by cryoscopic titration under argon that at a concentration of bis-$\widetilde{\pi}$-allylnickelbromide equal $1.56 \cdot 10^{-3}$ mole/l, the $\widetilde{\pi}$-allyl complex was dissociated by 37-40% and at a concentration of $1.54 \cdot 10^{-2}$ mole/l - by 18-20%. Hence, with a concentration increase of the $\widetilde{\pi}$-allyl complex in benzene the amount of the dimer form also increases. The addition of dimethylallene to bis-$\widetilde{\pi}$-allylnickelbromide results in the monomerization of the complex (the molecular weight of the interaction product of bis-$\widetilde{\pi}$-allylnickelbromide with dimethylallene - 240, calculated - 247).

In the IR-spectrum of the interaction product of bis-$\widetilde{\pi}$-allylnickelbromide with dimethylallene (Fig. 1) all the bands in the 1970 cm^{-1} region (ν_{as} for C=C=C) and in the 1020 cm^{-1} region (ν_{as} for C=C=C) disappear. The bands in the 915, 1000, 1300, 1635 and 3080 cm^{-1} regions may unambiguously testify to the presence of the vinyl group -CH=CH$_2$ in the complex, and the slight shift of the absorption band (ν_{as} for C=C) to the long wave region in comparison with its usual position (1640 cm^{-1}) may be the result of the co-ordination of this bond with Ni. In the 1800-1825 cm^{-1} region of the spectrum there are two absorption bands one of which is, apparently, the overtone of the 915 cm^{-1} band, and the other may be attributed to the vibrations of the Ni-H bond, which arises from the complex formation process. In the spectrum of the interaction product of bis-$\widetilde{\pi}$-allylnickelbromide with partially deuterated dimethylallene there is one band of weak intensity in the 1810 cm^{-1} region. Another band in this region disappears, and there appears a band of approximately the same intensity at 1285 cm^{-1}, where according to the rule of summation, an absorption band of the Ni-D bond must be present. In the IR-spectrum of the interaction product of bis-$\widetilde{\pi}$-allylnickelbromide with dimethylallene decomposed under air (Fig. 1) the 1800 cm^{-1} band disappears completely, the one that may be related to the Ni-H bond. The band in the 1670 cm^{-1} region and the two bands in the 970 and 740 cm^{-1} region appear simultaneously. These bands characterize internal double bonds -CH=CH-. Hydrolysis of the interaction product of bis-$\widetilde{\pi}$-allylnickelbromide with dimethylallene results in the evolution of hydrogen (established chromatographically).

It may be supposed that the appearance of the internal
double bond and the evolution of hydrogen is due to
the fact that bis-π -allylnickelbromide and dimethylallene interact to form an organometallic compound
of the carbenhydride type

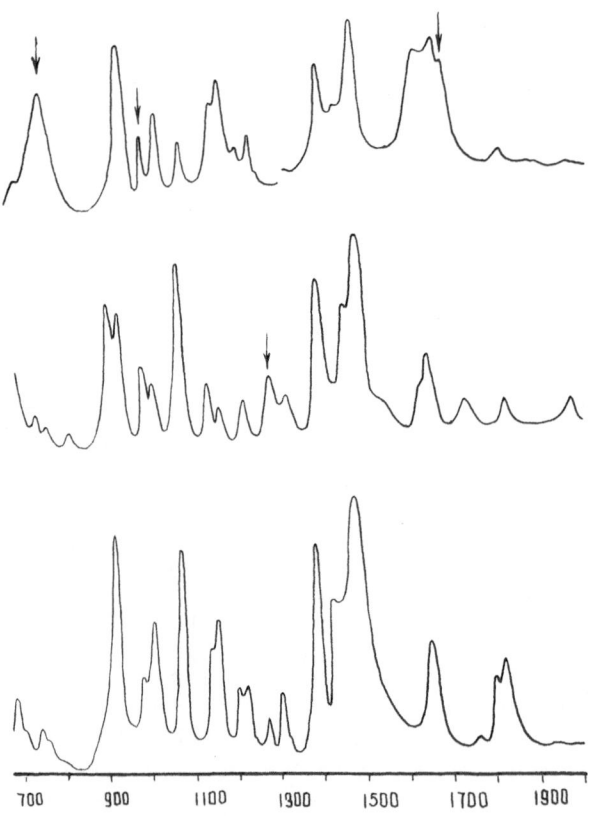

Fig. 1 IR-spectra of the interaction products of
         dimethylallene (a) and deuterated dimethyl-
         allene (b) with bis-π -C_3H_5NiBr (c) the
         decomposition product bis-π -C_3H_5NiBr + dime-
         thylallene

$$C_3H_5 + \!\!\!\!\!\!\left(\; C - CH_2 \; \right)\!\!\!\!\!\!\!_2 \quad C - CH = \!\!\!= NiBr$$

which arises as a result of hydrogen α-elimination in the complex. Decomposition of this organometallic compound may be accompanied by the formation of recombination products containing an internal double bond.

According to chromatomass-spectrometry, the decomposition of the interaction product of bis-π̃-allylnickelbromide with dimethylallene by sulphuric acid led to the formation of unsaturated hydrocarbons C_8H_{14}, $C_{13}H_{22}$, $C_{18}H_{30}$, $C_{23}H_{38}$:

$$CH_2 = \!\!\!= CH - CH_2 + \!\!\!\!\!\!\left(\; C \quad CH_2 \; \right)\!\!\!\!\!\!\!_n \quad C - CH_3$$

where n=0-3.

Hence, dimethylallene polymerization under the influence of bis-π̃ -allylnickelbromide takes place through the stage of monomer insertion into the Ni-C bond. In the process of dimethylallene polymerization an equilibrium is possible:

$$\left[\quad\underset{\overset{\textstyle C_3H_5-\underset{\overset{\displaystyle\Vert}{C}}{C}-CH_2\ -\ NiBr}{\underset{\textstyle H_3C\qquad CH_3}{\diagup\ \diagdown}}}{\underset{\textstyle H_3C}{\overset{\textstyle H_3C}{\diagdown}}\ \overset{\textstyle\diagup}{C} = C\ =CH_2}\quad\right]$$

$$C_3H_5-\underset{\overset{\displaystyle\Vert}{\underset{\textstyle\underset{H_3C\qquad CH_3}{\diagup\ \diagdown}}{C}}}{C}-CH_2\ -\ \underset{\overset{\displaystyle\Vert}{\underset{\textstyle\underset{H_3C\qquad CH_3}{\diagup\ \diagdown}}{C}}}{C}-CH\ =\underset{\overset{\displaystyle\mid}{H}}{NiBr}$$

        Polymerization of deuterated dimethylallene
/CH$_3$/$_2$ C=C=CD$_2$ revealed a kinetic isotopic effect
(Fig. 2). It may be assumed that hydrogen transfer
to the formation of a nickel hydride complex is the
limiting stage in the process of dimethylallene polymerization.

        To clarify the effect of the metal and ligand
nature on the process of allene polymerization we
investigated allene polymerization under the influence
of a catalyst on the basis of $(Co)_2/CO/_8$. This is an.
industrial catalyst cf oxosynthesis and it is readily
available.

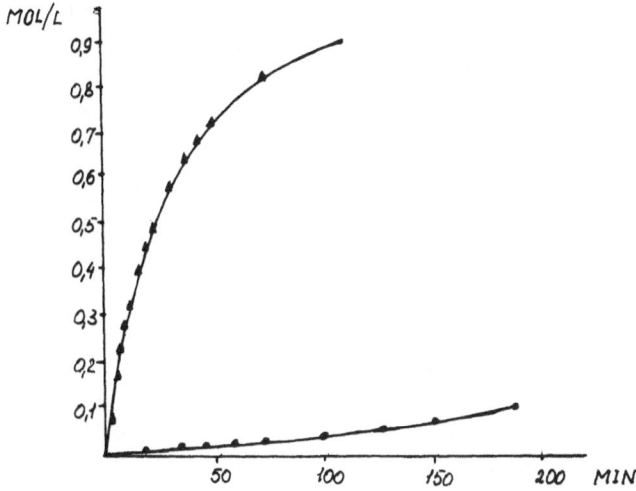

Fig. 2 Polymerization of dimethylallene (a) and
deuterated dimethylallene (b) on bis-$\widetilde{\pi}$-al-
lylnickelbromide at 30°C in benzene,
($\widetilde{\pi}$ -C$_3$H$_5$NiBr) = 1.8 10^{-3} mole/l;
[dimethylallene] = 1.25 mole/l

It is known that dicobaltoctacarbonyl contains
two types of carbonyl groups - bridge and terminal
ones and has the following structure:

During the interaction with acetylenes bridge carbonyl
groups are substituted and during the interaction with
butadiene - terminal ones. It was found that allene
polymerization under the influence of $(Co)_2/CO/_8$ proceeded with an induction period. To elucidate the nature of
the induction period the interation product of allene
with dicobaltoctacarbonyl at a molar ratio of allene
dicobaltoctacarbonyl = 3 was investigated. It was
established chromatographically that this interaction
led to the evolution of CO according to the equation:

$$C_3H_4 + Co_2(CO)_8 \longrightarrow Co_2(CO)_6 \cdot 2C_3H_4 + 2CO$$

The organo-cobalt compound obtained possessed high
activity in the allene polymerization reaction. The
structure of this compound is evidently analogous to
the π -allyl one.

We determined the kinetic parameters of allene polymerization of $Co_2(CO)_6 \cdot 2C_3H_4$ at various concentrations of the catalyst (3.0 to $8.4 \cdot 10^{-3}$ mole/l) (Fig. 3)
and of allene (0.9 to 2.1 mole/l) (Fig. 4) in a temperature range of 20 to 40°C. The reaction orders with respect
to the catalyst and monomer are close to I (Fig. 5)
and the activation energy is equal to about 48kJ/mole
(Fig. 6).

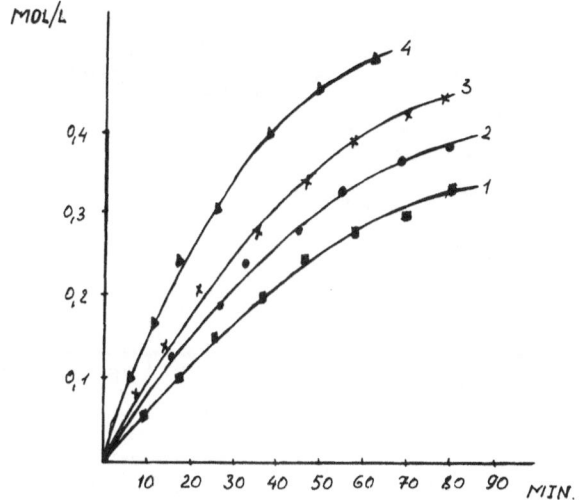

Fig. 3 The kinetic curves of allene polymerization on $Co_2(CO)_6 \cdot 2C_3H_4$ in benzene,
t = 30°C, [Allene] = 1.4 mole/l, the
catalyst concentration: (1) $3.0 \cdot 10^{-3}$ mole/l,
(2) $4.3 \cdot 10^{-3}$ mole/l, (3) $5.6 \cdot 10^{-3}$ mole/l,
(4) $8.4 \cdot 10^{-3}$ mole/l.

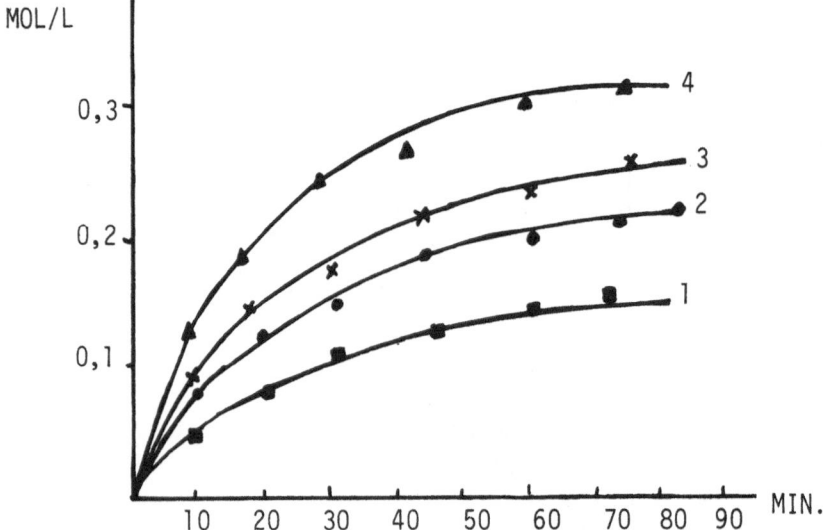

Fig. 4 The kinetic curves of allene polymerization on $Co_2(CO)_6 \cdot 2C_3H_4$ in benzene, at $t=30^{\circ}C$, [catalyst] $= 4.3 \cdot 10^{-3}$ mole/l, the concentration of allene: (1) 0.9 mole/l, (2) 1.4 mole/l, (3) 1.8 mole/l, (4) 2.1 mole/l.

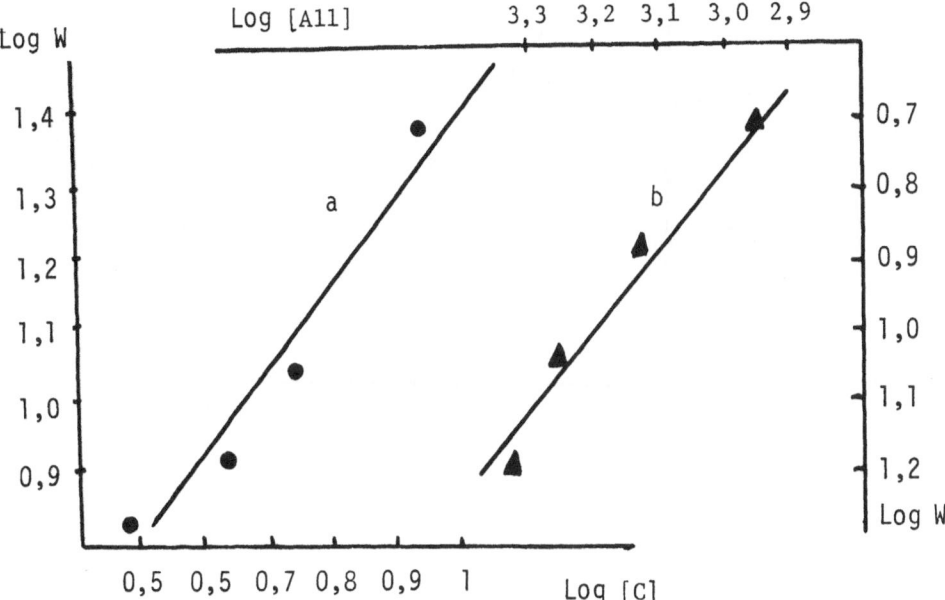

Fig. 5 Reaction orders with respect to the catalyst
(a) and monomer (b)

The polymer obtained was a white powder soluble
in hydrocarbons and chlorine-containing solvents. The
IR-spectrum testified to the vinylidene structure of
polyallene. An X-ray structural analysis revealed a
high crystallinity of the polymer ($\sim$65%). The melting
point of polyallene obtained on a Co-containing catalyst
is $\sim$164°.

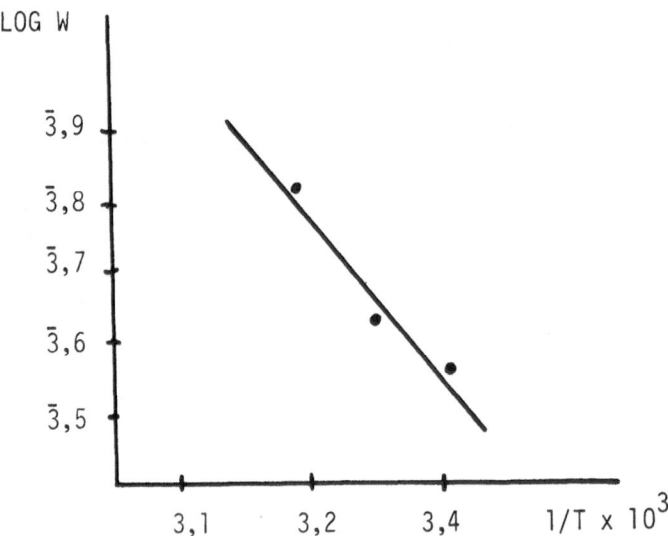

Fig.6 The dependence of the allene polymerization
      rate on temperature,
      [allene] = 1.4 mole/l,
      [catalyst] = 4.3 · 10^{-3} mole/l

The melting point of polyallene obtained on bis-
-allylnickelhaldide is 62°. This difference in the
properties of polyallene obtained on Ni- and Co-containing catalysts is apparently connected with the
nature of the metal and ligands in the co-ordination
sphere of the transition metal.

Polyallenes are susceptible to various polymer
transformations. In this connection they are of
particular interest as a new class of polyhydrocarbons
to obtain polymers, oligomers and intermediate products
for macromolecular and organic chemistry.

THE ROLE OF DONOR-ACCEPTOR INTERACTIONS IN
REACTIONS OF HOMO- AND COPOLYMERIZATION OF
CYCLIC AND ALLENE HYDROCARBONS

The conception "co-ordination" polymerization
includes another important aspect concerning the role
of weak intermolecular interactions that take place
during the formation of donor-acceptor (DA) complexes
either between the monomer and the initiator of
polymerization, or between the comonomers in the copolymerization process.

Our attention was drawn by the possibility of
using DA interactions between a monomer and organic
electron acceptors to initiate the process of polymerization. This possibility was studied on cyclic
sulphides and propargylamines[5,6].

The appearance of new absorption bands in the
UV-or visible part of the spectrum which are absent in
the spectra of the individual components testified to
the formation of DA complexes by the interaction of
S-containing monomers having various cycle size with
organic electron acceptors. The experimental data
obtained allowed us to establish that the donor capacity
and basicity of four-membered S-containing cycles
(trimethylene sulfide) connected with it was higher
than that of the three-membered (ethylenesulphide and
propylenesulphide) and the five-membered (thiophene and
tetrahydrothiophene) cycles.

The composition of the complexes was determined
by the method of "isomolar series of solutions", and
in the case of complexes of the complicated composition
"series of solutions with a variable concentration of
one of the components" were used. It was found that the
stronger donor - trimethylene sulfide (TMS) - formed
complexes of only 1:1 composition with all the acceptors
whereas propylenesulfide (PS) formed complexes of
variable composition. It will be shown later that the
composition of DA complexes is in good agreement with
the polymerization order with respect to the acceptor.

The equilibrium constants of the reactions of
complex formation in cases when the composition of
complexes was 1:1, were determined by Benesi-Hildebrand
and Scott equations. The value of k_e for TMS complex
with tetracyanoethylene (TCE) in a polar solvent is

2.4 ± 0.2 l/mole, and for the PS complex with TCE,
1.0 ± 0.3 l/mole, which also testifies to the fact
that the four-membered S-containing monomer is a
stronger donor than the three-membered PS one.

The formation of charge transfer complexes (CTC)
by the interaction of cyclic sulphides with organic
electron acceptors causes the shift of the electron
density on the sulphur atom and additional polarization
of the C-S bond, thus facilitating the fission of this
bond. However, only DA interaction is sufficient to
open the strained cycles. Under the action of CTC three-
and four-membered S-containing monomers undergo polymerization. The attempt to open the five-membered cycle
of tetrahydrothiophene and thiophene by DA interaction
failed. The opening reactions of these cycles in the
presence of other initiators are unknown. This is
apparently connected with the fact that the strain
energy in these cycles is very low.

The kinetic curves of PS and TMS polymerization
have various characters. In the case of PS polymerization in the presence of weak electron acceptors the
kinetic curve has a pronounced S-shaped character: the
induction period whose length depends on the strength
and amount of the acceptor; the period of reaction ac-
celeration; the period of maximum stationary rate and
gradual inhibition of the process. The nature of the
curve in the given case testifies to the commensurability
of the initiation and polymer chain propagation rates.
The data of IR- and UV-spectroscopy confirm experimentally that the process of CTC formation between PS and
weak electron acceptors occurs with time.

In the case of PS polymerization with a strong
electron acceptor - TCE and the TMS polymerization in
the presence of tetranitromethane and TCE, the kinetic
curves have a different character. There is no induction period, the reaction proceeds with a constant rate
up to small monomer conversions, the reaction then
slows down. This is in good agreement with UV spectroscopy data which show that the complexation process
occur instantaneously in these systems. The polymerization orders for both monomers with respect to the
initiator calculated from the kinetic curves, are in
accord with the composition of the corresponding DA
complexes: for the PS polymerization in the presence
of maleic anhydride (MA), the reaction is first order
with respect to the acceptor in nitrobenzene (the

composition of the complex in a polar medium is 1:1),
in chloroform the reaction is of the second order
with respect to the acceptor (the complex composition
in this case is 1:2); the TMS polymerization in the
presence of TCE proceeds according to the first order
with respect to the acceptor and the composition of
the complex formed in this case equals 1:1.

It was found that the polymerization rate in the
presence of organic electron acceptors was greatly
influenced by the medium polarity: an increase in the
solvent's dielectric permeability results in a sharp
increase in the reaction rate. This phenomenon is
evidently connected with an increase in the electron
transfer degree in the initiating complex and also
with the nature of the equilibrium of the complexation
reaction. It was shown experimentally for PS that the
equilibrium constant of the complexation reaction
increased with a rise of the solvent dielectric
permeability.

A comparison of the three- and four-membered
S-containing cyclic monomers behaviour in the polymerization under identical conditions (Table 5) shows that PS
is much more active despite the fact that it is a
weaker donor than TMS. PS undergoes polymerization in
the presence of such weak electron acceptors as MA and
trinitrobenzene (TNB). Even the nitrobenzene solvent
itself - a very weak electron acceptor, initiates
PS polymerization, but with a considerable induction
period (about 2 hours). TMS in the presence of weak
electron acceptors forms only CTC. The DA interaction
in these complexes is not, however, sufficient to
open the C-S bond in the less strained four-membered cycle.

Under the action of a strong acceptor - TCE - in
a polar solvent the initial rates of PS and TMS polymerization differ by almost two orders.

The experimental data presented graphically
show that the overall rate of cyclic sulfide polymerization initiated by DA interaction is most affected
by the cycle size and the opening capacity associated
with the latter; the donor capacity of the monomer
only affects the complexation and, evidently, the
initiation reaction.

Table 5. Comparison of PS and TMS behaviour in homopolymerization under the influence of CTC

Monomer	Electron acceptor	Monomer concent-ration, mole/l	Acceptor concen-tration, mole/l	Initial rate of polyme-riza-tion mole/l·sx10^4	Reac-tion time, hr	Polymer yields, %
PS	NB	4.3	-	-	12	60
PS	MA	3.0	1.00	0.018	-	-
TMS	MA	3.0	0.30	-	60	-
TMS	TNB	3.0	0.30	-	60	-
PS	TCE	2.0	0.05	4.86	0.5	18
TMS	TCE	2.0	0.05	0.050	3.0	2.4

Solvent - nitrobenzene (NB), temperature - 75°C

    This is in agreement with the proposed mechanism
of cyclic sulfide polymerization initiated by CTC:
1. Polymerization is realized in two stages:
    a) DA complex formation between the monomer and the
    electron acceptor

$$(CH_2)_n S + A \overset{Ke}{\rightleftharpoons} (CH_2)_n S^{+\delta} --- A^{-\delta}$$

where A is the electron acceptor
    b) a direct cycle opening in the complex-bonded mo-
nomer leading to the formation of an active sulfonium
ion

$$(CH_2)_n S^{+\delta} --- A^{-\delta} \underset{(CH_2)_n S}{\overset{Ki}{\longrightarrow}} \underset{A^-}{\overset{}{S}} - (CH_2)_n - S^+ - (CH_2)_n$$

2. Propagation of the polymer chain proceeds according
to the usual cationic mechanism.

    Apparently, CTC participate only in the initiation
stage, facilitating the breakage of the C-S bond, but
the active centre of the polymer chain propagation is

the sulphonium ion formed by the opening of the S-containing cycle. This is why in the case of cyclic sulphide homopolymerization the activity of the compound
will be determined by the strain energy of the cycle
and not by its ability to form a complex.

Quite a different picture is to be observed in
the reaction of the copolymerization of PS and TMS.
In this case, when even a small amount of more strained
PS is enough to form an active centre, the sulphonium
ion - the donor capacity of the monomer and its capacity to complexation with the active site may be a
decisive factor for the monomer insertion in the polymer
chain.

The experimental data obtained in the investigation of PS and TMS copolymerization employing the DA
interaction confirm the above assumption (Table 6).

Table 6. The dependence of the copolymerization
         rate and the composition of the PS
         copolymer with TMS on the composition
         of the initial mixture

TMS concentration, mole/l	PS concentration, mole/l	Composition of the monomer mixture, molar fraction of TMS	Copolymerization rate mole/l · s$\times 10^4$	Composition of the copolymer, molar fraction of TMS
2.0	0.0	1.00	0.050	–
1.8	0.2	0.90	0.092	–
1.7	0.3	0.85	0.116	0.95
1.4	0.6	0.70	0.125	0.90
1.0	1.0	0.50	0.080	0.74
0.5	1.5	0.25	0.083	0.51
0.0	2.0	0.00	4.860	–

[TCE] = 0.05 mole/l; solvent - nitrobenzene;
temperature - 75 °C

The experimental data presented testify unambiguously to the fact that it would be more correct to estimate the reactivity of polar monomers in ionic polymerization by their homopolymerization rates and not by their copolymerization constants.

Of particular interest was the possibility of using DA complexes to open the triple carbon-carbon bond, thus obtaining polymers with a conjugation system in the main chain.

We have experimentally discovered that propargylamines (primary, secondary and tertiary) during the interaction with organic electron acceptors form CTC which initiate the above amines polymerization, and the electron acceptors enter the polymer in equimolar amounts with respect to the monomer participating in the reaction. Hence, in this system DA complexes play the part of both the initiator of polymerization and the "activated" monomer.

This is the first case of the opening of the triple bond unconjugated with the heteroatom under the action of DA complexes. So far, initiation by DA interaction has been known mainly for monomers containing a vinyl group directly bonded with the heteroatom.

To clarify the effect of various substituents on the capacity of the triple bond to open under the action of DA complexes, the following propargylamines were synthesized: diethyl propargylamine, mono-, di- and tripropargylamines. DA complexes formed by the interaction of the above propargylamines with organic electron acceptors were studied by EPR, electron and PMR-spectroscopy.

It was established that the complexes formed were characterized by both a partial and a complete charge transfer and the relation between the two types of complexes depended on the value of the affinity energy to the electron of the acceptor and on the dielectric permeability of the solvent.

An analysis of the electron absorption spectra of various propargylamines DA complexes with TCE showed that the optical density of the absorption bands of complexes with both complete and partial charge transfer decreased in the following series:

diethyl propargylamine > monopropargylamine > dipropargylamine > tripropargylamine

The determination of the composition of the complexes
studied by the method of "a series of solutions with a
variable concentration of one of the components" and
"isomolar series of solutions" of the donor and acceptor
components revealed that their composition varied with
a predominance of 1:1.

The complex of diethyl propargylamine (DEPA) with
TCE was isolated in pure form. It appeared stable up
to 30-40°C which allowed its elemental composition to
be determined, IR-, EPR- and UV-spectra to be registered,
its electroconductivity to be ascertained and some
of its other properties to be investigated. Its composition corresponded exactly to the equimolar ratio of the
components.

To elucidate the participation of the triple bond
of propargylamines in the DA interaction with organic
electron acceptors, we synthesized diethyl propylamine
and investigated its complex with TCE and also registered the UV-spectrum of the propargyl alcohol - TCE
mixture. Comparison of these spectra with the absorption
spectrum of the DA complex DEPA - TCE showed that the
spectrum of diethyl propylamine with TCE had the same
absorption bands as the spectrum of DEPA - TCE, whereas
the spectrum of the propargyl alcohol - TCE mixture
revealed no absorption bands connected with charge
transfer. This experimental fact unambiguously indicates
that the dominant role in the forming of propargylamines
as electron donor partners belongs to the unshared
electron pair at the nitrogen atom, and not to the
electrons of the triple bond.

The analysis of the PMR-spectra of DEPA, mono-,
di-, and tripropargylamines, a number of model compounds
and their complexes with TCE also showed that mainly
the unshared electron pair at the nitrogen atom participated in the formation of DA complexes, but a considerable change in electron density does, however, occur
on the carbon triple bond, too, and the value of this
change depends on the nature of the substituent at nitrogen. If we recalculate the value of the chemical
shift due to DA interaction by one triple bond we shall
obtain the following data:
DEPA - 0.94 ppm, monopropargylamine - 0.38 ppm,
dipropargylamine - 0.28 ppm, tripropargylamine - 0.16 ppm.

The experimental data presented allow us to assure
that, despite the fact that it is separated from the
group which is responsible for the formation of charge
transfer, the triple bond in complex-bonded propargylamine may appear to possess a higher reactivity than
that of the conventional monomer, in particular, in
the polymerization reaction. Indeed, upon lighting
the above complexes with the visible or UV-region of
the spectrum or upon heating polymerization of the
complex-bonded monomers occurs with the formation of
polymers with a system of conjugated bonds, and as has
already been mentioned, the polymer involves the acceptor
in an equimolar amount with respect to propargylamine.
Among all the monomers considered DEPA, as was to be
expected, possessed the greatest activity in polymerization.

On the basis of the kinetic investigation results
of propargylamine polymerization by DA interaction we
have assumed the following mechanism of the triple
bond opening in nitrogen-containing monomers under the
influence of organic electron acceptors: at the first
stage the complex formation reaction occurs which leads
to the generation of the active sites and the "activated"
monomer; then in our opinion, complexes with a complete
charge transfer take part in the polymerization initiation (most likely, their ionic function does the
"work"); as for the reaction of polymer chain propagation, both complexes with complete charge transfer and
complexes with partial charge transfer, but in any
case, those having 1:1 composition participate in it.

The examples considered clearly illustrate quite
a new possibility for opening up heterocycles and
the triple bond by DA interactions, and if, in the
case of heterocyclic monomers polymerization, CTC
participates solely in the initiation stage in the
case of propargylamine polymerization,it does so both
at the stage of initiation and at the stage of polymer
chain propagation.

Now, let us consider the case when CTC takes part
solely in the activation of the monomer which then
polymerizes under the influence of the usual initiators.
This takes place, in particular, in the formation of
the $\pi\pi$ -type DA complexes necessary for the radical
polymerization of monomers which would not polymerize
according to the free-radical mechanism, as well as

for the formation of alternating copolymers possessing
a number of valuable technical and physiological properties.

The problems regarding the mechanism of alternating
copolymer formation, the role of the complexes with a
charge transfer between comonomers, and their influence
on the processes of polymer chain propagation, are the
subject of numerous discussions, and up to now there
is no single standpoint on this issue. In this connection, experimental data must be accumulated on the
copolymerization regularities of monomers with different
structures and different donor-acceptor properties.
The investigation of the propagation reaction mechanism
in such systems allows us to improve our knowledge of
the peculiarities of alternating copolymer formation
and to approach the creation of the alternating copolymerization theory.

Taking into account all said above, it is of
interest to consider the results of the kinetic investigation of the alternating copolymer formation of VCH,
indene (IN), 5-vinylnorbornene-2 (VNB) and 1,1-dimethyl-
allene (DMA) with maleic anhydride (MA) as well as a
simultaneous
comonomers[7],[8].

To prove the CTC formation between comonomers
and to investigate the properties of complexes the method
of electronic spectroscopy was selected. It turned out
that the maximum of optical density of the absorption
band of VNB and MA CTC laid in the 270 nm region, that
of IN and MA - in the 292 nm region, and that of VCH
and MA - in the 318 region.

The composition of CTC was determined using the
"isomolar series" technique. It was shown that the
maximum of optical density of CTC corresponded to the
equimolar relation of the donor and acceptor components
and the position of the maximum did not change in the
wavelength range studied which was typical of the
complexes with a 1:1 composition.

The equilibrium constants of the complex formation
were calculated according to the Benesi-Hildebrand and
Scott equations; they proved to be equal, at 20°C in
particular, for the VCH - MA system 0.033 $\pm$ 0.012 mole/l,
for the VNB - MA system 0.011 $\pm$ 0.03 mole/l and for the

IN - MA system 0.32 ± 0.14 mole/l. By the change in
the equilibrium constants with temperature the values
of the thermodynamic parameters of the complexation
process were estimated. Using an example of two systems,
VCH - MA and IN - MA, it is shown the order of these
values:
VCH - MA Δ H $= -13.6$ kJ/mole; Δ S $= -4.65$ J/(mole.K)
IN - MA Δ H $= -27.2$ kJ/mole; Δ S $= -100.5$ J/(mole·K)

The investigation of kinetic regularities of copolymerization showed that in the case of IN and VNB,
unlike VCH and DMA, the maximum rate of the reaction
occured at a comonomer ratio which did not correspond
to either the composition of comonomer CTC or that of
the copolymer formed.

To give reasons for this "anomalous" position of
the rate maximum, the kinetic features of copolymerization in solvents possessing different donor and acceptor
properties were studied. It was shown in special experiments that the initiation rate was independent of the
monomer feed composition and the "anomalous" position
of the copolymerization rate maximum was conditioned
by peculiarities of the chain propagation reaction.

Valuable information about the mechanism of the
chain propagation reaction during the alternating copolymer formation can be obtained by determining the reaction order with respect to the product of the monomer
concentrations. For the VCH-MA and DMA-MA systems,
observed the first order with respect to the product
of comonomer concentrations, in agreement with general
regularities of radical copolymerization, implies that
the polymer chain propagation proceeds predominantly
according to the mechanism of "homopolymerization" of
comonomer CTC. In the case of the VNB-MA and IN-MA
systems irrespective of the monomer mixture composition the reaction had half an order with respect to
the product of comonomer concentrations. Evidently,
for these pairs of comonomers polymer chain propagation is a consecutive addition of the free monomers
to the propagating chain end and the high values of
the chain cross-propagation rate constants due to the
formation of CTC between the adding monomer and the
propagating radical are responsible for the alternation of monomers units.

To elucidate the role of DA interactions in
alternating copolymerization, of certain interest was

the investigation of the copolymerization rate dependence on the temperature, initiation rate being constant.
It turned out that the copolymerization rate decreased
as the temperature rose. This testifies to a significant
contribution in DA interactions to the reactions under
study because only they can be responsible for the
decrease in the polymerization rate as the temperature
increased, since a rise in temperature causes a drop
in the concentration of DA complexes.

The investigation of copolymerization in solvents
with different donor and acceptor properties showed
that in most solvents, irrespective of their dielectric
constant, the maximum of copolymerization rate of IN
and VNB with MA was found at an MA content in the
initial monomer mixture of about 0.6 + 0.8 molar fraction. Only in solvents with strong donor capacities
did the rate maximum shift towards the excess of the
donor monomer. Thus, in dimethylformamide with a donor
number of 26.6 the rate maximum of copolymerization
is observed at an MA content in the monomer feed of
0.2 + 0.3 molar fraction.

For systems whose polymerization proceeds according to the mechanism of consecutive monomer addition
the molar fraction of the donor monomer in the mixture
(M_D), at which the maximum reaction rate is attained,
is related to the constants of the cross-propagation
rate by the following equation:

$$\left(\frac{1 - M_D}{M_D} \right)^2 = \frac{K_{AD}}{K_{DA}}$$

Proceeding from this the ratio of rate constants
of the chain cross-propagation in the IN and VNB
copolymerization with MA in various solvents was estimated. As was to be expected, in most solvents the
rate constant for addition of the donor monomer to
the acceptor end of the propagating chain K_{AD} is
larger than the rate constant for addition of the acceptor monomer to the donor end of the propagating
chain K_{DA} since the anhydride radical has a high reactivity due to the inductive effect of the carbonyl
groups in the α - and β - positions. Only in the
strongly donating solvent - dimethylformamide - due
to its competing DA interaction with the acceptor
macroradical does the "conversion" of the cross-propagation rate constants occur i.e. K_{AD} becomes lower than

K_{DA}. These experimental data give grounds for believing that the role of DA complexes in the process of chain propagation in the given comonomer systems is confined to the formation of CTC between the propagating radical and the adding monomer; the inequality of the chain cross-propagation rate constants is responsible for the "anomalous" position of the copolymerization rate maximum.

The competition of the solvent molecules with the comonomer molecules in the interaction reaction with the propagating macroradicals during copolymerization in solvents with high donor (dimethylformamide) and acceptor (acetonitrile) properties should lead not only to a change in the ratio of the cross-propagation rate constants, but also to a drop in the overall copolymerization rate. However, in the case of the IN-MA system rate drop was observed only in dimethylformamide, and in the case of the VNB-MA system in just these solvents, which are also characterized by higher dielectric permeability, the highest initial rate of the process was found. Evidently, the competitive interaction between the solvent and the propagating macroradical is compensated by a considerable increase in the CTC reactivity between the propagating radical and the adding monomer due to the high polarity of the medium.

A thorough analysis of the experimental data testifies to the somewhat different influence of the donor and acceptor properties of the solvent on the copolymerization the comonomer IN-MA and VNB-MA systems. In the case of the latter the donor (VNB$^{\cdot}$) and the acceptor (MA$^{\cdot}$) propagating radicals have, apparently, close activities. Indeed, when the activity of the macroradical with the end MA link considerably exceeded the activity of the donor propagating radical, the reaction in the donor solvent would be characterized by a significant decrease in the copolymerization rate, which was observed during the formation of the alternating copolymer in the IN-MA system. The cause of the greater activity of the VNB$^{\cdot}$ macroradical as compared to the IN$^{\cdot}$ radical might be the absence of radical stabilization due to the conjugation with the aromatic ring present in the IN molecule. Besides, the high activity of the VNB$^{\cdot}$ radical might be conditioned by the considerable strain energy of the norbornene bicycle.

Thus, the investigation of the regularities of the complexation reaction between comonomers, the study of copolymerization kinetics, as well as the study of the influence of the donor and acceptor properties of solvents on copolymerization showed that the chain propagation reaction in the monomer IN-MA and VNB-MA systems, unlike the VCH-MA and DMA-MA systems proceeded mainly through the consecutive addition of monomers with high cross-propagation rate constants due to the formation of CTC between the propagating macroradical and the adding monomer, i.e. on four different systems there was shown the dependence of the mechanism of the chain propagation reaction during the alternating copolymers formation on the properties of the donor and acceptor comonomers and the important part played by DA interactions in these processes.

CONCLUSION

The above results of the investigation of some aspects of coordination polymerization of a number of monomers made it possible to discover the peculiarities of polymerization of high α-olefins, to characterize the mechanism of polymerization of allene hydrocarbons and to determine the role of DA interactions in the polymerization and copolymerization of cyclic and allene compounds.

We have made an attempt at broader consideration of coordination polymerization including in this conception, not only coordination of monomers with transition metals, but also the bonding of monomers into DA complexes with acceptor molecules or with one another.

In our opinion, the data obtained contribute to improving and deepening modern knowledge of coordination polymerization in its various aspects.

REFERENCES

1. B.A. Krentsel and V.I. Kleiner,
   Some New Aspects in the Chemistry of Polyolefins,
   in: "Reviews of Science and Engineering. Chemistry
   and Technology of Macromolecular Compounds", v. 5,
   B.A. Dolgoplosk, ed., Nauka, Moscow (1974)

2. B.A. Krentsel, V.I. Kleiner, L.L. Stotskaya,
   V.V. Americ and D.V. Ivanukov,
   The Method of Polyvinylcycloalkanes Obtaining,
   Pat. 1581067, France; Pat. 1212558, England;
   Pat. 884855, Italy; Pat. 677014, Japan.

3. B.A. Krentsel, E.A. Mushina, E.M. Khar'kova and
   M.V. Shishkina,
   Polymerization of Allene and its Derivatives
   under the Influence of Bis -π- Allylnickel -
   bromide, Europ. Polym. J., 11, 865 (1975)

4. M.V. Shishkina, E.S. Juk, L.M. Zubrizky,
   E.A. Mushina and B.A. Krentsel,
   The Structure of Polyheptadiene,
   Vysokomol. Soedin., 18B, 2, 89 (1976)

5. L.L. Stotskaya, G.A. Oreshkina and B.A. Krentsel,
   The Method of S-containing Polymers Obtaining,
   Author Certification 325236, USSR

6. L.L. Stotskaya, V.S. Serebrjanikov, B.E. Davydov
   and B.A. Krentsel,
   The Method of Semiconducting Polymers Obtaining,
   Author Certification 368801, USSR

7. S.T. Bashkatova, V.I. Ojereliev, V.I. Kleiner,
   L.L. Stotskaya and B.A. Krentsel,
   Donor - Acceptor Complexes in the Copolymerization
   of Some Vinylcycloalkanes and Vinylcycloalkenes
   with Maleic and Chloromaleic Anhydride,
   Vysokomol. Soedin., 14A, 2640 (1972)

8. L.F. Kim, L.L. Stotskaya, B.A. Krentsel,
   V.P. Zubov, V.B. Golubev and I.L. Stoyachenko,
   Kinetic Features and Mechanism of Alternating
   Copolymerization of Indene with Maleic Anhydride,
   J. Macromol. Sci. - Chem., A12 (8), 1197 (1978)

CONTRIBUTORS

Bansleben, D.
Bukatov, G. D.
Burkhardt, T. J.
Buys, H. C. W. M.
Chiellini, E.
Fochi, G.
Gianinni, V.
Hagiwara, T.
Ishimori, M.
Kleiner, V. I.
Krentsel, B. A.
Langer, A. W.
Loeffler, P.
Masi, F.
Montaz, A.
Moser, G. A.
Muggee, J.
Mülhaupt, R.
Mushina, E. A.
Nakata, T.

Noltes, J. G.
Oschwald, A.
Overmars, H. G. J.
Peet, W. G.
Piccolo, O.
Pino, P.
Price, C. C.
Setterquist, R. A.
Sigwalt, P.
Solara, R.
Spassky, N.
Steger, J. J.
Stotskaya, L. L.
Tebbe, F. N.
Tsuruta, T
Vandenberg, E. J.
Vogl, O.
Yermakov, Y. I.
Zakarov, V. A.

INDEX

327